Polymeric Membranes for Water Purification and Gas Separation

Edited by

Rasel Das

Department of Chemistry, Stony Brook University, Stony Brook, NY 11794, United States

Published by **Materials Research Forum LLC**
Millersville, PA 17551, USA

Published as part of the book series
Materials Research Foundations
Volume 113 (2021)
ISSN 2471-8890 (Print)
ISSN 2471-8904 (Online)

Print ISBN 978-1-64490-162-5
eBook ISBN 978-1-64490-163-2

Distributed worldwide by

Materials Research Forum LLC
105 Springdale Lane
Millersville, PA 17551
USA
https://www.mrforum.com

Manufactured in the United States of America
10 9 8 7 6 5 4 3 2 1

Table of Contents

Preface

This book exquisitely conveys the fundamental science of different types of organic and synthetic polymers, which have been commonly used in industrial settings for water purification and gas separation. It includes the removal of organic and inorganic pollutants from the wastewater as well as the separation of important gases like O_2 and N_2, CO_2 and CH_4, and H_2 and N_2, etc., which are of paramount importance to the global development of clean energy and the environment. Every chapter of this book is written by well-known scientists from different research groups which have ensured a broad view of each topic, rather than reflecting the work of specific research groups. They discuss all the opportunities to optimize various types of membranes, such as microfiltration, ultrafiltration, nanofiltration, reverse osmosis, forward osmosis, and so on. It consists of overhauling classical membrane fabrication methods such as phase inversion, interfacial polymerization, electrospinning, layer-by-layer, etc. With the advent of nanotechnology, numerous nanomaterials such as carbon nanotube, two-dimensional nanosheets, metal-organic frameworks, covalent organic frameworks, porous organic cages, organic macrocyclic molecules, titanium dioxide, zinc oxide, mesoporous silica nanoparticles, and so on have raised tremendous public interest in separation science. The authors show how to use these nanomaterials effectively for the modification of polymeric membranes to secure desire productivity. It entails the good trade-off between water/gas permeability and solutes selectivity as well as efficient gas separation from mixture agents. Furthermore, the mechanism of pollutants retention and gas separation is beautifully illustrated in different chapters. Important steps of functionalization and grafting such as cross-linking and others are well corroborated. Factors affecting the membrane performance, such as operating condition, temperatures, operating pressure and flow rate, etc. are covered in detail. Some of the chapters, focusing on water purification, show how to control membrane fouling using different kinds of surface modification techniques. Each chapter is concluded by proving the authors' perspectives on key challenges revolving around membrane performance and stability issues. In summary, the present book is directed to experienced researchers in the fields of Polymer chemistry, Nanotechnology, Physics, General Chemistry, Organic chemistry, Inorganic chemistry, Environmental Science and Chemical Engineering, and the researchers who are actively working in the field, looking for polymer-based cross-disciplinary research thoughts. I am thankful to the authors who produced excellent chapters that will greatly benefit the potential readers interested in polymers-based water purification and gas separation topics. My very special thanks and gratitude would travel to Mr. Thomas Wohlbier, who agrees to publish the book on an international platform.

New York   Rasel Das

Polymeric Membranes for Water Purification and Gas Separation   Materials Research Forum LLC
Materials Research Foundations **113** (2021) 1-8   https://doi.org/10.21741/9781644901632-1

Chapter 1

Introduction

R. Das[1,2]

[1]Department of Biochemistry and Biotechnology, University of Science and Technology Chittagong, Chattogram-4202, Bangladesh

[2]Department of Chemistry, Stony Brook University, Stony Brook, NY 11794, United States

raseldas@daad-alumni.de

Abstract

In this chapter, a summary of different synthetic polymers-based water/gas separation techniques is given. It includes the separation of solutes from wastewater and also the purification of gases from various cocktails. Moreover, a summary of individual chapters of this book entitled 'Polymeric Membranes for Water Purification and Gas Separation' is shown in the conclusion section.

Keywords

Polymers, Membranes, Water Purification, Gas Separation

Contents

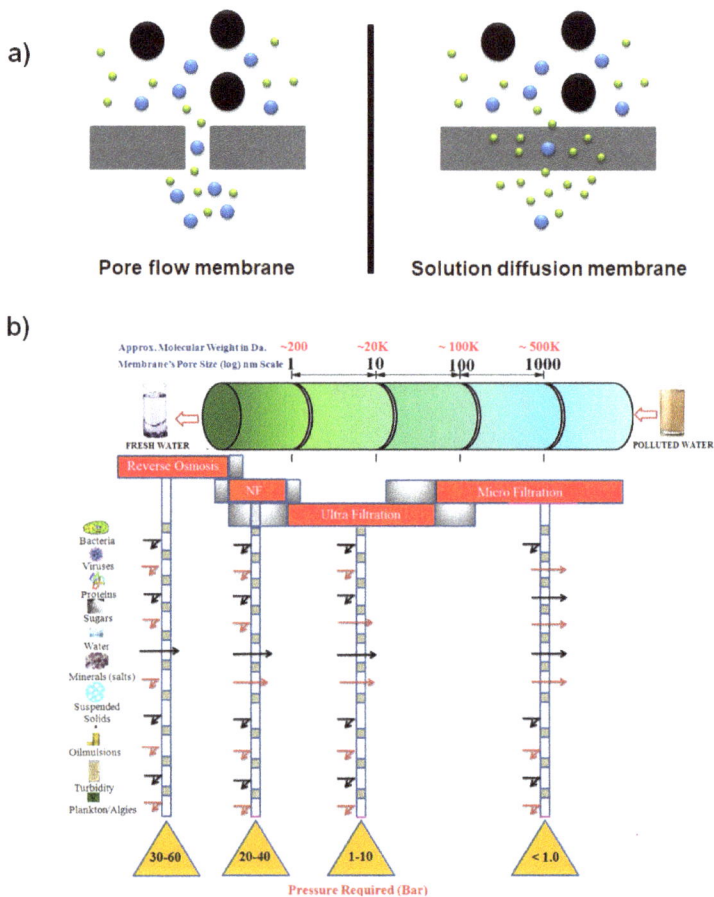

Figure 1. a) Schematic diagram of pore-flow membrane vs. solution-diffusion membrane transport. b) Diagrammatic representation of major membrane filtration methods.

1.   Background

Some areas of the world, especially the developing countries, have been severely affected by acute water shortages. This is due to freshwater pollution which has been caused by the addition of macro-, micro- and nano-pollutants. Advanced water purification technologies, which need less energy to purify contaminated water, are important to develop. Membrane technology is considered as one of the efficient methods to purify wastewater in several routes [1]. Four common types of polymeric membranes, such as microfiltration (MF), ultrafiltration (UF), nanofiltration (NF) and reverse osmosis (RO) are used to remove particles of desired sizes (See Fig. 1). Particles/solutes pass through MF and UF typically follow pore-flow mechanism, whereas RO membrane obeys solution-diffusion mechanism as revealed in Fig 1 (a). Meanwhile, the NF membrane goes along with a combination of pore-flow and solution-diffusion mechanisms. Based on pore types and their distribution, membranes are of two types, such as symmetrical/depth filter and asymmetric/screen filters as shown in Fig. 2. Symmetric membranes have a uniform pore structure throughout the thickness. This membrane retains solutes within the membrane (*i.e.* into the pores) and the mechanism of solutes capture follows simple size sieving, Brownian diffusion, adsorption, and electrostatic adsorption. On the other hand, asymmetric membranes consist of a thin active layer with fine pores overlaying a thicker support layer. This support layer typically has larger pores, which provide mechanical support to the active thin layer, but show little resistance to water flow. The asymmetric membranes separate solutes through the surface active layer of the membrane in a very thin film [2].

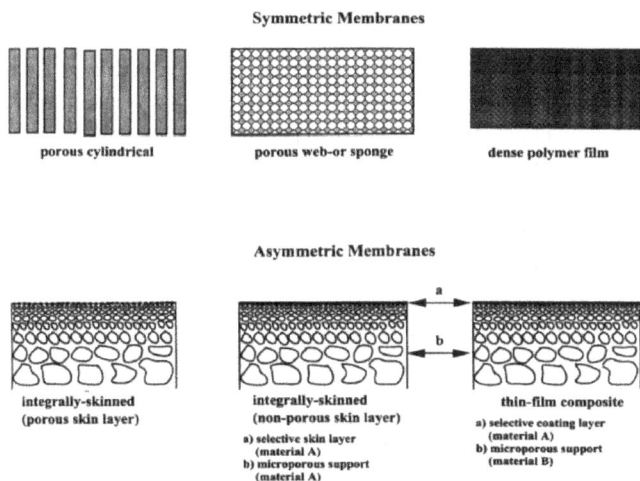

Figure 2. Structural representation of symmetric and asymmetric membrane structures. The figure is adapted with permission from the American Chemical Society [3].

1.1   Polymers for membrane fabrications

Many synthetic polymers are being used for membrane fabrication because of their high thermal, chemical and mechanical stabilities. To make a flat sheet or hollow fiber membranes, polymers should be flexible enough in such a way that one can easily fold them appropriately. Many synthetic polymers [4] are available for this purpose such as poly(phenylene oxide) polyesters, cellulose acetates, polypropylene, polyurethanes, polyetherimides, cellulose, polycarbonate esters, polyphenylene sulfide, ethyl celluloses, polyvinylidene fluorides, polyacetylenes, cellulose acetate butyrate, polyamide esters, polytetrafluoroethylene,   poly(trimethylsilylpropyne),   polyamide-hydrazide, polyacrylonitrile, cellulosic ester, polyetherketones, poly(phthalazine ether sulfone ketone),   polyamides,   cellulose   nitrates,   polycarbonates,   polyamideimides, polyestercarbonate, polyethersulfones, sulfonated poly (phthalazine ether sulfone ketone), polyureas,   sulfonated   polysulfones,   poly(trialkylsilylacetylenes),   polyvinylalcohol, polysiloxanes, polyimides, polyvinylchloride, polyvinylchloride, polyethylenes and polyvinylchloride. Such polymers have been used to fabricate different membrane processes [5]. However, some of these polymers have limitations and are not commonly practiced industrially. In general, we use very common polymers for membrane fabrication in our laboratory. In particular, one can use polysulfone, poly(acrylonitrile)-poly(vinyl chloride) copolymers, poly (vinylidene fluoride) and poly(acrylonitrile) for the synthesis of MF and UF membranes. Some other polymers such as nylons, cellulose acetate, poly(tetrafluoroethylene) and cellulose nitrate blends are also used in MF membranes preparation [6, 7]. In addition, poly (ether sulfone) has been extensively used for UF membranes. Polymers like polyester, polyether, polyamide and polyimide are useful for NF membranes. For fabricating the RO membranes, one can choose polysulfone coated with aromatic polyamides and cellulose acetate. However, all of these preceding polymers can be used to fabricate different types of membranes; their membrane performance could be drastically varied due to physicochemical differences as revealed in Table 1.

Table 1. Commonly used polymers for separation membranes in the membrane industry [8].

Polymer	Morphology			Membrane process
	Barrier type	**Cross-section**	**Barrier Thickness (mm)**	
Cellulose acetates	Nonporous	Anisotropic	~0.1	RO
	Mesoporous	Anisotropic	~0.1	UF
	Macroporous	Isotropic	50-300	MF
Cellulose nitrate	Macroporous	Isotropic	100-300	MF
Cellulose	Mesoporous	Anisotropic	~0.1	UF
Polyacrylonitrile	Mesoporous	Anisotropic	~0.1	UF
Polyetherimides	Mesoporous	Anisotropic	~0.1	UF
Polyethersulfones	Mesoporous	Anisotropic	~0.1	UF
	Macroporous	Isotropic	50-300	MF
Polyethyelene terephthalate	Macroporous	Isotropic track-etched	6-35	MF
Polyamide (aliphatic)	Macroporous	Isotropic	100-500	MF
Polyamide (aromatic)	Mesoporous	Anisotropic	~0.1	UF
Polyamide (aromatic)	Nonporous	Anisotropic/composite	~0.05	RO,NF
Polyether (aliphatic)	Nonporous	Anisotropic/composite	~0.05	RO,NF
Polyethylene	Macroporous	Isotropic	50-500	MF
Polyimides	Nonporous	Anisotropic	~0.1	GS,NF
Polypropylene	Macroporous	Isotropic	50-500	MF
Polysiloxanes	Nonporous	Anisotropic/composite	~0.1<1-10	NF
Polyvinylidenefluoride	Mesoporous	Anisotropic	~0.1	UF
	Macroporous	Isotropic	50-300	MF

Conclusion

Summary of different polymers for water purification and gas separation membranes, having macroporous, mesoporous and nonporous networks, is discussed. Chapter-2 depicts a detailed discussion on different types of synthetic polymers along with their structure and properties. These polymers are generally used for membrane fabrications. Different fabrication techniques, as well as membrane characterization tools, are extremely validated whether they are feasible for lab uses. At the end of this chapter, the authors discuss some membrane module testing systems, such as plate and frame, tubular module, capillary module, hollow fiber module spiral wound modules for separation processes. Chapter-3 covers the uses of polymers, which have been extensively used for MF and UF membrane preparation in water purification. Also, we show how to incorporate different materials as reinforcement agents to improve these membranes' productivity. Besides synthetic polymers, biopolymers are also highlighted in this chapter. At the end of this chapter, we discuss some factors which could influence the overall membrane performances during operation. Thin-film composite membrane is popular for NF-based application. This is well corroborated in Chapter-4. The authors first show the drawbacks of some classical membranes. The preparation methods of such membranes are highlighted too. Many nanomaterials such as metal and metal oxide nanoparticles, carbon-based materials, metal-organic frameworks, double hydroxides nanofillers and so on are discussed in such a way that one can overcome the current limitations of these membranes. In the last part of the chapter, they highlight the fouling types and show the strategy to mitigate the membrane fouling. RO and forward osmosis (FO) are very essential membranes for separation science. Chapter-5 highlights essential information such as the preparation, modification and uses of RO and FO membranes for solutes separation. The authors first show the desirable features of RO and FO membrane for enhanced performance. Then, they discuss the phase inversion, electrospinning, interfacial polymerization, layer by layer etc. for membrane synthesis. To achieve desirable properties, these membranes have to be modified and these topics such as coating, grafting and crosslinking are displayed in detail over there. At last, they draw their perspective to improve these polymeric membranes for water purification. In Chapter-6, we discuss the separation of important gases e.g. O_2/N_2 using polymeric membrane technology. Two important mechanisms of porous and non-porous membrane technology for gas separation are illustrated in the beginning of the chapter. In addition, the assemetric and symmetric membrane performances for gas separation are investigated. Some of the classical membrane technology's disadvantages are shown, and using various materials to overhaul these drawbacks is discussed. At the end of this chapter, we show the factors, which could directly affect the efficiency of polymeric membranes for O_2/N_2 separation. Salleh and her coworkers beautifully present the role of different polymers including polyimides, polysulfone, polyetherimides and so on for the separation of CO_2/CH_4 in Chapter-7. They highlight some drawbacks of classical membranes and discuss the strategies to solve these bottlenecks especially by using functionalization, composites preparation, introduction of new additives, and cross-linking methods, and so on. The performances of these membranes are discussed with their gas permeability cum selectivity. They reveal five main types of mechanisms of gas transport. At the end of the chapter, they show the research gaps and give their perspectives for future research design. Overview of membranes for H_2 and N_2 separations from various mixture gases is discussed in Chapter-8. The authors have beautifully written this chapter that is very comprehensive and substantial. They first discuss the history and major applications of both classical and advanced polymers. Performance check and mechanism of gas separation membranes are shown. Major headlines include H_2/CH_4, H_2/CO_2, CO_2/N_2, N_2/CH_4 and other kinds of separations. Challenges of purifying H_2 and N_2 are widely discussed as well as they highlight some membrane design principles. Some techniques of membrane separation optimization such as intrinsic microporosity control using different materials, such as metal-organic frameworks, covalent organic frameworks, porous organic cages, organic macrocyclic molecules, 2D nanosheets and so on are scrutinized. Different kinds of polymer modification techniques are given such as grafting, chain-functionalization, crosslinking and so on.

References

[1]   R. Das, P. Solís-Fernández, D. Breite, A. Prager, A. Lotnyk, A. Schulze, H. Ago, High flux and adsorption based non-functionalized hexagonal boron nitride lamellar membrane for ultrafast water purification, Chem. Eng. J. (2020) 127721. https://doi.org/10.1016/j.cej.2020.127721

[2]   G. M. Geise, H. S. Lee, D. J. Miller, B. D. Freeman, J. E. McGrath, D. R. Paul, Water purification by membranes: the role of polymer science. J. Polym. Sci. B Polym. Phys. 48 (2010) 1685-1718. https://doi.org/10.1002/polb.22037

[3]   I. Pinnau, B. Freeman, Formation and modification of polymeric membranes: overview, in, Membrane Formation and Modification, ACS Symposium Series, 744 (2000), pp. 1-22. https://doi.org/10.1021/bk-2000-0744.ch001

[4]   R. Das, Polymeric materials for clean water Springer Nature, Switzerland, 2019. https://doi.org/10.1007/978-3-030-00743-0

[5]   J. Ren, R. Wang, Preparation of polymeric membranes, in Membrane and desalination technologies, (2011), Springer. pp. 47-100. https://doi.org/10.1007/978-1-59745-278-6_2

[6]   Pinnau, I., et al., Advanced materials for membrane separations, American Chemical Society, Washington DC, 2004.

[7]   S. Nunes, K.V. Peinemann, Membrane Market. Membrane Technology: in the Chemical Industry, 2001, p. 4-5. https://doi.org/10.1002/3527600388.ch2

[8]   M. Ulbricht, Advanced functional polymer membranes, Polymer, 47 (2006), 2217-2262. https://doi.org/10.1016/j.polymer.2006.01.084

Polymeric Membranes for Water Purification and Gas Separation   Materials Research Forum LLC
Materials Research Foundations **113** (2021) 9-32   https://doi.org/10.21741/9781644901632-2

Chapter 2

Commonly used Polymers for Separation Science

Shubhalakshmi Sengupta[1], Anil Kumar Nallajarla[1], Aparajita Mukherjee[2] and Papita Das[3]*

[1]Department of Sciences and Humanities, Vignan's Foundation for Science, Technology & Research, Deemed to be University, Vadlamudi, 522213-Guntur, Andhra Pradesh, India

[2]Taradevi Harakhchand Kankaria Jain College, Kolkata-700002, West Bengal, India

[3]Department of Chemical Engineering, Jadavpur University, Kolkata-700032, West Bengal, India

* papitasaha@gmail.com

Abstract

Many synthetic and organic (bio-based) polymers have been used for membrane fabrications. In this chapter, we discuss the structure and properties of some commonly used polymers, which have been used for water purification and gas separation applications. To supplement that, we discuss some characterization tools and membrane module testing conditions for performance checks.

Keywords

Polymers, Membrane Separation, Water Purification, Filtration, Membrane Modules

Contents

1.   Introduction

One of the 21st century challenges is lack of freshwater due to increased world population, which is expected to reach 10 billion by 2050. Availability of pure drinking water in the developing world is another area of major concern. The innovation in the existing technologies, used for water purification, is very limited. In various separation processes, membranes are used as barriers which are semipermeable and can be used in separation and restriction of the transportation of substances. Thus, it involves concentration of components both in the retentate and permeate fractions. As of now, different types of membrane methods like membrane distillation (MD), microfiltration (MF), ultrafiltration (UF), nanofiltration (NF) and reverse osmosis (RO) have been used for this purpose. MF and UF techniques are used for water treatment, whereas NF and RO technique is mainly used for water softening and desalination purposes. Among this technique, MD is currently developing fast and it is an efficient technique for desalination. The membranes used in these techniques play a vital role in determining the economic and technological efficiency of the technique employed. The nature of material used and pore size of the membranes determine its application [1]. Membranes used for various applications can be varied depending on sectors such as biotechnological, chemical, and pharmaceutical and food industries. In majority of these cases, energy efficiency, better separation efficiency, reduction in steps of processing and improvement in product quality are the major lookouts for adopting a suitable technology [2].

In conventional filtration technologies whatever may be the fluid flow (gas or liquid), it is always perpendicular to the membrane surface so that the solutes get deposited on it and that requires a periodic interruption processing system for cleaning or replacing the filter. Alternatively, when the membrane filtration is tangential, the fluid flows parallel to the surface of the membrane. The solutes accumulated onto the surface are stripped away from the surface owing to high velocity. This increases the efficiency of the process. The application of hydraulic pressure for driving mass transport distinguishes the common separation processes like MF, UF, NF and RO [3]. The nature and properties of these membranes control the permeation of each component and their selective retention as they are separated with accordance to their molar masses and particle size. MF generally uses pressures lower than 0.2 MPa which is able to separate molecules in the range of 0.025 and 10 μm. Secondly, UF uses pressures which is greater than 1 MPa and is able to separate particles with molar masses in the range of 1 and 300 kDa. Thirdly, NF pressure ranges between 1 to 4 MPa and has a capability to separate particles in the range of 350 to 1000 Da. Finally, RO needs a pressure in the range of 4 to 10 MPa can accumulate particles below 350 Da and separate it. The membrane pore size is a crucial factor and is designated as molecular weight cut-off (MWCO). This indicates the molar mass of the component which is smallest that could be retained with an efficiency of around 95% [2, 4-8].

The solute's selectivity is determined by the dimension of the molecule to be separated and the pore size of the membrane. Solute diffusivity and the electric charges associated with it are also responsible. The separation properties of any membrane is also influenced by the material chemical composition and other factors like temperature, feed flow, pressure of flow and interaction between the molecules and also the surface morphology of the membrane. On the basis of the structure of a membrane, it can be asymmetrical or symmetrical having larger pore size and uniform pore size as respective characteristics. It can also be dense and porous. However, various materials are involved in fabricating these membranes for water purification and gas separations.

In this chapter, we highlight the materials, especially the polymers, which have been utilized for membranes fabrications. These polymeric membranes are fabricated from a wide array of polymers. There are various characterization tools for assessing their structures and also testing modules for their performance checks. Therefore, we also cover these aspects. They are required for various applications mainly discussed here in the purview of water purification [8].

2.   Polymers structure and properties

A substance which comprises of very large molecules each of which have many repeating subunits is known as a polymer. The definition of polymer says it is 'chemically made up of many repeating units'. The unit may be a single atom or a small group of atoms. The word "polymer" comes from the Greek word meaning many parts/multiple repeating units, and the repeating unit made up with carbon and hydrogen are called hydrocarbons. In some cases, the repeating unit made up with oxygen, nitrogen, sulfur, chlorine, fluorine, phosphorous and silicon. These larger molecules bring about some unique properties, which are not present in the smaller ones. They bring properties like viscoelasticity, formation of crystalline and semi-crystalline structures, enhanced mechanical properties like elasticity, toughness etc. Polymers may have natural origins or may be synthetically manufactured. The examples for naturally occurring polymers are Silk, Wool, DNA, Cellulose, Protein etc. Historically polymer was first coined by Jöns Jacob Berzelius in the year 1833. In 1920, the modern concept of polymers describes as "covalently bonded macromolecular structure" was proposed by Hermann Staudinger. In 1909, Bakelite which was created for manufacturing telephone components was the first synthetic polymer. In 1910, rayon was manufactured from cellulose and nylon was discovered in 1935.

There are different types in the synthetic polymers named as homopolymers, copolymers, thermoplastics, thermosets etc. Homopolymer is defined as polymer that has the same monomeric unit in the chain. Examples are likely polyvinylchloride (PVC) having vinyl chloride units and polypropylene with propylene units. Again, polymerization of more than one type of monomer results in formation of copolymers and the process is called copolymerization. Acrylonitrile butadiene styrene (ABS) and styrene-isoprene-styrene (SIS) are the common examples for the copolymers [9,10]. Depending upon the behaviour of a polymer on application of heat, it can be classified as thermoplastic and thermosetting polymers. A thermoplastic material become pliable and mouldable at certain elevated temperatures and consequently solidifies upon cooling. On the other hand, thermosetting polymers become irreversibly rigid when heated. Heat increases the cross linking between the chains and cures the polymer. Polyvinylchloride, polypropylene etc. are common examples of thermoplastics; whereas vulcanized rubber, polyester etc. are examples of thermosets [11].

3.   Classification of polymers

Polymers can be classified on the basis of various factors. Of them, classifications based on source, structure, mode of polymerization and molecular forces are of importance. As discussed earlier, depending upon sources the polymers can be classified as natural and synthetic polymers. Semi-synthetic ones are those which have origins from a natural source but have been modified artificially in a lab (e.g., cellulose acetate). Depending upon the structure, polymers can be of three types, such as linear, branch chained and crosslinked. Based on process through which they have been polymerized, it is possible to yield various types like additional polymers (e.g., polyethylene) and condensation (e.g., Nylon 66) polymers. Again, depending upon the molecular forces, polymers can be of different types like fibres and elastomers. Elastomers are solid rubber like polymers which can be easily stretched (e.g., rubber bands). The polymer chains can be easily stretched depending upon the application of forces. Fibres pose strong inter-molecular forces between its chains which result in less elasticity and high tensile properties. Nylon-66 is an example of this group of polymer [10].

4.   Uses of polymers

Polymers can be used in a wide variety of fields. Both synthetic and natural polymers have usage in our daily lives [9, 10]. They are used in manufacturing in the textile sector. For example, polyvinyl chlorideis are utilized in clothing; shoes, wetsuits, and spandex are produced from polymeric substances. Electronics like light emitting diodes, television parts, transistors etc. are polymeric products. Apart from these consumer goods, polymers are used widely in packaging, insulating and construction materials. They are also used for making automobile parts like rubber tires. Others have applications in paint, pharmaceutical, cosmetic and medical appliance industry. There are nowadays required in making polymeric currencies, bank cards, 3D printing and most importantly membrane separation techniques for water purification. This brief discussion on polymers and its properties in general is given in order to provide an impetus towards understanding the use of polymers as membranes for separation purposes [10]. Several polymers are used for membrane fabrication processes used for solutes separation. The chemical structures of some of these polymers are illustrated in Fig. 1.

Commercial membranes are composed of perfluorinated sulfonic acid polymers such as Nafion. It has good chemical, thermo-chemical and electrochemical properties [12]. Polyether ether ketone (PEEK) based membranes are widely being investigated as they have very high chemical and physical stability, and can be produced by simple economical methods [12]. Polyether sulfone (PES) is a hydrophilic membrane, which is easy to wet and has a fast filtration rate along with a low protein binding affinity that could minimise the membrane biofouling [13]. Polysulfone can be manufactured easily having reproducibility with pore sizes as small as 40 nm. These membranes have potential to be used in processes like waste water treatment, haemodialysis, food products processing and gas separation [13]. Polyvinylidene difluoride (PVDF) based flat membrane is commercially available which has been extensively used for water and gas purification. Polytetrafluoroethylene (PTFE) membrane is the best for its hydrophobicity, chemical and thermal stability. Moreover, its good chemical resistance and excellent mechanical strength have attracted researchers' attention [14]. PTFE based membranes are popularly used for membrane distillation application because of its hydrophobicity. This type of polymers have various properties ranging from high chemical and temperature resistance to good electrical insulation capacity along with superior resistance to environmental factors such as UV, weathering etc. It also possess flexibility, resistance to fatigue, low moisture absorption coupled with its availability in purified food and medical grades [15]. Again, linear polymers like polycarboxylates having high molecular mass, polymers of acrylic acid or copolymers of acrylic and maleic acid, biodegradable, non-conventional polymers like poly lactic acid are also used for making polymeric membranes nowadays [16].

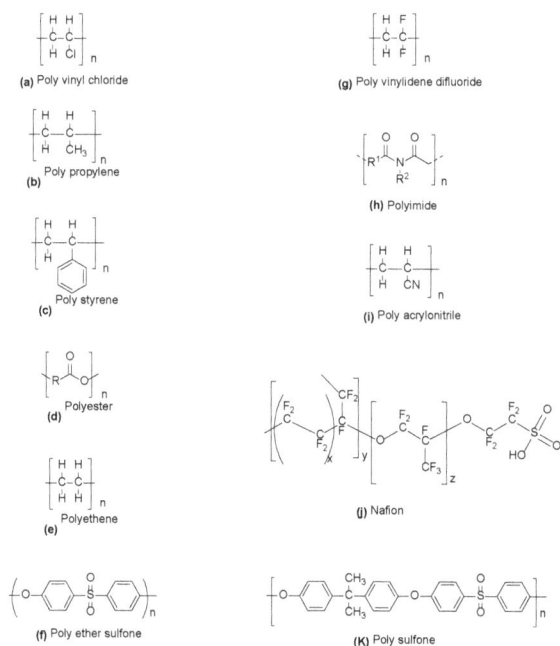

Figure 1. Chemical structures of commonly used polymers for membrane fabrication: a) Polyvinyl chloride, b) Polypropylene, c) Polystyrene, d) Polyester, e) Polyether sulfone, g) Polyvinylidine difluoride, h) Polyamide, i) Polyacrylonitrile, j) Nafion, and k) Polysulfone.

Table 1. Commonly used polymers used in separation science and their properties (Reproduced with permission from [20]).

Material	Technique Used in	Advantages	Disadvantages	Mechanical Strength and durability	Hydrophilicity and Contact Angle	pH range	Chlorine resistance
Polysulfone	Microfiltration/Ultrafiltration	Good Mechanical Strength, Chemically resistant		√√	√ ~75	1–13[a]	√√
Polyethersulfone	Microfiltration/Ultrafiltration	Rigid, compaction resistant, very permeable		√√	√ ~70	1–13[a]	√√
Polyacrylonitrile	Microfiltration/Ultrafiltration	Oxidant tolerant, narrow pore size distribution		√	√√ ~60		
Polyvinylidene fluoride	Microfiltration/Ultrafiltration	Very oxidant tolerant, Chlorine resistant	Broader pore size distribution	√√	× 100	2–11[a]	√
Polyethylene	Microfiltration/Ultrafiltration (uncommon)	High resistant to organic solvents, low cost, oxidant tolerant	Poor thermal properties, Weaker fouling resistance	√	×		×
Polypropylene	Microfiltration/Ultrafiltration (uncommon)	High resistance to organic solvents, decent mechanical strength	Low fouling resistance, not oxidant tolerant	√	√	2–13	×
Polyvinyl chloride	Microfiltration/U		Poor thermal stability, not	×	×		

15

	Process	Properties	Low permeability (RO), oxidant tolerant		pH	
Cellulose acetate	Ultrafiltration (occasionally) Reverse Osmosis, Microfiltration/Ultrafiltration	Renewable resource	×	√√	5–8.5	√√
Polyamide	Reverse Osmosis (TFC active layer), Nanofiltration, Microfiltration/Ultrafiltration (occasionally)	Small pores, excellent retention, selectivity	Relatively impermeable/dense √	√√ -55	1–13[b]	×

Symbols are used as follows: √√√= excellent, √√= good √= fair, and X = poor. [a]Poor long-term stability in basic conditions. [b]Poor long-term stability in acidic or basic conditions.

5.   Polymeric membrane fabrication

In recent times, different techniques are available for membrane fabrication. The selection of the process depends on the polymer and the structure required. Phase inversion, stretching, interfacial polymerization and electrospinning are the commonly used techniques. In phase inversion, a homogeneous polymeric solution in a controlled manner is converted towards solid state by immersion-precipitation, thermally induced phase separation, evaporation-induced phase separation and vapour induced phase separation processes. Interfacial polymerization (IP) is a commercial method of fabrication of thin film composites for RO and nanofiltration (NF) membranes. In the basic IP method, a microporous PSf support is usually immersed in a polymeric amine solution and then the amine-incorporated membrane is again immersed in di-isocyanate in hexane. The membranes are then cross-linked by applying heat (sometimes not necessary) to form the porous membrane. Stretching is also a technique applied for microporous membranes used for MF, and UF processes. In this process, the polymer is melt-extruded and then stretched to get the porous structure. In track-etching, non-porous membranes are irradiated with heavy ions to form damage tracks across irradiated polymeric films. In recent times, electrospinning also emerges as a one of sophisticated techniques to fabricate porous membranes [1] In the following Table 2, some commonly used polymers for separation processes and their fabrication techniques are given.

Table 2: Commonly used polymers and their membrane fabrication techniques (Reproduced with permissions from [1]).

Water treatment Process	Polymers used for membrane fabrication	Fabrication techniques	Average pore size of the membrane
Reverse Osmosis	Cellulose acetate/triacetate	Phase inversion, Solution casting	3–5 Å
	Aromatic polyamide Polypiperzine Polybenziimidazoline		3–5 Å
Nanofiltration	Polyamides	Interfacial polymerization Layer-by-layer deposition Phase inversion	0.001–0.01 μm
	Polysulfones Polyols Polyphenols	Phase inversion Solution wet-spinning	
Ultrafiltration	Polyacrylonitrile (PAN)		0.001 – 0.1 μm

	Polyethersulfone (PES)		
	Polysulfone (PS)		
	Polyethersulfone (PES)		
	Poly(phthazine ether sulfone ketone) (PPESK)		
	Poly(vinyl butyral)		
Microfiltration	Polyvinylidene fluoride (PVDF)	Phase inversion Stretching Track-etching	0.1–10 μm
	Poly(tetrafluorethylene) (PTFE)	Phase inversion Stretching Track-etching	
	Polypropylene (PP)	Phase inversion Stretching Track-etching	
	Polyethylene (PE)	Phase inversion Stretching Track-etching	
Membrane distillation	PTFE PVDF	Phase inversion Stretching Electrospinning	0.1–1 μm

5.1   Types of polymeric membranes

Membranes are semi permeable which can separate two phases of matter and can selectively allow certain materials to pass while others to retain. Few controlling factors of membrane efficiency are its chemical composition, pressure, temperature, internal reactions etc. [19]. Selectivity of the membrane depends upon particle dimension, pore size, the diffusivity of solute in the matrix and electric charges associated [21]. Separation performance of the membrane depends upon pressure, temperature, feed flow, chemical composition of the membrane and its surface [22]. Most important membrane characteristics include thickness, porosity, pore diameter, solvent permeability, density, symmetry or asymmetry, mechanical and chemical resistance etc. [23]. Symmetrical membranes have uniform cross sectional pore sizes unlike asymmetric membranes which have larger pores [24, 25]. Morphologically, membranes can be of two categories i.e. dense and porous. In case of the dense membranes, the components undergo both dissolution and diffusion across the membrane while in porous membranes, the permeate transport in continuous fluid phase filling the pores of the membrane [26]. The other membranes are of two types: organic (e.g., cellulose acetate, polyamide etc.) or inorganic, like the metallic or ceramic membranes [25].

Membranes are of several generations:

•   Membranes derived from cellulose acetate belong to first generation. These are pH, heat, microbe and disinfectant sensitive [21].

•   Membranes derived from synthetic polymers belong to second generation. Such membranes are resistant to strong acids and bases, high temperatures and hydrolysis [21]. However, they are susceptible to mechanical compacting.

•   Membranes of ceramic materials deposited on graphite surfaces belong to third generation. They are tolerant to high temperature, total pH range, high pressure and mechanical compacting [21].

•   A hybrid of conventional electrodialysis and membranes with different pore sizes belong to fourth generation [21]. The process is termed as continuous electrophoresis with porous membranes (CEPM), and the driving force in here is a directly applied electric field [21, 27-28].

6.   Membranes characterization tools

Nowadays, polymeric membranes are widely used technologies in gas separation, recovery, water purification and wastewater treatment. In order to ensure better and desirable membrane performances, these membranes need extensive characterization using various techniques. This help to understand their structure and chemical properties of membranes and confirm the quality and purity of the prepared membranes [29]. There are a lot of important techniques to understand the physical structure of polymeric membranes which are hereby discussed.

6.1   Scanning electron microscopy (SEM)

SEM is a technique almost similar to an optical microscope but only consists of electrons instead of light [30]. Prepared membranes are morphologically and topographically characterized by SEM. This technique is capable of determining the pore size of the porous polymeric membranes as well as measuring the thickness of layers in dense membranes [31]. In order to protect the polymers from burning, generally, polymeric membranes are thin-coated by gold or platinum which makes the membrane conductive. In SEM, data are presented as surface image which is expected to be free of cracks and holes. It is a surface of uniform structure having small holes in case of porous membranes. Average pore size is calculated with MATLAB software. Composite membranes made of two or more than two polymers can also be analyzed by SEM [32]. A porous support holds a thin layer of selective polymer and cross-section surface examination gives structural information of two polymers. In case of polymeric gas separation membrane, the transport mechanism is based on solution-diffusion model. In this case, the fully dense surface is free from fully penetrating holes and cross-section analysis is done in several reports [33]. SEM detects the thickness of the dense layer to calculate gas permeability. For undergoing SEM technique, the sample must be solid in form and free of moisture as SEM is operated at vacuum and it can cause sample loss by evaporation [34, 35]. The samples are either to be electrically conductive or they must be coated. SEM does not measure membrane pore size in gas operation membrane. It measures the films of thickness more than 10 nm [36]. A SEM micrograph is given in Fig. 2 which shows a porous membrane structure.

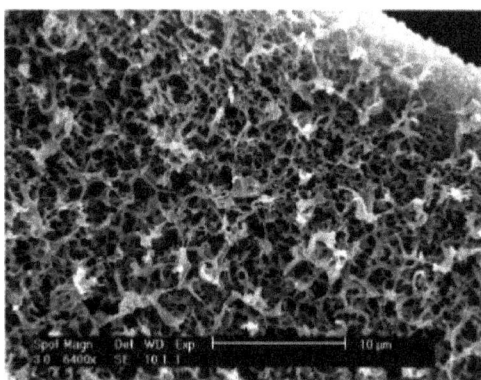

Figure 2. SEM of polypropylene MF membrane where porous membrane structure is visible. Reproduced with permission from [20].

6.2   Transmission electron Microscopy (TEM)

TEM is another technique with magnification reaching 50 million and making it suitable for nano scale measurements [37]. In TEM, electrons are shot from an electron gun pointed at the sample. Electrons pass through the sample and these transmitted electrons create an image in a fluorescent screen that could give the details of internal membrane structure [37]. TEM is extremely useful in analysis of mixed-matrix membranes [38]. The nanoparticles to be uniformly distributed through the membrane can also be characterized. Agglomeration of particles as a result of molecular electrostatic forces could be analysed by TEM [39]. TEM requires special preparation of sample unlike SEM. Sample should be within 2.5 mm in diameter and up to 100 micron thick. Volatile samples cannot be used as it too causes sample evaporation and damage to the instrument. TEM only provides 2D image and does not give topographical data of the sample surface.

6.3   X-ray diffraction (XRD)

XRD is a unique technique which helps in the study of crystal structure of the polymeric membrane and the distance between the polymeric chains [39]. Mechanical properties can also be elicited by XRD. Besides, XRD identifies the chemical compounds, which help to confirm the sample purity [40]. In this technique, a tungsten filament generates X-ray beams, which is directed to the sample and the scattered X-ray is thereafter detected and analysed [40]. Bragg's law is used to calculate the diffraction angle [41, 42]. XRD represents data by intensity versus diffraction peak angle.   Membranes having filler content can be analysed for their structural properties through XRD [43]. These structural properties are important to characterize in determining gas permeation through these membranes [44, 45]. XRD accurately measures large crystalline structures as compared with small structures [46] and does not strongly interact with lighter elements [47].

6.4   Fourier-transform infrared spectroscopy (FTIR)

This is a unique technique which helps to determine the functional groups of polymers. This method is applicable on gases, liquids and solids [48] and it is a quantitative method. In FTIR, an infrared radiation is sent to the sample using a reference solvent and some of the radiation from the sample will absorb while passing through or transmitting the rests. The Fourier transform function is a mathematical model which converts the data into spectrum. Another method to generate FTIR spectrum without using reference material is attenuated total reflection (ATR) [49]. FTIR gives data as the transmittance versus the wavelength. When the sample absorbs the entire radiation, the transmittance is zero and consequently 100% transmittance represents that the sample and the reference absorbed equal amounts of radiation. FTIR is very unique for tracing the hydroxyl, carboxylic acids and others. FTIR is capable of differentiating such chemical groups based on their wavelength [49]. This technique also determines the compatibility between the polymer and the solvent [50]. For mixed matrix membranes, FTIR determines the interaction of polymer and fillers and it is studied by the detection of functional groups of the fillers within the FTIR spectrum [51-53].

7.   Modules for polymeric membrane testing

The arrangement of a membrane into devices and hardware is called "membrane module", and it is used for the separation of the feed stream into retentate and permeate stream. The retentate means the remaining product after the filtration and the solution or compound which passes through the membrane in the process of filtration is called permeate. Module design is an important operation in the membrane separation process. Two types of membranes will be used for the design of a module: a) flat sheet membrane and b) tubular membrane. There are two important things which will affect or lead to the range of the module, one will be the size of the membrane and second one is the method which is used for the packaging of a membrane into device. Nowadays, the membrane separation has been installed from small laboratories to big scale industries also. Here some of the separation techniques use e.g. membrane modules (RO), pervaporation (PV), (MF), (UF) and electrodialysis (ED) [54]. There are five basic designs that will be available in the membrane modules namely a) spiral wound, b) hollow fiber, c) capillary, d) plate and frame and e) tubular membrane.

Every membrane module has its own applications. For example, in RO separation and (PV) techniques; spiral wound, hollow fiber and plate and frame membrane modules are used. In UF and MF separation process, plate and frame, tubular, capillary membrane modules are used. Particularly in ED separation process, the plate and frame membrane modules are used. Even if the same membrane module design is using for different separation process, which means the design for the both separation processes is not same. For example, the plate and frame module designed for (ED) is quite different to the design make for pervaporation. The details of different membrane modules uses are highlighted in Table 3 and a schematic representation is given in Fig. 3.

Table 3. Membrane module testing for separation processes.

Types	Spiral Wound	Hollow fiber	Plate and Frame	Tubular	Capillary
RO	Yes	Yes	Yes	Yes	No
PV	Yes	Yes	Yes	No	No
ED	No	No	Yes	No	No
UF	Yes	No	Yes	Yes	Yes
MF	No	No	Yes	Yes	Yes

7.1   Spiral wound modules

The spiral wound module is manufactured by winding the permeate spacers and feed spacers along with the flat sheet membranes around a perforated central pipe. As mentioned earlier that this module is mainly used for RO, pervaporation and UF operations. The replacement of module elements is easily possible and further scaled up for bulk scale operations which are the major advantages in spiral wound module. Spiral wound module cannot be mechanically cleaned as hollow fibre module and because of that the particular material has a low tolerance. The spiral wound modules has limited applications in UF because of low tolerance by increasing the width space between the membranes and the tolerance will be high for the UF [54].

7.2   Hollow fibre module

The hollow fibre module is manufactured by packing the many membranes in cylindrical shape. It is of two types either out-side (shell side) feed types or in-side (Bore side) feed types as shown in Fig. 3. The out-side-feed-type is used in case of high-pressure conditions and in-side-feed-type is used in case of low and medium pressure conditions [55]. Hollow fibre module consists of two layers, one is an active layer having the thickness of greater than 40 nm and another one is a porous non-selective support layer with thickness of 200 micro meters [56]. Hollow fiber modules have largest applications in membrane science. The main and important thing about hollow fiber module is that the feed stream needs to be very clean.

7.3   Capillary module

The capillary module is also like hollow fibre module. In capillary module, the fiber has larger diameter than for the RO. In this modular design, the feed flow will be arranged inside the feed flow as like active layer skin, which gives good control of concentration polarisation. Modules are fixed in shells with 0.8-1.0 meter in diameter and approximately 1.0 meter length. The packing densities are lower than RO unit because of its high diameter size. The capillary modules are mainly used in UF method. The capillary membranes for UF utilize polymers including polyacrylonitrile, polysulphone, chlorinated poly olefins etc. [54]. In certain cases, capillary membrane modules are also used in MF and UF. The fiber size (approximately 600 micron) is greater than the fiber size used in UF.

7.4   Plate and frame module

The plate and frame module is one of the earliest modules in membrane science. This module contains the membranes, feed and product spacers that are combined together in a metallic frame [57]. The main applications of plate and frame module are ED and UF, and the membrane arranged in flat sheet form. The spacers will not stick together with the membrane and they provide channels for the free flow of feed and product. But this is not using regularly in the present cases; this will specially be employed for the purposes of waste water treatment with higher levels of suspended solids like landfill leachate [55].

7.5   Tubular module

The tubular module consists of a straight and long membrane tube surrounded by a support tube and a porous support layer. The flow of feed is from internally along with the tube and the permeate passes through the membrane then into the porous layer and through appropriate holes in the support tube [54]. The main applications of the tubular module are UF and MF. The development of inorganic membrane materials in tubular form module is a particular significance in UF applications [54, 58].

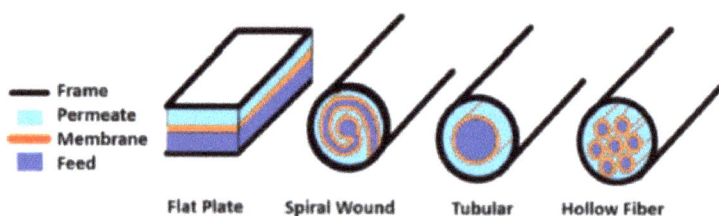

Figure 3. A schematic representation of some polymeric membrane module structures. (Reproduced with permission from [20].

Conclusion

Membrane technologies are used worldwide for purification purposes. These membranes are processed from varied types of polymeric materials. Among the materials used, polymers are groups of widely sought materials for membrane fabrications [59-67]. These membrane structures are determined by various morphological techniques. The applications of these membranes are determined by various membrane testing modules. Thus, these polymeric membranes have a huge potential for membrane separation processes especially in reclaiming wastewater and purification of drinking water [68-74]. In order to achieve sustainability in our technologies application of such novel processes is required worldwide.

Acknowledgement

SS would like to that DST (SEED) division (Sanction No: SP/YO/2019/1283G) for providing financial support during the writing tenure of this chapter

References

[1]   B.S. Lalia, V. Kochkodan, R. Hashaikeh, N. Hilal, A review on membrane fabrication: Structure, properties and performance relationship, Desalination 326 (2013) 77-95. https://doi.org/10.1016/j.desal.2013.06.016

[2]   C. de Morais Coutinho, R.C. Chiu, R.C. Basso, A.P. Ribeiro, L.A. Gonçalves, L.A. Viotto, State of art of the application of membrane technology to vegetable oils: A review, Food. Res. Int. 42 (2009) 536-550. https://doi.org/10.1016/j.foodres.2009.02.010

[3]   R. Das, M. Khayet, Nanotechnology based platforms for efficient water desalination, Desalination 451 (2019) 1-1. https://doi.org/10.1016/j.desal.2018.11.011

[4]   R. Das, M. Kuehnert, A. Sadat Kazemi, Y. Abdi, A. Schulze, Water softening using a light responsive, spiropyran-modified nanofiltration membrane, Polymers 11 (2019) 344 (1-10). https://doi.org/10.3390/polym11020344

[5]   R. Das, Nanohybrid catalyst based on carbon nanotube, Springer, New York, USA, 2017. https://doi.org/10.1007/978-3-319-58151-4

[6]   C. Y. Pan, G.R. Xu, K. Xu, H.L. Zhao, Y. Q. Wu, H.C. Su, J.M. Xu, R. Das, Electrospun nanofibrous membranes in membrane distillation: Recent developments and future perspectives, Sep. Purif. Technol. 221 (2019) 44-63. https://doi.org/10.1016/j.seppur.2019.03.080

[7]   S. B. Abd Hamid, S. K. Zain, R. Das, G. Centi, Synergic effect of tungstophosphoric acid and sonication for rapid synthesis of crystalline nanocellulose, Carbohyd. Polym. 138 (2016) 349-335. https://doi.org/10.1016/j.carbpol.2015.10.023

[8]   G.R. Xu, J.M. Xu, H. C. Su, X. Y. Liu, H. L. Zhao, H. J. Feng, R. Das, Two-dimensional (2D) nanoporous membranes with sub-nanopores in reverse osmosis desalination: Latest developments and future directions, Desalination 451 (2019) 18-34. https://doi.org/10.1016/j.desal.2017.09.024

[9]   M.R. Kessler, Polymer matrix composites: A perspective for a special issue of polymer reviews, Polym. Rev. 52.3 (2012) 229-233. https://doi.org/10.1080/15583724.2012.708004

[10]   T. X. Mei, D. Rodrigue, A review on porous polymeric membrane preparation. Part II: Production techniques with polyethylene, polydimethylsiloxane, polypropylene, polyimide, and polytetrafluoroethylene, Polymers 11.8 (2019) 1310 (1-35). https://doi.org/10.3390/polym11081310

[11] Information on https://ncert.nic.in/ncerts/l/lech206.pdf.

[12] S. Ayaz, H.Y. Yu, Investigation of thermo-mechanical behavior, proton transfer and methanol permeation of polymer electrolyte membrane in low sulfonated state modified with thermally stable surface functionalized graphene oxide nanosheets, Polym. Test. 93 (2021) 106941 (1-9). https://doi.org/10.1016/j.polymertesting.2020.106941

[13] N.I.M. Nawi, M.R. Bilad, N. Zolkhiflee, N.A.H. Nordin, W.J. Lau, T. Narkkun, K. Faungnawakij, N. Arahman, T.M.I. Mahlia, Development of A Novel Corrugated Polyvinylidene difluoride Membrane via Improved Imprinting Technique for Membrane Distillation, Polymers 11 (2019) 865 (1-13). https://doi.org/10.3390/polym11050865

[14] S. Tungrapa, T. Pungparn, M. Weerasombut, I. J. Jangchud, P. Fakum, S. Semongkhol, C. Meechaisue, P. Supaphol, Cellulose acetate fibers: effect of solvent system on morphology and fiber diameter. Cellulose 14 (2007) 563-575. https://doi.org/10.1007/s10570-007-9113-4

[15] P. Yadav, N. Ismail, M. Essalhi, M. Tysklind, D. Athanassiadis, N. Tavajohi, Assessment of the environmental impact of polymeric membrane production, J. Membr. Sci. 622 (2021) 118987 (1-8). https://doi.org/10.1016/j.memsci.2020.118987

[16] E. Kianfar, V. Cao, Polymeric membranes on base of PolyMethyl methacrylate for air separation: a review, J. Mater. Res. Technol. 10 (2021) 1437-1461. https://doi.org/10.1016/j.jmrt.2020.12.061

[17] K. Amulya, R. Katakojwala, S. Ramakrishna, S.V. Mohan, Low carbon biodegradable polymer matrices for sustainable future, Composites Part C: Open Access 4 (2021) 100111 (1-13). https://doi.org/10.1016/j.jcomc.2021.100111

[18] Z. Gao, Y. Wang, H. Wu, Y. Ren, Z. Guo, X. Liang, Y. Wu, Y. Liu, Z. Jiang, Surface Functionalization of Polymers of Intrinsic Microporosity (PIMs) Membrane by Polyphenol for Efficient CO2 Separation, Green Chem. Eng. 2 (2021) 70-76. https://doi.org/10.1016/j.gce.2020.12.003

[19] H. Strathmann, The use of membranes in downstream processing, Food Biotechnol. 4 (1990) 253-272. https://doi.org/10.1080/08905439009549739

[20] D. M. Warsinger, S. Chakraborty, E.W. Tow, M.H. Plumlee, C. Bellona, S. Loutatidoi, H. A. Arafat, A review of polymeric membranes and processes for potable

water reuse, Prog. Polym. Sci. 81 (2018) 209-237.
https://doi.org/10.1016/j.progpolymsci.2018.01.004

[21]   M. Cheryan, Ultrafiltration and microfiltration handbook, CRC press, USA, 1998.
https://doi.org/10.1201/9781482278743

[22]   L. Lin, K.C. Rhee, S.S. Koseoglu, Bench-scale membrane degumming of crude
vegetable oil: Process optimization, J. Membr. Sci. 134 (1997) 101–108.
https://doi.org/10.1016/S0376-7388(97)00098-7

[23]   B. Ostegaard, Applications of membrane processing in the dairy industry. In D.
MacCarthy (Ed.), Concentration and drying of foods, Oxford: Elsevier Applied
Science Publishers, 1989, pp. 133–145.

[24]   Q. Liu, G. R. Xu, R. Das, Inorganic scaling in reverse osmosis (RO) desalination:
Mechanisms, monitoring, and inhibition strategies, Desalination 468 (2019) 114065
(1-17). https://doi.org/10.1016/j.desal.2019.07.005

[25]   B. Pan, X. Zhang, Z. Jiang, Z. Li, Q. Zhang, J. Chen, Polymer and polymer based
nanocomposite adsorbents for water treatment, in R. Das (Ed.), Polymeric materials
for clean water, Springer, Cham, 2019, pp 93-120. https://doi.org/10.1007/978-3-030-
00743-0_5

[26]   C. C. Pereira, A. C. Habert, R. Nobrega, C. P. Borges. New insights in the removal
of diluted volatile organic compounds from dilute aqueous solution by pervaporation
process, J. Membr. Sci. 138 (1998) 227-235. https://doi.org/10.1016/S0376-
7388(97)00225-1

[27]   S.I. Nakao, Determination of pore size and pore size distribution: Filtration
membranes, J. Membr. Sci. 96 (1994) 131-165. https://doi.org/10.1016/0376-
7388(94)00128-6

[28]   W. H. Modler, Milk processing, in: N. Shuryo and W. H. Modler (Eds), Food
protein processing applications, Wiley: VCH inc., 2000, pp 1-88.

[29]   B.D. Freeman, Basis of permeability/selectivity tradeoff relations in polymeric gas
separation membranes, Macromolecules 32 (1999) 375–380.
https://doi.org/10.1021/ma9814548

[30]   M. R. Scheinfein, J. Unguris, H. K. Michael, D. T. Pierce, R.J. Celotta, Scannng
electron microscopy with polarization analysis (SEMPA), Rev. Sci. Instrum. 61(1990)
2501-2527. https://doi.org/10.1063/1.1141908

[31]   R. Ziel, A. Haus, A. Tulke, Quantification of the pore size distribution (porosity profiles) in microfiltration membranes by SEM, TEM and computer image analysis, J. Membr. Sci. 323 (2008) 241–246. https://doi.org/10.1016/j.memsci.2008.05.057

[32]   Y. Ren, J. Zhu, S. Cong, J. Wang, B. Van der Bruggen, J. Liu, Y. Zhang, High flux thin film nanocomposite membranes based on porous organic polymers for nanofiltration, J. Membr. Sci. 585 (2019) 19–28. https://doi.org/10.1016/j.memsci.2019.05.022

[33]   S. Taheri, M. Ams, H. Bustamante, L. Vorreiter, M. Withford, S M. Clark, A practical methodology to assess corrosion in concrete sewer pipes, in: MATEC Web of Conferences, EDP Sciences, 2018, 199, pp 6010 (1-4). https://doi.org/10.1051/matecconf/201819906010

[34]   X. Wencheng, J. Yang, C. Liang, Investigation of changes in surface properties of bituminous coal during natural weathering processes by XPS and SEM, Appl. Surf. Sci. 293 (2014) 293-298. https://doi.org/10.1016/j.apsusc.2013.12.151

[35]   R. Ziel, A. Haus, A. Tulke, Quantification of the pore size distribution (porosity profiles) in microfiltration membranes by SEM, TEM and computer image analysis, J. Membr. Sci. 323, 2 (2008) 241-246. https://doi.org/10.1016/j.memsci.2008.05.057

[36]   B. Chakrabarty, A. K. Ghosal, M. K. Purkait. SEM analysis and gas permeability test to characterize polysulfone membrane prepared with polyethylene glycol as additive, J. Colloid Interface Sci. 320 (2008) 245-253. https://doi.org/10.1016/j.jcis.2008.01.002

[37]   P.S. Goh, A.F. Ismail, S.M. Sanip, B.C. Ng, M. Aziz, Recent advances of inorganic fillers in mixed matrix membrane for gas separation, Sep. Purif. Technol. 81 (2011) 243–264. https://doi.org/10.1016/j.seppur.2011.07.042

[38]   F.H. Akhtar, M. Kumar, L.F. Villalobos, H. Vovusha, R. Shevate, U. Schwingenschlogl, K.V. Peinemann, Polybenzimidazole-based mixed membranes with exceptional high water vapor permeability and selectivity, J. Mater. Chem. A. 5.41 (2017) 21807–21819. https://doi.org/10.1039/C7TA05081J

[39]   C.Y. Tang, Z. Yang, Transmission electron microscopy, in: N. Hilal, A. F. Ismail, T. Matsuura., D. A. Radcliffe (Eds.), Membrane Characterization, Elsevier, 2017, pp. 145-159. https://doi.org/10.1016/B978-0-444-63776-5.00008-5

[40] M. Saberi, P. Rouhi, M. Teimoori, Estimation of dual mode sorption parameters for CO2 in the glassy polymers using group contribution approach, J. Membr. Sci. 595 (2020) 117481. https://doi.org/10.1016/j.memsci.2019.117481

[41] M. T. Weller, Inorganic Materials Chemistry; Oxford University Press: Oxford, UK, 1995.

[42] C. Suryanarayana, M.G. Norton, X-ray Diffraction: A Practical Approach; Springer: New York, NY, USA, 2013.

[43] Y. Alqaheem, A. Alomair, A. Alhendi, S. Alkandari, N. Tanoli, N. Alnajdi, A. Quesada-Peréz, Preparation of polyetherimide membrane from non-toxic solvents for the separation of hydrogen from methane, Chem. Cent. J. 12 (2018) 1-8. https://doi.org/10.1186/s13065-018-0449-7

[44] Q. Shen, S. Cong, R. He, Z. Wang, Y. Jin, H. Li, X. Cao, J. Wang, B. Van der Bruggen, Y. Zhang, SIFSIX-3-Zn/PIM-1 mixed matrix membranes with enhanced permeability for propylene/propane separation, J. Membr. Sci. 588(2019) 117201 (1-7). https://doi.org/10.1016/j.memsci.2019.117201

[45] S. Thomas, D. Rouxel, D. Ponnamma, Spectroscopy of Polymer Nanocomposites, William Andrew Publishing:Oxford, UK, 2016.

[46] A. Ali, Failure Analysis and Prevention, IntechOpen: London, UK, 2017. https://doi.org/10.5772/65149

[47] J.H. Fendler, Nanoparticles and nanostructured films: Preparation, Characterization, and Applications, Wiley: Hoboken, NJ, USA, 2008

[48] T. Xiao, H. Yuan, Q. Ma, X. Guo, Y. Wu, An approach for in situ qualitative and quantitative analysis of moisture adsorption in nanogram-scaled lignin by using micro-FTIR spectroscopy and partial least squares regression, Int. J. Biol. Macromol. 132 (2019) 1106–1111. https://doi.org/10.1016/j.ijbiomac.2019.04.043

[49] D.W. Sun, Modern Techniques for food Authentication, Elsevier Science: Amsterdam, The Netherlands, 2008.

[50] S. Hansen, S Pedersen-Bjergaard, K. Rasmussen, Introduction to Pharmaceutical Chemical Analysis, Wiley: Hoboken, NJ, USA, 2011. https://doi.org/10.1002/9781119953647

[51] J.K. Adewole, A.L. Ahmad, S. Ismail, C.P. Leo, A.S. Sultan, Comparative studies on the effects of casting solvent on physico-chemical and gas transport properties of

dense polysulfone membrane used for CO2/CH4 separation, J. Appl. Polym. Sci. 132 (2015) 42205 (1-10). https://doi.org/10.1002/app.42205

[52]   H. Zhu, L. Wang, X. Jie, D. Liu, Y. Cao, Improved interfacial affnity and CO2 separation performance of asymmetric mixed matrix membranes by incorporating postmodified MIL-53(Al), ACS Appl. Mater. Interfaces 8 (2016) 22696–22704. https://doi.org/10.1021/acsami.6b07686

[53]   J. Grdadolnik, ATR-FTIR spectroscopy: Its advantages and limitations, Acta Chim. Slov. 49 (2002) 631–642.

[54]   B C. Smith, Fundamentals of Fourier Transform Infrared Spectroscopy, Taylor & Francis: Abingdon-on-Thames, UK, 1995.

[55]   K. Scott, Hand Book of Industrial membranes, First ed., Elsevier, Netherlands, 1998.

[56]   R.W. Baker, Membrane technology and applications. Wiley online Library: Hoboken, NJ, USA, 2012. https://doi.org/10.1002/9781118359686

[57]   J. Balster, E. Drioli, L. Giorno, Hollow fiber membrane module, in: E. Drioli, L. Giorno (Eds.), Encyclopedia of membranes, Springer, 2016, pp. 959-957. https://doi.org/10.1007/978-3-662-44324-8_1583

[58]   B. Gu, D.Y. Kim, J.H. Kim, D.R. Yang, Mathematical model of flat sheet membrane modules for FO process: Plate-and-frame module and spiral-wound module, J. Membr. Sci. 379 (2011) 403-415. https://doi.org/10.1016/j.memsci.2011.06.012

[59]   E. O. Ezugbe, S. Rathilal, Membrane Technologies in Wastewater Treatment: A Review, Membranes 10 (2020) 89 (1-28). https://doi.org/10.3390/membranes10050089

[60]   Y. Zhu, D. Wang, L. Jiang, J. Jin, Recent progress in developing advanced membranes for emulsified oil/water separation, NPG Asia Mater. 6 (2014) e101 (1-11). https://doi.org/10.1038/am.2014.23

[61]   M.A. Aroon, A.F. Ismail, T. Matsuura, M.M. Montazer-rahmati, Performance studies of mixed matrix membranes for gas separation: a review, Sep. Purif. Technol. 75 (2010) 229-242. https://doi.org/10.1016/j.seppur.2010.08.023

[62]   D. Bastani, N. Esmaeili, M. Asadollahi, Polymeric mixed matrix membranes containing zeolites as a filler for gas separation applications: A review, J. Ind. Eng. Chem. 19 (2013) 375-393. https://doi.org/10.1016/j.jiec.2012.09.019

[63]   R. Faiz, K. Li, Polymeric membranes for light olefin/paraffin separation, Desalination 287 (2012) 82-97. https://doi.org/10.1016/j.desal.2011.11.019

[64]   S. Sengupta, D. Ray, Vegetable oil-based polymer composites: synthesis, properties and their applications, in: V. K. Thakur, M. K. Thakur, M. R. Kessler (Eds.), Handbook of composites from renewable materials, polymeric composites, Wiley, 2017, pp. 441-466. https://doi.org/10.1002/9781119441632.ch120

[65]   C.A. Scholes, S.E. Kentish, G.W. Stevens, Effects of minor components in carbon dioxide capture using polymeric gas separation membranes, Sep. Purif. Rev. 38 (2009) 1-44. https://doi.org/10.1080/15422110802411442

[66]   A.P. Ribeiro, J.M. de Moura, L.A. Gonçalves, J.C. Petrus, L.A. Viotto, Solvent recovery from soybean oil/hexane miscella by polymeric membranes, J. Membr. Sci. 282 (2006) 328-336. https://doi.org/10.1016/j.memsci.2006.05.036

[67]   S. Sridhar, B. Smitha, T.M. Aminabhavi, Separation of carbon dioxide from natural gas mixtures through polymeric membranes-a review, Sep. Purif. Rev. 36 (2007) 113-174. https://doi.org/10.1080/15422110601165967

[68]   Z.Y. Yeo, T.L. Chew, P.W. Zhu, A.R. Mohamed, S.P. Chai, Conventional processes and membrane technology for carbon dioxide removal from natural gas: a review, J. Nat. Gas. Chem. 21 (2012) 282-298. https://doi.org/10.1016/S1003-9953(11)60366-6

[69]   M.M. Aijumaily, M.A. Alsaadi, N.A. Hashim, Q.F. Alsalhy, R. Das, F. Mjalli, Embedded high-hydrophobic CNMs prepared by CVD technique with PVDF-co-HFP membrane for application in water desalination by DCMD, Desalin. Water Treat. 142 (2019) 37-48. https://doi.org/10.5004/dwt.2019.23431

[70]   Y. Kodaira, T. Miura, S. Ito, K. Emori, A. Yonezu, H. Nagatsuka, Evaluation of Crack Propagation Behavior of Porous Polymer Membranes, Polym. Test. 96 (2021), 107124 (1-14). https://doi.org/10.1016/j.polymertesting.2021.107124

[71]   P.C. DeLeo, H. Summers, K. Stanton, M.W. Lam, Environmental risk assessment of polycarboxylate polymers used in cleaning products in the United States, Chemosphere 258 (2020) 127242 (1-9). https://doi.org/10.1016/j.chemosphere.2020.127242

[72] S. Cong, J. Wang, Z. Wang, X. Liu, Polybenzimidazole (PBI) and benzimidazole-linked polymer (BILP) membranes, Green Chem. Eng. 2 (2021) 44-56. https://doi.org/10.1016/j.gce.2020.11.007

[73] F.G. Torres, O.P. Troncoso, F. Piaggio, A. Hijar, Structure–property relationships of a biopolymer network: The eggshell membrane, Acta. Biomater. 6 (2010) 3687-3693. https://doi.org/10.1016/j.actbio.2010.03.014

[74] Z. F. Cui, Y. Jiang, R. W. Field, Fundamentals of Pressure-Driven Membrane Separation Processes, Membr. Technol. (2010) 1-18. https://doi.org/10.1016/B978-1-85617-632-3.00001-X

Polymeric Membranes for Water Purification and Gas Separation   Materials Research Forum LLC
Materials Research Foundations **113** (2021) 33-68   https://doi.org/10.21741/9781644901632-3

Chapter 3

Microfiltration and Ultrafiltration Membranes for Water Purification

Priya Banerjee[1*], Sandipan Bhattacharya[2], Rasel Das[3,4], Papita Das [2,5],
Aniruddha Mukhopadhyay[6]

[1] Department of Environmental Studies, Centre for Distance and Online Education, Rabindra Bharati University, Kolkata 700091, India

[2] Department of Chemical Engineering, Jadavpur University, Kolkata 700032, India

[3] Department of Biochemistry and Biotechnology, University of Science and Technology Chittagong, Foy's lake-4202, Chattogram, Bangladesh

[4] Department of Chemistry, Stony Brook University, Stony Brook, NY 11794, United States

[5] School of Advanced Studies in Industrial Pollution Control Engineering, Jadavpur University, Kolkata 700032, India

[6] Department of Environmental Science, University of Calcutta, Kolkata 700019, India

*prya_bnrje@yahoo.com

Abstract

Random and rampant urbanization have rendered one third of the global population vulnerable to water scarcity. For mitigating this problem, membrane technology has been widely investigated for desalination as well as reuse and reclamation of wastewater. Polymeric membranes have reportedly displayed significantly lower consumption of energy with potential self-cleansing and antifouling. Application of microporous and ultraporous polymeric membranes for water purification has been vividly discussed herein. The diverse facets of water treatment technologies using polymeric membranes compiled herein will facilitate all potential readers like academicians, environmentalists, industrialists and membrane technologists, and to address the water scarcity challenges of our society.

Keywords

Polymers, Biopolymers, Microfiltration, Ultrafiltration, Emerging Pollutants, Desalination

Contents

1.   Introduction

In the last century, water consumption rates had nearly doubled in comparison to the growth of population, thereby making water scarcity a reality for the vast majority of population of the 21st century. The Earth has limited supplies of fresh water which are being rapidly depleted every day. According to a study conducted by the United Nations, two third of the global population of the world will be experiencing severe water crisis by 2025 [1]. The rapid and unplanned urbanization, which is rampant nowadays, often results in the destruction of adjacent water bodies, further worsening the water crisis situation. Other than manmade causes, climate change induced by global warming and other weather extremities have led to the rising of the sea level, causing submergence of land and freshwater resources and increased rates of evaporation. These events have collectively increased the salinity of both fresh and marine water resources thereby resulting in a sharp decline in the availability of freshwater resources. Also, due to the complexities associated with the monitoring and regulation of the components responsible for water pollution, there is a lack of effective water treatment technologies. This in turn results in a dearth of clean potable water for the rapidly growing population [2]. This impending scarcity of water is expected to affect all the aspects of social progress including the health, economy and sustainable development of the country or region under concern. Thus, in keeping with the requirements of the present times, it has been considered imperative to determine economic and efficient technologies for the purpose of treating wastewater in compliance with the regulatory guidelines for reuse in domestic and industrial sectors with the aim of restoring safe environment.

Generally, water treatment is carried out as primary secondary and tertiary processes which include methods of physical separation, microbial treatments and chemical treatments like electrocoagulation, photocatalysis, adsorption, etc. respectively [3].

Nonetheless, majority of the conventional technologies of water treatment has resulted in inefficient removal of pollutants from different streams of wastewater [3].

The drawbacks of each process of wastewater treatment have been shown in Fig. 1. Out of all the water treatment processes discussed so far, water treatment by membrane technologies have been widely applied owing to its chemical and thermal inertness and expedient techniques of regeneration. Due to the prospect of performing water purification up to the level of ionic filtration and many other added advantages, membrane filtration has emerged as one of the more popular methods of water treatment practiced in recent times. As membranes only permit the passage of aqueous solvents, membrane filtration efficiently separates different particles, solutes, gases and other pollutants present in effluents.

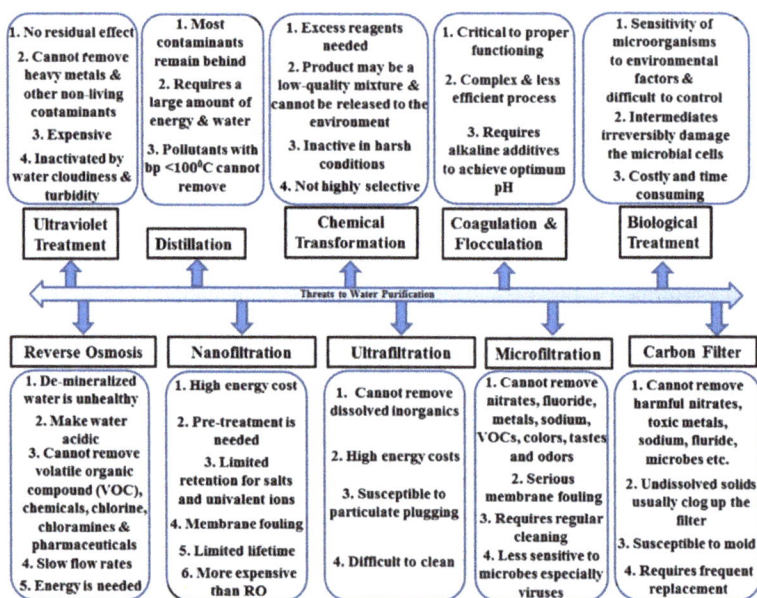

Figure 1. Schematic depiction of major limitations of conventional processes of water purification (adapted from Das et al. [3]).

The membrane separation process that are generally applied for wastewater treatment include distillation, different types of dialysis, micro (MF), ultra (UF) and nano (NF) filtration, osmosis (reverse (RO) and forward (FO)) and vacuum filtration. MF, UF, NF and RO are pressure driven processes that are widely applied for water treatment carried out on large scale. However, the high energy requirement of these processes is a drawback for their wide scale commercial application. Other than that, the precipitation of the pollutants results in the fouling and pore blockage of the membranes, thus reducing their efficiency and shortening their lifetime [3]. The membranes also need periodic cleaning and substitution as they are incapable of self-cleaning.

As per recent studies, the FO membranes used for capacitive deionization and distillation are capable of efficient desalination but are not obtainable for commercial scale applications [4]. Due to these limitations, the scientists are impelled to find alternative technologies for cost effective treatment of wastewater. The success in the field of adsorptive removal of water pollutants has prompted the researchers to explore different biomaterials for the purpose of synthesizing the membranes. Biomaterials generally have low carbon footprint and can be used individually or in the form of composites for manufacturing the membranes. Composites are considered to be advantageous over individual membranes as because on one hand they possess the favorable properties of their parent components and on the other hand they are able to overcome limitations of their individual components due to their composite nature. With reference to these points, inexpensive and easily available biopolymers like polysaccharides, lignin, alginate etc. can be considered as ideal components for the membrane-based separation of pollutants [5]. Furthermore, the rapid advancement in the fields of nanoscience and nanotechnology has prompted the researchers to develop highly functionalized nanocomposite with biomaterials as their base for the removal of emerging pollutants via adsorption. Different polymers reported for synthesis of MF and UF membranes and subsequent implementation of these membranes for water purification and desalination have been extensively reviewed herein.

2.   Polymers for membrane preparation

A polymer is substance or material consisting of repeating subunits of very large molecules or macromolecules or a mixture of macromolecules having a molecular weight higher than 1000 gmol^{-1} [6]. Polymers are assembled in the form of chain(s) containing thousands to millions of repeating units of monomers bearing positive, negative or both charges. The polymers chains may be 400 to 8000 ft in length [6]. A polymer may be cationic, anionic, amphoteric or nonionic according to the charge of its monomers [6]. The overall charge density of the polymer depends upon the charge of its monomers and is usually expressed as percentage. Polymers may be classified as linear, branched or cross linked (as shown in Fig. 2) on the basis of its chain orientation.

Polymers can be classified into biopolymers or chemical polymers on the basis of their parent material. Some of the chemical polymers that have been widely mentioned in recent studies include cellulose acetate (CA), polyacrylonitrile, polyaniline nanoparticles (NPs), poly(arylene ether ketone), polyamide, poly(etherimide), polyethersulfone (PES), polysulfone, polyethylene glycol (PEG), polyimide, poly(methacrylic acid), polyvinyl alcohol (PVA), polyvinyl chloride, and polyvinylidene fluoride (PVDF) [6]. Biopolymers like cellulose, chitin, chitosan etc. have been widely used for membrane synthesis. A large number of studies conducted in recent times have emphasized on improvement of polymeric membranes in terms of antifouling properties, chemical resistance, energy consumption, salt rejection and water flux, and of by blending and surface modification.

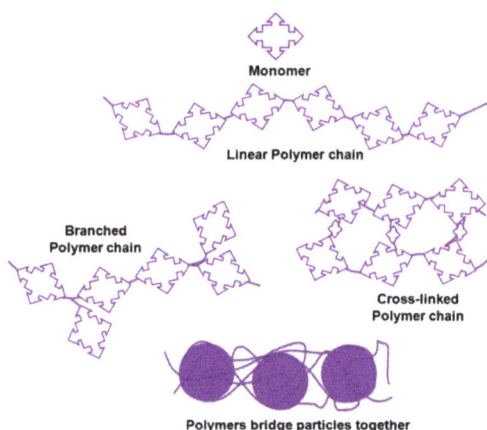

Figure 2. Classification of polymers (adapted from Das [6])

Polymeric membranes are cheaper compared to those made from inorganic materials and ceramics [7]. Also, a higher level of water production can be achieved through polymeric membranes. Polymeric membranes easily fabricated and conveniently aligned as hollow fiber and spiral wounds for achieving optimum efficiency [7,8]. This chapter reviews the application of polymeric materials for water treatment and desalination reported in recent studies. More precisely, the recent developments in the synthesis of polymeric membrane for MF and UF have been discussed herein. The selection of an appropriate polymer for filtration is critical as it is responsible for the permeate quality and cost incurred by the concerned process. Thus, selection of an appropriate polymer ensures the avoidance of issues like frequent replacement of membranes and unjustified consumption of energy. The current limitations of newly synthesized polymers have also been discussed in this chapter.

3.   Polymeric membrane for microfiltration (MF)

MF membranes have been mainly applied for wastewater treatment processes, membrane distillation (MD) and membrane bioreactors (MBRs) [6,9,10,11,12,13]. Industrial and domestic effluents contain toxic organic pollutants (like pharmaceuticals) which exert detrimental impacts on the aquatic species and the general environment alike. Modern technologies like advanced oxidation processes (AOP) have demonstrated high efficiency for the purpose of treating recalcitrant wastewater [6]. Photocatalysis by TiO_2 is one the most renowned AOPs capable of completely degrading the organic molecules [6]. The direct integration of TiO_2 nanoparticles within PES and PVDF MF membranes has been found to enhance the efficiency of the same [6]. Though addition of TiO_2 decreased membrane porosity, it enhanced membrane efficiency for degrading different pollutants like dyes, pharmaceuticals, pesticides, etc.

MD is a temperature driven process used for separating pollutants and water molecules by passing the fluid through a porous membrane [10,14]. MD is carried out on the basis of vapor permeation. Therefore, MD is carried out with hydrophobic membranes. MD is driven by the difference of vapor pressure developing across a hydrophobic membrane. However even though MD is conveniently implemented, it has still not gained wide scale industrial application due to several limitations like fouling and declining flux that are most prominent and difficult to address. However, direct contact MD (DCMD) membranes have recently exhibited an increase in flux when treated by tetrafluromethane (CF_4) plasma surface modification [10,14]. The overall flux of the plasma modified membrane exceeded that of the untreated membranes. CF_4 plasma treatment converted the PVDF membranes to super hydrophobic plasma membranes having increased flux and greater capacity for salt rejection [10,14].

Membrane bioreactor (MBR) has recently gained widescale popularity as a mode of water treatment because of its ability to produce permeates of high quality, noticeable decrease in the generation of sludge, ability to control solids and reducing the operational costs [15,16]. Among its many advantages, a significant disadvantage of MBR operation is membrane fouling [17,18]. In an effort to overcome this, MF membranes are being prepared with GO to produce membranes having antifouling properties [6]. This development is influenced by the unique hydrophilicity and large negative zeta potential of GO which can be ascribed to functional group present on GO surfaces. These properties of GO facilitate permeation of water through the membrane and reduces biofouling [6]. Higher percentage of GO in the membrane increased the negative zeta potential induced by GO, decreased the thickness of the biofilm layer formed on the GO incorporated membrane and significantly increased pure water flux [6]. The high energy demand incurred for avoiding fouling of membrane is an unintended disadvantage of the MBR technology. Thus, in order to alleviate this problem, investigations on osmotic MBRs (OMBRs) have escalated recently [19,20,21]. The OMBRs, are driven by the osmotic pressure gradient formed between the feed and draw sectors of the membrane. Nevertheless, in contrast to conventional MBR, which has a microporous membrane, OMBRs possess a high rejection semipermeable membrane. In comparison to other membrane technologies, OMBRs have lower fouling, but still experience membrane fouling. The increased salinity and the resultant accumulation of salt, the interaction between organic foulants and inorganic ions and the gradient of low soluble salts formed under high ionic strength might give rise to more complicated fouling scenarios [22]. However, the compaction of foulants is milder in OMBRs due to the absence of hydraulic pressure. Hence it may be considered that fouling is effectively reduced in presence of hydrodynamic shear.

4.   Polymeric membranes for ultrafiltration (UF)

Polymeric materials have been widely used for the preparation of UF, PS and PES membranes due to their robust mechanical properties, a very wide operational range of pH and high chemical stability [23,24,25,26,27,28,29]. However, their implementation in the area of water purification is restricted due to their hydrophobic nature which results in reduced membrane permeability. Recent studies have reportedly used naturally hydrophobic polymers like PMAA, PVC and PVDF for fabrication of UF membranes [30,31,32,33,34,35]. Membrane hydrophobicity poses many problems like decline in water flux during an ongoing process as a result of accumulation of organic compounds. These organic compounds in turn facilitate the attachment and growth of microorganisms on the membrane surface causing membrane fouling and resultant membrane failure. Therefore, necessary modifications are needed to be made in the polymeric materials in order to improve their ability for greater performance in the water treatment methods. These modifications increase membrane in turn reducing the probability of membrane fouling in case of liquid-based filtration.

4.1   Incorporation of TiO$_2$ NPs into polysulfone (PSF)

For increasing the hydrophilicity of the UF membranes, hydrophilic materials like NPs and amphiphilic copolymers are incorporated into the membranes by blending and surface modification [6]. According to recent studies, incorporation of modified TiO$_2$

particles in hybrid PSF membranes increase the hydrophilicity of the latter [6,7]. Polymer chains of (2-hydroxyethyl methacrylate) (P(HEMA)) on TiO_2 NPs were grafted via atom transfer radical polymerization for synthesis of these membranes. The reason behind the impregnation of PSF membranes with the modified TiO_2 NPs is enhancing the membrane performance, controlling the NP agglomeration on membrane surface and preventing the NP leakage from membrane during the process of filtration. Modified TiO_2 reportedly demonstrates better dispersibility within the polymer than unmodified TiO_2 [6]. TiO_2-HEMA modified PSF membrane displayed enhanced hydrophilicity, greater water flux and superior antifouling property in comparison with PSF membrane impregnated with unmodified TiO_2 particles.

4.2   Integration of mesoporous silica particles (MSP-1) within polysulfone (PSF)

Addition of inorganic particles into the casting mixture of the membrane before phase inversion is a simple and effective method by which additional particle-based functionalities can be incorporated into a membrane. As such this method of membrane modification is widely studied [36,37]. With the aim of improving the properties of the PSF membrane they are impregnated with surfactant templated MSP-1 with or without PEG as a molecular porogen [6]. The terminal silanol (Si-OH) groups present on the pore walls and external surfaces of MSP-1 particles increased the hydrophilicity of the PSF membranes. Integration of MSP-1 and PEG modified the archetypal morphology of the phase inversion PSF membrane. The PSF-MSP membranes demonstrate mechanical properties and they are equivalent to those sans MSP. Porogen free membranes made from casting solutions having low polymer content demonstrated significantly higher permeate flux when modified with MSP-1. A 5 % increase in the MSP -1 loading to the membrane has a highly damaging effect on the flux, whereas a 10 % increase yielded permeate flux equivalent to MSP-free control. Nonetheless, the inclusion of PEG porogen in the casting mixture yielded no statistically significant changes in terms of flux or rejection. The composite membrane also exhibited superior dextran rejection than MSP-free membrane. Moreover, the composite membrane exhibited lower decline of flux and increased rejection in comparison with its MSP free membranes during the fouling tests carried out with humic acid solutions.

4.3   Integration of silica and zinc oxide (ZnO) NPs within polyvinyl chloride (PVC) and poly(methacrylic acid) (PMAA) matrices

Recent studies have reported an increase in hydrophilicity of PVC UF membranes as a result of integration of NPs into PVC matrices [39]. Five PVC membranes were fabricated with variable ZnO percentages by phase inversion method using water (coagulant) and PEG (pore forming additive). The membranes impregnated with ZnO

NPs exhibited higher hydrophilicity in comparison to pristine PVC membrane. Results revealed an increase in water flux with a corresponding increase in ZnO loading up to 3 wt %. However, any further increase in ZnO percentages resulted in clustering of ZnO particles on the membrane surface leading to a decline of water flux (Fig. 3). Increase in ZnO loadings (upto 3 wt%) also lead to a parallel increase in porosity. However, a further increase in ZnO percentage resulted in a decline of membrane porosity. Pristine PVC membranes recovered 69 % of water flux while membranes impregnated with ZnO (3 wt %) recovered 92 % water flux post bovine serum albumin (BSA) permeation. Impregnation of PMAA with NPs has also been examined recently for enhancing PMAA based UF membrane efficiency. Grafting of superhydrophilic silica NPs into PMAA membrane matrices resulted in an increase in membrane wettability [6].

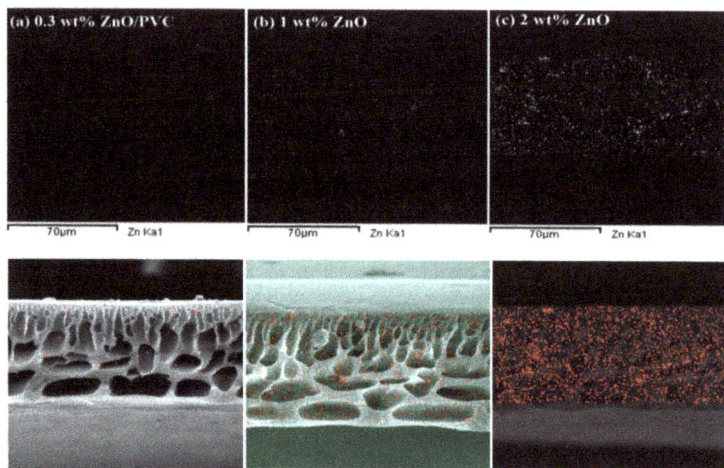

Figure 3. Agglomeration of ZnO NPs in PVC-ZnO casting solution as revealed by Energy dispersive X-ray (EDAX) spectroscopy. Both top and bottom panel have scale of 70 μm (Adapted from Rabiee et al. [39]).

4.4   Polyethersulfone (PES)

PES has been frequently utilized for synthesis of UF membranes [6]. In a recent study, carboxylic and amine group bearing mesostructured silica particles were incorporated into PES membranes for improving hydrophilicity of the latter [40]. Integration of ordered mesoporous silica particles significantly modified membrane pore size, porosity and morphology and also increased membrane hydrophilicity. Enhanced membrane hydrophilicity and surface porosity in turn increased water permeation. The antifouling property of the membrane also increased due to this modification, especially against irreversible fouling without any negative effect against the protein rejection potential of the membrane. The modified membrane yielded stable permeation performance throughout repeated stability tests. Another recent study has proposed polyether sulfone amide (PESA), a PES based hydrophilic polymer for membrane synthesis [41]. PESA was subjected to interfacial polymerization using gallic acid (GA) and 3,5-diaminobenzoic acid (DBA). PESA membranes demonstrated more hydrophilicity in comparison to pure PES membrane. Functionalization of PESA membranes using DBA and GA made the same more hydrophilic. PESA and functionalized PESA membranes possessed greater roughness than pure PES membranes. Moreover, PESA membranes were also more efficient than PES membranes in terms of antifouling properties, pure water flux and humic acid rejection [41].

Incorporation of polyaniline (PANI) NPs was found to enhance the hydrophilicity of PES UF membranes [6]. Incorporation of PANI NPs also enhanced the fouling resistance, flux recovery, pollutant rejection and antimicrobial potential of the PES membranes. Moreover, PANI synthesized indigenously by authors was found to yield superior membrane properties in comparison to commercial PANI. PES UF membranes can also be modified using polyethylene glycol and silver NPs [42]. For some separation process, hollow polymeric membranes were favored due to convenient module fabrication, excellent flexibility, high surface area and self-mechanical support [42]. The modified membrane demonstrated enhanced antimicrobial properties, hydrophilicity and water flux. Dextran-grafted halloysite nanotubes (HNTs) have also been used for modification of PES UF membranes [43]. HNT modified membranes reportedly demonstrate enhanced antifouling property, hydrophilicity and water flux in comparison to unmodified PES membranes [43]. Moreover, HNT modified membranes were found to exhibit slightly lower porosity but greater pore size in comparison to that of PES membranes.

4.5   Polyvinylidene fluoride (PVDF)

PVDF also enhances membrane hydrophilicity. PVDF UF membranes have also been modified with dopamine solution via self-polymerization [44]. Any unreacted polydopamine was rinsed off with water. The dopamine coated PVDF membrane was further dipped in a TiO_2 NP solution whereby Dopamine expedited the uniform distribution and attachment and prevented agglomeration of TiO_2 NPs on PVDF membrane surface. The notable reduction in the water contact angle of the membrane denoted an increase in its hydrophilicity due to its modification. Both the pure water flux and the BSA rejection capacity of the membrane increased significantly due to the modifications. A significant improvement of antifouling properties of the membrane was indicated by the enhanced flux recovery ratio (> 90 %) and reduced irreversible fouling ratio [44].

4.6   Poly(arylene ether ketone) (PAEK) and poly(ether imide) (PEI)

PAEK and PEI have also been investigated for preparation of UF membranes [45, 46]. PAEK membranes functionalized with carboxylic acid groups (PAEK-COOH) have been found to be more hydrophilic than traditional PAEK membranes [46]. PAEK-COOH was then used for the synthesis of tight UF membranes via the process of nonsolvent induced phase inversion. The resultant membrane displayed high water permeation flux and dye rejection. The membrane was also thermally stable and was suitable for high temperature filtration application. These membranes exhibited a flux recovery rate of 91.5% for BSA. They also demonstrated impressive antifouling potential. The adsorption rate of these membranes was found to be lower than 5.0 % for all tested dyes.

In a recent study, PEI was blended with N-phthaloylated chitosan (NPHC) in order to enhance the hydrophilicity and antifouling properties of the resultant UF membranes [45]. The modified membranes also demonstrated higher surface roughness as compared with the unmodified membranes. Pure water and permeate flux as well as heavy metal ion separation, of the PEI–NPHC membranes were increased with a corresponding increase in NPHC concentration of blended membranes. An NPHC concentration of 2.0 wt % yielded maximum flux recovery [45].

5.   Biopolymer based membranes

Biopolymers are widely synthesized from vegetable (alginate, cellulose, polyisoprene, starch, etc.) or animal (polybutylene succinate, polyhydroxyalcanoates, polylactic acid, etc.) derivatives [47]. Biopolymers like chitin, chitosan, collagen and sericin are obtained via bacterial fermentation [47]. The application of biopolymers for the "green" synthesis of membranes have been well studied and well documented in scientific literature. However, utilization of biopolymers on a commercial scale is still a challenge. The advantages of using bio-based solvents include low consumption of energy sources, biodegradability and reduced toxicity [48]. Nevertheless, the cost of replacing the standard technology and the corresponding production lines in use for green synthesis of plastics from petrochemical compounds are extremely high. Moreover, the existing rate of biopolymer production is insufficient for meeting the large requirement of the membrane industry for the same. However, synthetic polymers, synthesized from petroleum materials, cause a massive surge in pollution of the environment which in turn prompts the researchers to actively pursue an interest in biopolymer-based membranes due to their sustainable nature. Traditional plastics generally require an immensely long time to degrade, resulting in secondary pollution and high cost of treatment. Therefore, in order to solve these problems, an effort must be taken to substitute petroleum-based membrane with more biocompatible alternatives. Natural materials can be used as efficient substitutes of petroleum-based compounds. These materials occur in abundance and are renewable. Further research is required for harnessing the true potential of these resources. Few such biopolymers investigated for membrane synthesis has been discussed as follows.

5.1   Biopolymers obtained via bacterial fermentation

5.1.1   Polylactic acid (PLA)

PLA is a widely investigated bioplastic with commercial applications. This biodegradable polyester is largely derived from lactic acid. The amorphous and crystalline structure of PLA is due to the result of copolymerization of L and D-lactic acid which impacts the properties of resultant PLA [49]. PLA could be synthesized by ring opening polymerization of cyclic lactide monomer or direct polycondensation of hydroxyl acid. The utilization of PLA as a constituent for the synthesis of membrane synthesis is a comparatively recent idea. PLA-MF membranes prepared by thermally induced phase separation (TIPS) had pore sizes ranging from 0.6 to 4.4 μm [47]. The membrane performance was tested in terms of bacteria cell retention and protein molecule permeation [50]. The synthesis and drying of the polymer solution prior to quenching caused variations in the structures of pores present near the surface of the membranes which in turn enhanced bacterial cell rejection by the membrane. These membranes are synthesized from recyclable and compostable polymers. Hence, they are considered as justifiable alternatives to conventional synthetic polymer membranes usually employed for water treatment. Porous PLA membranes with PEG as a pore former were produced by Chitarattha and Phaechamud [51]. Authors also investigated the effect of non-solvents like ethanol, glycerin and isopropanol on synthesis of membranes in terms of the mechanism of pore formation in the PLA matrices [52]. They successfully tailored the membrane morphology and efficiency as required for the final application of the membranes by altering the production parameters of the same. Interestingly, Moriya et al. [53] exhibited an interesting option for synthesizing PLA having high hydrophilicity by utilizing PEG as a porogen. Modification of poly(ε-caprolactone) (PCL) with PLA showed lower membrane resistance towards yeast cells than unmodified ones [47]. Minbu et al. [54] used Tween 80 (surfactant) and 1,4-dioxane (solvent) for synthesis of MF-PLA membranes via the NIPS technique. A recent study proposed the synthesis of porous PLA membranes by hot water droplet induced phase separation [55].

5.1.2 Polyhydroxyalkanoates (PHA)

PHA are linear bioplyesters of 3,4,5 and 6 hydroxyalkanoic acids which are produced from the fermentation of different substrates used as carbon sources by both Gram positive and Gram–negative bacteria. The most well-known polymers of PHA family are polyhydroxybutyrate (PHB) and poly(3-hydroxybutyrate-co-3- hydroxyvalerate) (PHBV) [56]. PHB is the most common PHA produced by fermentation by bacteria like *Alcaligenes, Azobacter, Bacillus* and *Pseudomonas* [47]. PLA and PHBV have been combined in a recently developed procedure to synthesize novel blended membranes [57,58]. The membranes synthesized by the mixture of PLA and PHBV had increased strength and ductility when compared to membranes synthesized from only PHBV. Along with this, the time required for the natural degradation of this hybrid membrane is much shorter when compared to the the membranes made from only PHBV [47]. Biodegradable membrane for the purpose of water treatment was synthesized by Keawsupsak et al. [58] by blending PLA with PBAT, PHBV or PBS via non solvent (water) induced phase separation. PHB has restrictive use as a polymer owing to its high crystallinity and brittleness. According to Miguel et al. [47], PHB acts as a high barrier for CO_2 with a good separation of CO_2 from water. Dense PV membranes using PHB was synthesized by Villegas et al. [59, 60] for the separation of methanol/water and methanol/methyl-tert-butyl ether (MTBE) mixtures. In case of both mixtures, PHB membranes offered good structural stability in response to swelling, and yielded high flux as well as moderate selectivity.

5.1.3 Poly(butylene succinate) (PBS)

PBS is obtained as a semi crystalline biopolymer as a result of polymerization of succinic acid and butanediol [47]. Both succinic acid and butanediol are easily synthesized from bio-based renewable resources [47]. Succinic acid is synthesized from microbial fermentation of renewable stocks like glucose and starch. Butanediol may be synthesized via catalytic reduction of succinic acid or direct fermentation of sugar [47]. Esterification of succinic acid and butanediol and subsequent polycondensation of the resultant oligomers lead to the formation of PBS. In recent studies, PBS has been combined with PES and CA for synthesis of blended membranes [61,62,63]. The blended membranes possessed lower crystallinity and demonstrated larger mechanical resistance than unmodified PBS membrane. Moreover, the PES-PBS systems exhibited higher tensile strength and lower membrane crystallinity than CA-PBS systems prepared with equivalent blend ratios. PBS-PES systems were further blended with PEG for the synthesis of membranes having enhanced efficiency for wastewater treatment.

5.2 Biopolymers derived from vegetable sources

5.2.1 Cellulose-based polymers

Cellulose is a polysaccharide synthesized by plants. It is one of the primary polymers present in nature and it is composed of long macromolecular β-D-glucose chains [47]. Cellulose is primarily obtained from lignocellulosic biomass made up of lignin and hemicellulose. Cellulose derivatives are synthesized via different methodologies. Cellulose acetate (CA), for example is generally prepared by treating cellulose with acetic acid in presence of sulfuric acid and acetic anhydride as catalysts. Conversely, hydroxyethyl cellulose (HEC) is prepared by treating alkali modified cellulose with ethylene oxide [64]. One of the first commercially implemented polymeric membranes was prepared in the 1960s using asymmetric CA fabrics [47]. These membranes had high flux and high salt rejection capacity. CA membranes synthesized by the acetyl substitution of cellulose, a naturally abundant polymer, have been utilized widely at the commercial level to control the microorganism load in untreated water subjected to RO and UF [65,47]. CA has been widely synthesized from cotton, recycled paper or wood cellulose [47]. CA membranes have lower aversion to chloride in comparison to PA membranes and have largely replaced the same in desalination processes. CA also has a lower membrane fouling tendency than conventional PA membranes used for water treatment processes [66]. Further investigations on CA could enhance membrane technology and replace traditional synthetic polymers in commercial applications. In order to improve antifouling properties of CA in a sustainable manner, authors have impregnated electrospun CA fiber matrix with chitin nanocrystals [67,68]. As a result of which, the membrane surface became highly hydrophilic thereby reducing the membrane fouling. Previous studies have investigated the effect of PVP addition and alterations in coagulation bath temperature on asymmetric CA membranes synthesis via phase inversion method [47]. According to the results obtained, hydrophilicity and flux of the membranes improved with a corresponding increase in the amount of PVP blended in the membrane. Previous studies have also reported CA-PSF UF membranes prepared via phase inversion [47]. Kanagaraj et al., [69] investigated the antifouling capacity, selectivity and flux of membranes prepared by combining CA with hydrophilic surface modifying macromolecules (SMMs). The resultant membranes displayed better flux recovery and antifouling properties. Varanasi et al [70] synthesized biodegradable UF membranes from P-amine-epiclorohydrin (PAE) cellulose nanofibers and silica NPs (as shown in Fig. 4). The NPs functioned as spacers which controls the pore size of the nanofibers. The used membranes were recycled for traditional paper production. Interestingly, Medina-Gonzalez et al. [71] substituted highly toxic NMP (conventional solvent) with methyl lactate (bio-solvent) for synthesis of CA. The synthesized membrane exhibited a sponge-like matrix having a pore size similar to that of commercial UF membranes. Results indicated that the fabricated membranes could be used for water treatment upto a transmembrane pressure of 5 bar. In another study, NF cellulose membranes were fabricated via phase inversion using 1-ethyl-3-methylimidazolium acetate [72]. The synthesized membrane was stable in different solvents. These membranes also displayed dye rejection similar to the available organic solvent-based NF membranes with a maximum Brilliant Blu R rejection capacity of ~100 %.

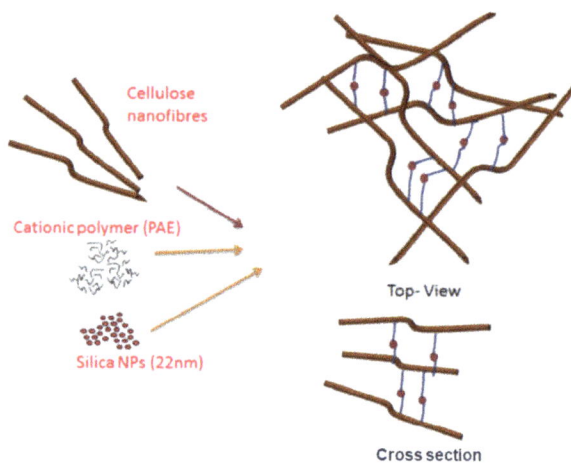

Figure 4. Interconnected cellulose nanofibers and silica NPs (Adapted from Galiano et al. [47]).

5.2.2 Alginate

Alginate is a natural polysaccharide extracted from brown algae (*Phaeophyceae*) [73]. Alginates have been widely investigated and applied in food and biomedical sectors [47]. It is composed of straight chains of *L*-gluronic acid and 1-4 linked mannuronic acid [47]. Algal extracts obtained by treating algae with NaOH are filtered and precipitated using Na/Ca chloride to yield alginates [73]. Alginates are mainly blended with chitosan (CS), for the synthesis of multilayer membranes for water treatment. Layers of this membrane are held together by the electrostatic interactions occurring between poly-cations of the protonated amines of CS chains and the poly-anions of the carboxylate groups of alginates [47,74]. Alginate-CS membranes have been previously reported for herbicide adsorption [74]. The synthesized membrane had numerous sites on their surface which was exposed to the positive/negative charged particles of the pesticide molecules to bind to. Authors also described the method for the preparation of bilayer membranes for the purpose of absorbing Paraquat from water and eventually, herbicides like Clomazone, Difenzoquat and Diquat from contaminated water [74,75]. The blended biopolymeric membranes simultaneously adsorbed negatively charged herbicides and positively charged compounds on the CS and alginate layer respectively. Paiva et al [76] synthesized uniform and homogenous alginate-CS layers alternate layers of the same in a Petri dish. Incorporation of metal in the membrane composition primarily depends upon solution pH and the metal's affinity for the polymer being used. These organic-inorganic membranes may be utilized as potential antimicrobial systems for inhibiting bacterial growth.

5.2.3 Starch

Starch is a plant-based polymer synthesized via photosynthesis. Starch consists of amylase (30 %), amylopectin (70 %) and other lipids and proteins (<1 %) from plants. Starch is a cost effective, completely biodegradable material having good renewability [47]. Starch based polymers have been widely applied as edible films and food packaging in the food industry. However, these materials have been rarely investigated as a component of polymeric membranes. A recent study had reported an antibacterial membrane synthesized using a blend of hydrolyzed starch and CA in presence of glycerol as a plasticizer via phase inversion [77]. The antibacterial activity of these membranes was tested with *E. coli*. Complete inhibition of bacterial growth on the membrane surface proved the applicability of these membranes in medical and food sectors. The presence of starch strongly contributed to the biodegradability of the membrane. Moreover, the total starch and polymer content was found to influence the hydrophilicity of the membrane. In another study, montmorillonite (MMT) and carboxymethyl cellulose (CMC) were blended as reinforcing fillers in a starch matrix for synthesis of biocomposite films [78]. Water resistance of starch was seen to improve by blending CMC with nano clay. Diffusion of water molecules through this composite was found to decrease with a simultaneous reduction of membrane hydrophilicity caused by the saturation of hydrogen bonds of the hydroxyl groups present in MMT layers and CMC starch chains.

5.3   Biopolymers derived from animal sources

5.3.1   Chitosan (CS)

Chitin, poly(β-(1-4)-N-acetyl-D-glucosamine) was identified for the first time in 1884 [47]. Chitin, a natural, biocompatible and nontoxic polysaccharide extracted from shells of crustaceans, has attracted significant scientific interest in recent times. Nearly 50% deacetylation of chitin results in the formation of CS, a biorenewable material soluble in acidic aqueous solvents [79]. Raval et al. [80] reported the synthesis of a high flux thin-film composite (TFC) RO membrane modified with CS. The CS modified membrane reportedly exhibited higher water flux in comparison to unmodified TFC RO membranes. Moreover, an increase in water flux of the CS modified membranes was recorded with a parallel increase in temperature without any loss of their salt rejection efficiency. This process is more promising than solar powered RO, where low-grade thermal energy is utilized for heating the feed water. In another study, Egusa et al. [81] investigated the nano-fibrillation of CS. They synthesized nanofiber sheets from nanofibrillated CS under neutral conditions. The modified CS nanofiber sheets strongly inhibited mycelial growth of *Microsporum* and *Trichophyton* and were resistant to fungal degradation. In another study, Cooper et al. [82] synthesized CS-based membranes via electrospinning for application in antibacterial filters. Incorporation of CS (25 %) within the nanofibrous membrane resulted in 50 % inhibition of *S. aureus* colonization in comparison to unmodified membranes. In another study, authors have investigated the antibacterial activity of the CS membranes [47]. Results obtained in this study indicated that antibacterial activity of CS membranes were dependent upon the degree of deacetylation of CS [47]. The CS membrane reported in this study had potential bactericidal properties. Hence, CS may be considered as an efficient material for synthesis of membranes for water treatment.

5.3.2 Collagen and sericin

Collagen is the most abundant protein present in the bones, cartilages, skin and tendons [47]. Collagen consists of several polypeptides and is extensively used in the biomedical industry. Collagen may be extracted from different animal species. However, it is primarily derived from byproducts of slaughter. Sericin, the major component of silk, is a macromolecular protein made up of 17–18 different amino acids [83]. Collagen and sericin are usually extracted using enzymatic, chemical, and ultrasound processes [84]. Previous studies have reported the synthesis of cross-linked hydrophilic membranes using collagen fibers and fibrils obtained from bovines [47]. This membrane offered an advantage of yielding high flux even at low temperatures. However, due to the proteinaceous nature of the membrane, it is sensitive to microbial degradation, high temperature and extreme pH. The collagen membranes reported in this study exhibited good water selectivity and separation efficiency in comparison to hydrophilic membranes. Nevertheless, rapid degradation is a major disadvantage of this collagen-based membrane [47]. For overcoming this limitation, collagen is often cross-linked to other polymers for membrane synthesis.

6. Factors affecting membrane efficacy

Membrane efficiency (in terms of permeate quality and flux) is reportedly guided by the parameters discussed as follows [85].

6.1 Operating condition

The membrane efficiency is firmly related to process parameters like temperature, operating pressure and flow rate.

6.1.1 Temperature

With the increase in temperature, the permeate flux also increases as rise in temperature causes a decline of viscosity and level of concentration polarization of oily wastewater [86]. Moreover, increase of temperature also decreases the removal of chemical oxygen demand (COD) and the electrical conductivity of oily wastewater [87]. In another study, membrane permeability was found to decline at temperatures above 24 °C [30]. However, in the same study it was observed that the effect of temperature was non-significant when the flux permeate was maintained at $11\text{-}12 \text{ L m}^{-2} \text{ h}^{-1}$. However, in presence of fouling, flux decrease irrespective of any rise in temperature [88]. Change in temperature causes the alteration of diffusion coefficient and rate of adsorption and in turn influences the flux.

6.1.2 Operating pressure

Flux reportedly increases with a parallel increase in pressure. However, does not occur in case of concentration polarization phenomenon and membrane fouling [86]. In another study, it was reported that an increase in operation pressure had resulted in a corresponding increase in dye rejection and the transport rate of solvent [89].

6.1.3 Flow rate

The increase in flow rate results in a corresponding increase of mass transport and permeates flux as well as a reduction of possible concentration polarization [89].

6.2 Membrane characteristics

According to a previous study, ployacrylonitrile membranes with greater hydrophilicity can remove surfactants more efficiently than fluoropolymers with comparatively lower hydrophilicity though the latter exhibits higher resistance to temperature and chemicals [85]. Moreover, the membranes reported in this study yielded higher permeate flux when used for crossflow or vibratory-enhanced filtration than in a dead-end mode. In another study, tubular polyamide loose -NF membranes reportedly yielded a higher permeate flux

(75 L m^{-2}) but lower removal of organics (98%) and conductivity (70%) in comparison to tubular tight-NF membranes that generated lower flux (49 L m^{-2}) and less efficient removal of organics (99.7%) and conductivity (95%) [85]. However, almost no significant difference was noted in stable permeate fluxes of metal membranes differing in pore sizes (0.5, 1, and 5 μm) when used for direct filtration of greywater [85]. Moreover, permeate yielded by metal membranes having a pore size of 0.5 μm was of comparatively higher quality (in terms of color, COD, conductivity, and turbidity than those generated by metal membranes with pore sizes of 1 and 5 μm. Thus, from this result, it can be inferred that the selection of appropriate membrane is of significant importance for achieving maximum efficiency in case of wastewater treatment.

Membrane charge, hydrophilicity and pore size are closely associated with its performance efficiency [85]. The charge of membrane and solute must be considered as significant parameters for guiding separation processes. In scenarios where charge of the solutes differs from charge of the membrane, a larger gravitational force operates and increases the probability of membrane fouling. Membrane fouling eventually causes a decline in flux and alters membrane selectivity [85]. The membrane pore size is also a determining factor guiding rejection of uncharged contaminants. Moreover, hydrophilicity of a membrane is more implicative of higher interaction with water than lower interactions with solutes. Nevertheless, hydrophobic membranes have higher antifouling potential and lower flux in comparison to hydrophilic ones [86].

6.2.1 Feed characteristics

Charge, chemical nature, geometry, hydrophilicity and pH of feed solutions exert significant effects on membrane efficiency. pH of the feed solution can influence and even alter membrane charge. At a pH exceeding the isoelectric point, the membrane will acquire negative charge and vice versa [86]. The isoelectric point differs with membrane type [86]. The change in pH also influences membrane hydrophilicity [86]. Moreover, temperature of the feed solutions accelerates the process of mass transfer through the membrane [87].

6.2.2 Fouling

In wastewater, the major fouling components are particulate and dissolved matter (both inorganic and organic), salts (both multivalent and monovalent), pathogens and surfactant [86]. According to a previous study, hydraulic resistance of the fouling cake layers is guided by the amount of organic matter and calcium present in greywater and the formed cake may enhance micropollutant removal by adsorption of the same within its layers [86]. However, surfactants (like sodium dodecyl sulfate) strongly interact with calcium present in effluents to form micelles that may accelerate membrane fouling. This often leads to unintended accumulation of organic substance in permeates as the surfactant outcompetes organics and occupies the active adsorption sites of calcium. Moreover, shower gels and conditioners present in greywater can reportedly cause more serious fouling of the membrane in comparison to shampoos and hand soaps [86].

6.3   Pretreatment of membrane

MF (0.2 μm) based pretreatment is more effective than cartridge (10 μm) based ones for enhancing the water flux and permeate quality of RO processes [90]. Another study reported that membrane filtration integrated with photocatalytic processes exhibit higher efficiency for degradation of organic pollutant present in greywater [85]. Moreover, presence of TiO_2 (photocatalyst) significantly reduced membrane fouling through the formation of large organo-TiO_2 aggregates. However, photocatalytic efficiency of TiO_2 may be affected adversely by these aggregates. Results obtained in this study indicated a reduction in membrane fouling with a corresponding increase in duration of UV irradiation [85].

7.   Application of polymeric membranes for water purification

An extensive review of recent literature suggested that polymer nanocomposite based UF membranes (as evident from both Table 1 and Table 2) have been most widely investigated for desalination and wastewater treatment. Incorporation of NPs in polymer matrices was found to enhance membrane porosity and hydrophilicity as well as antifouling properties. Composite UF membranes have been most widely investigated and applied for desalination and wastewater treatment. These membranes have been provided high flux and solutes retention of good quality. Incorporation of photocatalytic NPs into polymeric membranes has further facilitated efficient degradation of pollutants in solution besides rejection. However, membrane efficiency and reusability should be determined using real effluents for elucidating the feasibility of using these membranes on an industrial scale.

Table 1. Polymer nanocomposite membranes for desalination

Polymer type	Membrane	Major findings	Ref.
2-hydroxyethyl metha-crylate, glycidyl metha-crylate and glycerol metha-crylate were crosslinked with poly (ethylene glycol) dimethacrylate	NF	• Chemical modification of a series of water content equivalent copolymers increased water/salt permeability selectivity; • Distributed hydrophilic functional groups may lead to increased selectivity and may represent a strategy for improving water/salt selectivity of advanced membrane materials	[91]
Cellulose acetate / poly-ethylene glycol with integrated candle soot NPs	RO	•CS NPs contributed in improving the salt rejection % with a slight reduction in the water flux behavior; •93% salt rejection using the membranes;	[92]
TiO_2-sulfonated polymer embedded polyetherimide	UF	• Three different electrolytes (NaCl, Na_2SO_4 and $MgSO_4$) with varying concentrations from 50 - 500 mg L^{-1} and groundwater were filtered; • Fabricated membranes are lower MWCO UF membranes (8kDa); • Higher flux was recorded for groundwater filtration.	[93]
Negatively charged polyaniline (PANI) nanoparticles incorporated within polysulfone (PSF)	UF	• 2 wt% PANI loading leads to a 25 fold increase in the molecular weight cut-off (from 0.2 to 4.8 kDa); • Improvement in porosity (from 20% to 64%), and a 2-fold increase in permeability (from 8×10^{-12} to 16×10^{-12} m^3 m^{-2} Pa^{-1} s^{-1}); • 2.5 times enhancement in the permeate flux; • Salt rejection between 40–53%, equivalent to the NF performance.	[94]
PVDF hollow fiber membrane	UF	• Treatment of shale gas fracturing wastewater; • Significantly reduced fouling of UF membranes	[95]

Hydrolyzed polyacrylonitrile (PAN) modified with PEI	NF	• Treatment of textile effluent; • Membrane prepared at optimized condition achieved a high flux and dye rejection (~97%) and lowest irreversible fouling ratio (8.1%) for humic acid	[96]
Polyethersulfone	UF	• Lower MWCO membranes exhibited a negligible relative flux decline; • 90% retention of humic acid; • Good antifouling property	[97]
Monolayer graphene (G) coated polypropylene (PP) and PVDF	MF	• G – PP membranes exhibited better desalination efficiency than G – PVDF ones; • Both membranes had defects that improved with Nylon 6,6 sealing; • Optimum sealing resulted in 84% blockage of KCl ions	[98]
Cellulose/graphene oxide composite	MF	• High water flux (334.7 ± 10.4 Lm^{-2}h^{-1}); • Good antifouling property; • Suitable for pretreatment of seawater prior to RO	[99]
Polyamide – graphene oxide composite	RO-MF	• High salt rejection (99.7%); • High permeance (3.0 L m^{-2}h^{-1}bar^{-1}); • Membranes exhibited good stability and anti-fouling property	[100]

Table 2. Polymer nanocomposite membranes for wastewater treatment

Polymer type	Membrane	Major findings	Ref.
1-Methyl-2 pyrrolidinone, polyvinylpyrrolidone and polysulfone with graphene oxide and carboxylic-functionalized carbon nanotube fillers	UF	• Rejection of methyl orange ($17.8 \pm 6.7\%$) is observed to be always higher than that of methylene blue ($1.9 \pm 1.5\%$); • > 90% rejection of bovine serum albumin (BSA); • Addition of carbon nanofillers enhanced the rejection of negatively charged molecules with increased permeability at low nanofiller mass loading	[101]

Mesoporous carbon (MPC) bearing Ag NP loaded polyethersulfone	UF	• Bacterial attachment on the membrane surface was reduced dramatically; • Damage of *Bacillus subtilis* (92.94%) and *Escherichia coli* (93.21%); The polymeric mixed matrix membrane entirely mitigated biofouling over 99% by the combination of the bactericidal effect of silver and the anti-adhesion properties of MPC.	[102]
Poly(m-phenylene isophthalamide)	UF	• Membrane exhibited a high porosity, narrow pore radius distribution and excellent hydrophilicity, leading to its high performance in the filtration process; • Exhibit good antifouling properties; • Efficient separation of BSA and phosphates.	[103]
Polydopamine NP coated polyethersulfone	UF	• Membrane exhibited 92.9 % BSA rejection and 166 $Lm^{-2}h^{-1}$ pure water flux; • Membrane also demonstrated static adsorption capacities of 20.23, 17.01 and 10.42 mg g^{-1} of Pb^{2+}, Cd^{2+} and Cu^{2+} respectively; • Adsorption data could be fitted better with the Langmuir isotherm model and pseudo 2nd order kinetics	[104]
Alumina (Al_2O_3) NP embedded in PES membranes	UF	• Thinner skin layer, higher porosity and hydrophilicity enhanced water permeation almost by three times; • Composite formation enhanced uptake of Cu from aqueous solutions	[105]
Cellulose acetate composite with organically modified montmorillonite	UF	• Enhanced membrane hydrophilicity; • Enhanced adsorption of Humic acid	[106]

Electrospun PES nanofibers	MF	• Enhanced flux and anti-fouling properties; • Improved water flux; • High retention of TiO_2 microparticles (99%)	[107]
PVDF nanofibers	MF	• Improved membrane hydrophilicity; • High flux rates under low pressure; • High retention of polystyrene microparticles (99%)	[108]
TiO_2 NP coated PES and PVDF	MF	• Enhanced flux and anti-fouling properties; • Membranes exhibited high photocatalytic degradation of methylene blue and anti-inflammatory drugs (ibuprofen and diclofenac); • Membranes exhibited good reuse potential without any loss of photocatalytic activity	[109]
Carboxylated MWCNT/ PES nanocomposite	MF	• Increased membrane porosity and hydrophilicity; • High pure water permeance (20 $Lm^{-2}h^{-1}bar^{-1}$); • Enhanced removal of bromothymol blue and methyl orange (95%)	[110]

Conclusion

Polymer membrane-based desalination and wastewater treatment processes have significantly improved over the last few years. However, certain initiatives can promote these processes further. Only a fraction of these membranes has been synthesized using "green" materials or processes. Further studies are required for reducing the thickness of membrane layers and defects present in the membrane. Investigations should also include life cycle assessment methodologies for ensuring complete green synthesis of membranes. More utilization of biopolymers is encourage able for synthesis of nanocomposites for preparing membranes. These membranes should be tested for performance efficiency using real effluents for assessment of their true potential for desalination and water treatment.

References

[1] Coping with water scarcity. A strategic issue and priority for system-wide action. UN-water (2006).

[2] R. Das, M.E. Ali, S.B.A. Hamid, S. Ramakrishna, Z.Z. Chowdhury, Carbon nanotube membranes for water purification: a bright future in water desalination, Desalination 336 (2014) 97–109. https://doi.org/10.1016/j.desal.2013.12.026

[3] P. Banerjee, R. Das, P. Das, A. Mukhopadhyay, Membrane Technology, in: R. Das (Ed.), Carbon Nanotubes for Clean Water, Springer, Cham, 2018, pp. 127-150. https://doi.org/10.1007/978-3-319-95603-9_6

[4] P.S. Goh, A.F. Ismail, B.C. Ng, Carbon nanotubes for desalination: performance evaluation and current hurdles, Desalination 308 (2013) 2–14. https://doi.org/10.1016/j.desal.2012.07.040

[5] Z. Zia, A. Hartland, M.R. Mucalo, Use of low-cost biopolymers and biopolymeric composite systems for heavy metal removal from water, Int. J. Env. Sci. Technol. 17 (2020) 4389–4406. https://doi.org/10.1007/s13762-020-02764-3

[6] R. Das, Polymeric Materials for Clean Water, Springer, Berlin, Germany, 2019. https://doi.org/10.1007/978-3-030-00743-0

[7] L.Y. Ng, A.W. Mohammad, C.P. Leo, N. Hilal, Polymeric membranes incorporated with metal/metal oxide nanoparticles: a comprehensive review, Desalination 308 (2013) 15–33. https://doi.org/10.1016/j.desal.2010.11.033

[8] N.L. Le, S.P. Nunes, Materials and membrane technologies for water and energy sustainability, Sustain. Mater. Technol. 7 (2016) 1–28. https://doi.org/10.1016/j.susmat.2016.02.001

[9] A. Abdel-Karim, T.A. Gad-Allah, A.S. El-Kalliny, S.I.A. Ahmed, E.R. Souaya, M.I. Badawy, M. Ulbricht, Fabrication of modified polyethersulfone membranes for wastewater treatment by submerged membrane bioreactor, Sep. Purif. Technol. 175 (2017) 36–46. https://doi.org/10.1016/j.seppur.2016.10.060

[10] B.B. Ashoor, S. Mansour, A. Giwa, V. Dufour, S.W. Hasan, Principles and applications of direct contact membrane distillation (DCMD): a comprehensive review, Desalination 398 (2016) 222–246. https://doi.org/10.1016/j.desal.2016.07.043

[11] A. Giwa, S. Daer, I. Ahmed, P. Marpu, S. Hasan, Experimental investigation and artificial neural networks ANNs modeling of electrically-enhanced membrane bioreactor for wastewater treatment, J. Water Proc. Eng. 11 (2016) 88–97. https://doi.org/10.1016/j.jwpe.2016.03.011

[12] A. Giwa, S. Hasan, Theoretical investigation of the influence of operating conditions on the treatment performance of an electrically-induced membrane

bioreactor, J. Water Proc. Eng. 6 (2015) 72–82.
https://doi.org/10.1016/j.jwpe.2015.03.004

[13]   X. Wang, C. Wang, C.Y. Tang, T. Hu, X. Li, Y. Ren, Development of a novel anaerobic membrane bioreactor simultaneously integrating microfiltration and forward osmosis membranes for low-strength wastewater treatment, J. Mem. Sci. 527 (2017) 1–7. https://doi.org/10.1016/j.memsci.2016.12.062

[14]   B.B. Ashoor, H. Fath, W. Marquardt, A. Mhamdi, Dynamic modeling of direct contact membrane distillation processes, in: I.A. Karimi, R. Srinivasan (Eds.), Computer Aided Chemical Engineering, Elsevier, Netherlands, 2012, pp. 170–174. https://doi.org/10.1016/B978-0-444-59507-2.50026-3

[15]   S.W. Hasan, M. Elektorowicz, J.A. Oleszkiewicz, Start-up period investigation of pilot-scale submerged membrane electro-bioreactor (SMEBR) treating raw municipal wastewater, Chemosphere 97 (2014) 71–77.
https://doi.org/10.1016/j.chemosphere.2013.11.009

[16]   C.H. Neoh, Z.Z. Noor, N.S.A. Mutamim, C.K. Lim, Green technology in wastewater treatment technologies: integration of membrane bioreactor with various wastewater treatment systems, Chem. Eng. J. 283 (2016) 582–594.
https://doi.org/10.1016/j.cej.2015.07.060

[17]   J. Lee, H.R. Chae, Y.J. Won, K. Lee, C.H. Lee, H.H. Lee, I.C. Kim, J.M. Lee, Graphene oxide nanoplatelets composite membrane with hydrophilic and antifouling properties for wastewater treatment, J. Mem. Sci. 448 (2013) 223–230.
https://doi.org/10.1016/j.memsci.2013.08.017

[18]   M. Aslam, A. Charfi, G. Lesage, M. Heran, J. Kim, Membrane bioreactors for wastewater treatment: a review of mechanical cleaning by scouring agents to control membrane fouling, Chem. Eng. J. 307 (2017) 897-913.
https://doi.org/10.1016/j.cej.2016.08.144

[19]   P. Krzeminski, L. Leverette, S. Malamis, E. Katsou, Membrane bioreactors—a review on recent developments in energy reduction, fouling control, novel configurations, LCA and market prospects, J. Mem. Sci. 527 (2017) 207–227.
https://doi.org/10.1016/j.memsci.2016.12.010

[20]   W. Luo, H.V. Phan, M. Xie, F.I. Hai, W.E. Price, M. Elimelech, L.D. Nghiem, Osmotic versus conventional membrane bioreactors integrated with reverse osmosis for water reuse: biological stability, membrane fouling, and contaminant removal, Water Res. 109 (2017) 122–134. https://doi.org/10.1016/j.watres.2016.11.036

[21]   X. Wang, V.W.C. Chang, C.Y. Tang, Osmotic membrane bioreactor (OMBR) technology for wastewater treatment and reclamation: advances, challenges, and

prospects for the future, J. Mem. Sci. 504 (2016) 113-132.
https://doi.org/10.1016/j.memsci.2016.01.010

[22]   G. Qiu, Y.P. Ting, Short-term fouling propensity and flux behavior in an osmotic membrane bioreactor for wastewater treatment, Desalination 332 (2014) 91–99. https://doi.org/10.1016/j.desal.2013.11.010

[23]   B. Díez, N. Roldán, A. Martín, A. Sotto, J.A. Perdigón-Melón, J. Arsuaga, R. Rosal, Fouling and biofouling resistance of metal-doped mesostructured silica/polyethersulfone ultrafiltration membranes, J. Mem. Sci. 526 (2017) 252–263. https://doi.org/10.1016/j.memsci.2016.12.051

[24]   S. Mokhtari, A. Rahimpour, A.A. Shamsabadi, S. Habibzadeh, M. Soroush, Enhancing performance and surface antifouling properties of polysulfone ultrafiltration membranes with salicylate-alumoxane nanoparticles, Appl. Surf. Sci. 393 (2017) 93–102. https://doi.org/10.1016/j.apsusc.2016.10.005

[25]   Y. Orooji, M. Faghih, A. Razmjou, J. Hou, P. Moazzam, N. Emami, M. Aghababaie, F. Nourisfa, V. Chen, W. Jin, Nanostructured mesoporous carbon polyethersulfone composite ultrafiltration membrane with significantly low protein adsorption and bacterial adhesion, Carbon 111 (2017) 689–704. https://doi.org/10.1016/j.carbon.2016.10.055

[26]   N. Sharma, M.K. Purkait, Impact of synthesized amino alcohol plasticizer on the morphology and hydrophilicity of polysulfone ultrafiltration membrane, J. Mem. Sci. 522 (2017) 202–215. https://doi.org/10.1016/j.memsci.2016.08.068

[27]   M. Son, H. Kim, J. Jung, S. Jo, H. Choi, Influence of extreme concentrations of hydrophilic pore-former on reinforced polyethersulfone ultrafiltration membranes for reduction of humic acid fouling, Chemosphere 179 (2017) 194–201. https://doi.org/10.1016/j.chemosphere.2017.03.101

[28]   S. Velu, G. Arthanareeswaran, H. Lade, Removal of organic and inorganic substances from industry wastewaters using modified aluminosilicate-based polyethersulfone ultrafiltration membranes, Environ. Prog. Sustain. Energy 36 (2017) 1612-1620. https://doi.org/10.1002/ep.12614

[29]   H. Wu, Y. Liu, L. Mao, C. Jiang, J. Ang, X. Lu, Doping polysulfone ultrafiltration membrane with TiO_2-PDA nanohybrid for simultaneous self-cleaning and self-protection, J. Mem. Sci. 532 (2017) 20–29. https://doi.org/10.1016/j.memsci.2017.03.010

[30]   A. Behboudi, Y. Jafarzadeh, R. Yegani, Polyvinyl chloride/polycarbonate blend ultrafiltration membranes for water treatment, J. Mem. Sci. 534 (2017) 18–24. https://doi.org/10.1016/j.memsci.2017.04.011

[31] Y.W. Huang, Z.M. Wang, X. Yan, J. Chen, Y.J. Guo, W.Z. Lang, Versatile polyvinylidene fluoride hybrid ultrafiltration membranes with superior antifouling, antibacterial and self-cleaning properties for water treatment, J Colloid Interf. Sci. 505 (2017) 38–48. https://doi.org/10.1016/j.jcis.2017.05.076

[32] Z. Maghsoud, M. Pakbaz, M.H.N. Famili, S.S. Madaeni, New polyvinyl chloride/thermoplastic polyurethane membranes with potential application in nanofiltration, J. Mem. Sci. 541 (2017) 271–280. https://doi.org/10.1016/j.memsci.2017.07.001

[33] H. Wang, Z.-M. Wang, X. Yan, J. Chen, W.-Z. Lang, Y.-J. Guo, Novel organic-inorganic hybrid polyvinylidene fluoride ultrafiltration membranes with antifouling and antibacterial properties by embedding N-halamine functionalized silica nanospheres, J. Ind. Eng. Chem. 52 (2017) 295–304. https://doi.org/10.1016/j.jiec.2017.03.059

[34] B. Yang, X. Yang, B. Liu, Z. Chen, C. Chen, S. Liang, L.-Y. Chu, J. Crittenden, PVDF blended PVDF-g-PMAA pH-responsive membrane: effect of additives and solvents on membrane properties and performance, J. Mem. Sci. 541 (2017) 558-566. https://doi.org/10.1016/j.memsci.2017.07.045

[35] C. Zhao, J. Lv, X. Xu, G. Zhang, Y. Yang, F. Yang, Highly antifouling and antibacterial performance of poly (vinylidene fluoride) ultrafiltration membranes blending with copper oxide and graphene oxide nanofillers for effective wastewater treatment, J. Colloid Interf. Sci. 505 (2017) 341–351. https://doi.org/10.1016/j.jcis.2017.05.074

[36] S. Daer, J. Kharraz, A. Giwa, S.W. Hasan, Recent applications of nanomaterials in water desalination: a critical review and future opportunities, Desalination 367 (2015) 37–48. https://doi.org/10.1016/j.desal.2015.03.030

[37] A. Giwa, N. Akther, A. Al Housani, S. Haris, S.W. Hasan, Recent advances in humidification dehumidification (HDH) desalination processes: improved designs and productivity, Renew. Sustain. Energy Rev. 57 (2016) 929–944. https://doi.org/10.1016/j.rser.2015.12.108

[38] D. Qadir, H. Mukhtar, L.K. Keong, Mixed matrix membranes for water purification applications, Sep. Purif. Rev. 46 (2017) 62–80. https://doi.org/10.1080/15422119.2016.1196460

[39] H. Rabiee, V. Vatanpour, M.H.D.A. Farahani, H. Zarrabi, Improvement in flux and antifouling properties of PVC ultrafiltration membranes by incorporation of zinc oxide (ZnO) nanoparticles, Sep. Purif. Technol. 156 (2015) 299–310. https://doi.org/10.1016/j.seppur.2015.10.015

[40]   A. Martín, J.M. Arsuaga, N. Roldán, J. de Abajo, A. Martínez, A. Sotto, Enhanced ultrafiltration PES membranes doped with mesostructured functionalized silica particles, Desalination 357 (2015)16–25. https://doi.org/10.1016/j.desal.2014.10.046

[41]   K.L. Mercer, 2017 State of the water industry: strengthening our connections, J. American Water Works Assoc. 109 (2017) 56–65. https://doi.org/10.5942/jawwa.2017.109.0090

[42]   J.A. Prince, S. Bhuvana, K.V.K. Boodhoo, V. Anbharasi, G. Singh, Synthesis and characterization of PEG-Ag immobilized PES hollow fiber ultrafiltration membranes with long lasting antifouling properties, J. Mem. Sci. 454 (2014) 538–548. https://doi.org/10.1016/j.memsci.2013.12.050

[43]   H. Yu, Y. Zhang, X. Sun, J. Liu, H. Zhang, Improving the antifouling property of polyethersulfone ultrafiltration membrane by incorporation of dextran grafted halloysite nanotubes, Chem. Eng. J. 237 (2014) 322–328. https://doi.org/10.1016/j.cej.2013.09.094

[44]   L. Shao, Z.X. Wang, Y.L. Zhang, Z.X. Jiang, Y.Y. Liu, A facile strategy to enhance PVDF ultrafiltration membrane performance via self-polymerized polydopamine followed by hydrolysis of ammonium fluotitanate, J. Mem. Sci. 461 (2014) 10–21. https://doi.org/10.1016/j.memsci.2014.03.006

[45]   P. Kanagaraj, A. Nagendran, D. Rana, T. Matsuura, S. Neelakandan, T. Karthikkumar, A. Muthumeenal, Influence of N-phthaloyl chitosan on poly (ether imide) ultrafiltration membranes and its application in biomolecules and toxic heavy metal ion separation and their antifouling properties, Appl. Surf. Sci. 329 (2015) 165–173. https://doi.org/10.1016/j.apsusc.2014.12.082

[46]   C. Liu, H. Mao, J. Zheng, S. Zhang, Tight ultrafiltration membrane: preparation and characterization of thermally resistant carboxylated cardo poly (arylene ether ketone)s (PAEK-COOH) tight ultrafiltration membrane for dye removal, J. Mem. Sci. 530 (2017)1–10. https://doi.org/10.1016/j.memsci.2017.02.005

[47]   F. Galiano, K. Briceno, T. Marino, A. Molino, K.V. Christensen, A. Figoli, Advances in biopolymer-based membrane preparation and applications, J. Mem. Sci. 564 (2018) 562-586. https://doi.org/10.1016/j.memsci.2018.07.059

[48]   Y. Gu, J. François, Bio-based solvents: an emerging generation of fluids for the design of eco-efficient processes in catalysis and organic chemistry, Chem. Soc. Rev. 42 (2013) 9550-9570. https://doi.org/10.1039/c3cs60241a

[49]   M. Niaounakis, Biopolymers: applications and trends, first ed., William Andrew, New York, 2015.

[50]   T. Tanaka, M. Ueno, Y. Watanabe, T. Kouya, M. Taniguchi, D.R. Lloyd, Poly(l-lactic acid) microfiltration membrane formation via thermally induced phase separation with drying, J. Chem. Eng. Japan 44 (2011) 467–475. https://doi.org/10.1252/jcej.11we030

[51]   S. Chitrattha, T. Phaechamud, Modifying poly(L-lactic acid) matrix film properties with high loaded poly(ethylene glycol), Key Eng. Mater. 545 (2013) 57–62. https://doi.org/10.4028/www.scientific.net/KEM.545.57

[52]   T. Phaechamud, S. Chitrattha, Pore formation mechanism of porous poly(dl-lactic acid) matrix membrane, Mater. Sci. Eng. C 61 (2016) 744–752. https://doi.org/10.1016/j.msec.2016.01.014

[53]   A. Moriya, P. Shen, Y. Ohmukai, T. Maruyama, H. Matsuyama, Reduction of fouling on poly(lactic acid) hollow fiber membranes by blending with poly(lactic acid)-polyethylene glycol-poly(lactic acid) triblock copolymers, J. Mem. Sci. 415–416 (2012) 712–717. https://doi.org/10.1016/j.memsci.2012.05.059

[54]   H. Minbu, A. Ochiai, T. Kawase, M. Taniguchi, D.R. Lloyd, T. Tanaka, Preparation of poly(L-lactic acid) microfiltration membranes by a nonsolvent-induced phase separation method with the aid of surfactants, J. Mem. Sci. 479 (2015) 85–94. https://doi.org/10.1016/j.memsci.2015.01.021

[55]   A.C. Chinyerenwa, W. Han, Q. Zhang, Y. Zhuang, K.H. Munna, C. Ying, H. Yang, W. Xu, Structure and thermal properties of porous polylactic acid membranes prepared via phase inversion induced by hot water droplets, Polymer 141 (2018): 62-69. https://doi.org/10.1016/j.polymer.2018.03.011

[56]   M.M. Reddy, S. Vivekanandhan, M. Misra, S.K. Bhatia, A.K. Mohanty, Biobased plastics and bionanocomposites: current status and future opportunities, Prog. Polym. Sci. 38 (2013) 1653–1689. https://doi.org/10.1016/j.progpolymsci.2013.05.006

[57]   H.C. Chang, T. Sun, N. Sultana, M.M. Lim, T.H. Khan, A.F. Ismail, Conductive PEDOT:PSS coated polylactide (PLA) and poly(3-hydroxybutyrate-co-3-hydroxyvalerate) (PHBV) electrospun membranes: Fabrication and characterization, Mater. Sci. Eng. C 61 (2016) 396–410. https://doi.org/10.1016/j.msec.2015.12.074

[58]   K. Keawsupsak, A. Jaiyu, J. Pannoi, P. Somwongsa, N. Wanthausk, P. Sueprasita, C. Eamchotchawalit, Poly(lactic acid)/biodegradable polymer blend for the preparation of flat-sheet membrane, J. Teknol. Sci. Eng. 69 (2014) 99–102. https://doi.org/10.11113/jt.v69.3405

[59]   M. Villegas, E.F. Castro Vidaurre, J.C. Gottifredi, Sorption and pervaporation of methanol/water mixtures with poly(3-hydroxybutyrate) membranes, Chem. Eng. Res. Des. 94 (2015) 254–265. https://doi.org/10.1016/j.cherd.2014.07.030

[60]   M. Villegas, A.I. Romero, M.L. Parentis, E.F. Castro Vidaurre, J.C. Gottifredi, Acrylic acid plasma polymerized poly(3-hydroxybutyrate) membranes for methanol/MTBE separation by pervaporation, Chem. Eng. Res. Des. 109 (2016) 234–248. https://doi.org/10.1016/j.cherd.2016.01.018

[61]   V. Ghaffarian, S.M. Mousavi, M. Bahreini, H. Jalaei, Polyethersulfone/poly (butylene succinate) membrane: effect of preparation conditions on properties and performance, J. Ind. Eng. Chem. 20 (2014) 1359–1366. https://doi.org/10.1016/j.jiec.2013.07.019

[62]   V. Ghaffarian, S.M. Mousavi, M. Bahreini, Effect of blend ratio and coagulation bath temperature on the morphology, tensile strength and performance of cellulose acetate/poly(butylene succinate) membranes, Desalin. Water Treat. 54 (2015) 473–480. https://doi.org/10.1080/19443994.2014.883329

[63]   V. Ghaffarian, S.M. Mousavi, M. Bahreini, M. Afifi, Preparation and characterization of biodegradable blend membranes of PBS/CA, J. Poly. Environ. 21 (2013) 1150-1157. https://doi.org/10.1007/s10924-012-0551-1

[64]   M.A. El-Sheikh, S.M. El-Rafie, E.S. Abdel-Halim, M.H. El-Rafie, Green synthesis of hydroxyethyl cellulose-stabilized silver nanoparticles, J. Polym. 2013 (2013) 1–11. https://doi.org/10.1155/2013/650837

[65]   A.K. Mishra, Smart Materials for Waste Water Applications, Scrivener Publishing, Wiley, 2016. https://doi.org/10.1002/9781119041214

[66]   M.I. Baoxia, M. Elimelech, Gypsum scaling and cleaning in forward osmosis: measurements and mechanisms, Environ. Sci. Technol. 44 (2010) 2022–2028. https://doi.org/10.1021/es903623r

[67]   L.A. Goetz, B. Jalvo, R. Rosal, A.P. Mathew, Superhydrophilic anti-fouling electrospun cellulose acetate membranes coated with chitin nanocrystals for water filtration, J. Membr. Sci. 510 (2016) 238–248. https://doi.org/10.1016/j.memsci.2016.02.069

[68]   N. Naseri, A.P. Mathew, L. Girandon, M. Fröhlich, K. Oksman, Porous electrospun nanocomposite mats based on chitosan–cellulose nanocrystals for wound dressing: effect of surface characteristics of nanocrystals, Cellulose 22 (2015) 521–534. https://doi.org/10.1007/s10570-014-0493-y

[69]   P. Kanagaraj, A. Nagendran, D. Rana, T. Matsuura, Separation of macromolecular proteins and removal of humic acid by cellulose acetate modified UF membranes, Int. J. Biol. Macromol. 89 (2016) 81–88. https://doi.org/10.1016/j.ijbiomac.2016.04.054

[70] S. Varanasi, Z.X. Low, W. Batchelor, Cellulose nanofibre composite membranes – biodegradable and recyclable UF membranes, Chem. Eng. J. 265 (2015) 138–146. https://doi.org/10.1016/j.cej.2014.11.085

[71] Y. Medina-Gonzalez, P. Aimar, J.F. Lahitte, J.C. Remigy, Towards green membranes: preparation of cellulose acetate ultrafiltration membranes using methyl lactate as a biosolvent, Int. J. Sustain. Eng. 4 (2011) 75–83. https://doi.org/10.1080/19397038.2010.497230

[72] F.M. Sukma, P.Z. Çulfaz-Emecen, Cellulose membranes for organic solvent nano filtration, J. Membr. Sci. 545 (2018) 329–336. https://doi.org/10.1016/j.memsci.2017.09.080

[73] K.Y. Lee, D.J. Mooney, Alginate: properties and biomedical applications, Prog. Polym. Sci. 37 (2012) 106–126. https://doi.org/10.1016/j.progpolymsci.2011.06.003

[74] M. Agostini de Moraes, D.S. Cocenza, F. da Cruz Vasconcellos, L.F. Fraceto, M.M. Beppu, Chitosan and alginate biopolymer membranes for remediation of contaminated water with herbicides, J. Environ. Manag. 131 (2013) 222–227. https://doi.org/10.1016/j.jenvman.2013.09.028

[75] D.S. Cocenza, M.A. De Moraes, M.M. Beppu, L.F. Fraceto, Use of biopolymeric membranes for adsorption of paraquat herbicide from water, Water Air. Soil Pollut. 223 (2012) 3093–3104. https://doi.org/10.1007/s11270-012-1092-x

[76] R.G. de Paiva, M.A. de Moraes, F.C. de Godoi, M.M. Beppu, Multilayer biopolymer membranes containing copper for antibacterial applications, J. Appl. Polym. Sci. 126 (2012): E17-E24. https://doi.org/10.1002/app.36666

[77] A. Zarei, V. Ghaffarian, Preparation and characterization of biodegradable cellulose acetate-starch membrane, Polym. Plast. Technol. Eng. 52 (2013) 387–392. https://doi.org/10.1080/03602559.2012.752831

[78] H. Almasi, B. Ghanbarzadeh, A.A. Entezami, Physicochemical properties of starch-CMC-nanoclay biodegradable films, Int. J. Biol. Macromol. 46 (2010) 1–5. https://doi.org/10.1016/j.ijbiomac.2009.10.001

[79] P. Wu, M. Imai, Novel biopolymer composite membrane involved with selective mass transfer and excellent water permeability, in: R.Y. Ning (Ed.), Advancing Desalination, InTech, Rijeka, Croatia, 2012, pp. 57–81. https://doi.org/10.5772/50697

[80] H.D. Raval, P.S. Rana, S. Maiti, A novel high-flux, thin-film composite reverse osmosis membrane modified by chitosan for advanced water treatment, RSC Adv. 5 (2015) 6687–6694. https://doi.org/10.1039/C4RA12610F

[81]   M. Egusa, R. Iwamoto, H. Izawa, M. Morimoto, H. Saimoto, H. Kaminaka, S. Ifuku, Characterization of chitosan nanofiber sheets for antifungal application, Int. J. Mol. Sci. 16 (2015) 26202–26210. https://doi.org/10.3390/ijms161125947

[82]   A. Cooper, R. Oldinski, H. Ma, J.D. Bryers, M. Zhang, Chitosan-based nanofibrous membranes for antibacterial filter applications, Carbohydr. Polym. 92 (2013) 254–259. https://doi.org/10.1016/j.carbpol.2012.08.114

[83]   Z. Wang, Y. Zhang, J. Zhang, L. Huang, J. Liu, Y. Li, G. Zhang, S.C. Kundu, L. Wang, Exploring natural silk protein sericin for regenerative medicine: an injectable, photoluminescent, cell-adhesive 3D hydrogel, Scientific Reports 4 (2014) 7064. https://doi.org/10.1038/srep07064

[84]   M.M. Schmidt, R.C.P. Dornelles, R.O. Mello, E.H. Kubota, M.A. Mazutti, A.P. Kempka, I.M. Demiate, Collagen extraction process, Int. Food Res. J. 23 (2016) 913–922.

[85]   B. Wu, Membrane-based technology in greywater reclamation: A review, Sci. Total Environ. 656 (2019) 184-200. https://doi.org/10.1016/j.scitotenv.2018.11.347

[86]   R. Mulyanti, H. Susanto, Wastewater treatment by nanofiltration membranes, IOP Conf. Series: Earth Environ. Sci. 142 (2018) 012017. https://doi.org/10.1088/1755-1315/142/1/012017

[87]   A. Rahimpour, B. Rajaeian, A. Hosienzadeh, S.S. Madaeni, F. Ghoreishi, Treatment of oily wastewater produced by washing of gasoline reserving tanks using self-made and commercial nanofiltration membranes, Desalination 265 (2011) 190-198. https://doi.org/10.1016/j.desal.2010.07.051

[88]   Z. Gönder, S.A. Beril, B. Hulusi, Advanced treatment of pulp and paper mill wastewater by nanofiltration process: Effects of operating conditions on membrane fouling, Sep. Purif. Technol. 76 (2011) 292-302. https://doi.org/10.1016/j.seppur.2010.10.018

[89]   A.M.F. Shaaban, A.I. Hafez, M.A. Abdel-Fatah, N.M. Abdel-Monem, M.H. Mahmoud, Process engineering optimization of nanofiltration unit for the treatment of textile plant effluent in view of solution diffusion model, Egyptian J. Pet. 25 (2016) 79-90. https://doi.org/10.1016/j.ejpe.2015.03.018

[90]   V.M. Boddu, T. Paul, M.A. Page, C. Byl, L. Ward, J. Ruan, Gray water recycle: Effect of pretreatment technologies on low pressure reverse osmosis treatment, J. Environ. Chem. Eng. 4 (2016) 4435-4443. https://doi.org/10.1016/j.jece.2016.09.031

[91]   H. Luo, K. Chang, K. Bahati, G.M. Geise, Engineering selective desalination membranes via molecular control of polymer functional groups, Environ. Sci. Technol. Lett. 6 (2019) 462-466. https://doi.org/10.1021/acs.estlett.9b00351

[92]   A.E. Abdelhamid, A.M. Khalil, Polymeric membranes based on cellulose acetate loaded with candle soot nanoparticles for water desalination, J. Macromol. Sci. A 56 (2019) 153-161. https://doi.org/10.1080/10601325.2018.1559698

[93]   Y. L. Thuyavan, N. Anantharaman, G. Arthanareeswaran, A.F. Ismail, R. V. Mangalaraja, Preparation and characterization of TiO_2-sulfonated polymer embedded polyetherimide membranes for effective desalination application, Desalination 365 (2015) 355-364. https://doi.org/10.1016/j.desal.2015.03.004

[94]   R. Mukherjee, R. Sharma, P. Saini, S. De., Nanostructured polyaniline incorporated ultrafiltration membrane for desalination of brackish water, Environ. Sci. Water Res. Technol. 1 (2015) 893-904. https://doi.org/10.1039/C5EW00163C

[95]   H. Chang, T. Li, B. Liu, C. Chen, Q. He, J.C. Crittenden, Smart ultrafiltration membrane fouling control as desalination pretreatment of shale gas fracturing wastewater: The effects of backwash water, Environ. Int. 130 (2019) 104869. https://doi.org/10.1016/j.envint.2019.05.063

[96]   S. Zhao, Z. Wang, A loose nano-filtration membrane prepared by coating HPAN UF membrane with modified PEI for dye reuse and desalination, J. Mem. Sci. 524 (2017) 214-224. https://doi.org/10.1016/j.memsci.2016.11.035

[97]   A.A., Saif, C.J. Wright, N. Hilal, Investigation of UF membranes fouling and potentials as pre-treatment step in desalination and surface water applications, Desalination 432 (2018) 115-127. https://doi.org/10.1016/j.desal.2018.01.017

[98]   F.M. Kafiah, Z. Khan, A. Ibrahim, R. Karnik, M. Atieh, T. Laoui, Monolayer graphene transfer onto polypropylene and polyvinylidene difluoride microfiltration membranes for water desalination, Desalination 388 (2016) 29-37. https://doi.org/10.1016/j.desal.2016.02.027

[99]   Y. Ibrahim, F. Banat, A.F. Yousef, D. Bahamon, L.F. Vega, S.W. Hasan, Surface modification of anti-fouling novel cellulose/graphene oxide (GO) nanosheets (NS) microfiltration membranes for seawater desalination applications, J. Chem. Technol. Biotechnol. 95 (2020) 1915-1925. https://doi.org/10.1002/jctb.6341

[100] J. Shi, W. Wu, Y. Xia, Z. Li, W. Li, Confined interfacial polymerization of polyamide-graphene oxide composite membranes for water desalination, Desalination 441 (2018) 77-86. https://doi.org/10.1016/j.desal.2018.04.030

[101] Z. Wan, Y. Jiang, Synthesis-structure-performance relationships of nanocomposite polymeric ultrafiltration membranes: A comparative study of two carbon nanofillers, J. Mem. Sci. (2020) 118847. https://doi.org/10.1016/j.memsci.2020.118847

[102] Y. Orooji, F. Liang, A. Razmjou, G. Liu, W. Jin, Preparation of anti-adhesion and bacterial destructive polymeric ultrafiltration membranes using modified mesoporous

carbon, Sep. Purif. Technol. 205 (2018) 273-283.
https://doi.org/10.1016/j.seppur.2018.05.006

[103] C.E. Lin, J. Wang, M.-Y. Zhou, B.-K. Zhu, L.-P. Zhu, C.-J. Gao, Poly (m-phenylene isophthalamide) (PMIA): A potential polymer for breaking through the selectivity-permeability trade-off for ultrafiltration membranes, J. Mem. Sci. 518 (2016) 72-78. https://doi.org/10.1016/j.memsci.2016.06.042

[104] X. Fang, J. Li, X. Li, S. Pan, X. Zhang, X. Sun, J. Shen, W. Han, L. Wang, Internal pore decoration with polydopamine nanoparticle on polymeric ultrafiltration membrane for enhanced heavy metal removal, Chem. Eng. J. 314 (2017) 38-49. https://doi.org/10.1016/j.cej.2016.12.125

[105] N. Ghaemi, A new approach to copper ion removal from water by polymeric nanocomposite membrane embedded with γ-alumina nanoparticles, Appl. Surf. Sci. 364 (2016) 221-228. https://doi.org/10.1016/j.apsusc.2015.12.109

[106] F.S. Dehkordi, M. Pakizeh, M. Namvar-Mahboub, Properties and ultrafiltration efficiency of cellulose acetate/organically modified Mt (CA/OMMt) nanocomposite membrane for humic acid removal, Appl. Clay Sci. 105 (2015) 178-185. https://doi.org/10.1016/j.clay.2014.11.042

[107] S. Khezli, M. Zandi, J. Barzin, Fabrication of electrospun nanocomposite polyethersulfone membrane for microfiltration, Polym. Bull. 73 (2016) 2265-2286. https://doi.org/10.1007/s00289-016-1607-5

[108] Z. Li, W. Kang, H. Zhao, M. Hu, N. Wei, J. Qiu, B. Cheng, A novel polyvinylidene fluoride tree-like nanofiber membrane for microfiltration, Nanomaterials 6 (2016) 152. https://doi.org/10.3390/nano6080152

[109] K. Fischer, M. Grimm, J. Meyers, C. Dietrich, R. Gläser, A. Schulze, Photoactive microfiltration membranes via directed synthesis of TiO_2 nanoparticles on the polymer surface for removal of drugs from water, J. Mem. Sci. 478 (2015) 49-57. https://doi.org/10.1016/j.memsci.2015.01.009

[110] A.K. Shukla, J. Alam, M. Rahaman, A. Alrehaili, M. Alhoshan, Ali Aldalbahi, A facile approach for elimination of electroneutral/anionic organic dyes from water using a developed carbon-based polymer nanocomposite membrane, Water Air Soil Pollut. 231 (2020) 1-16. https://doi.org/10.1007/s11270-020-04483-4

Polymeric Membranes for Water Purification and Gas Separation Materials Research Forum LLC
Materials Research Foundations **113** (2021) 69-112 https://doi.org/10.21741/9781644901632-4

Chapter 4

Thin Film Nanocomposite of Nanofiltration Membrane for Water Softening and Desalination

Muhammad Hanis Tajuddin[1,2], Muhammad Faris Hamid[1,2], Norhaniza Yusof[1,2]*,
Ahmad Fauzi Ismail[1,2]*

[1]Advanced Membrane Technology Research Centre (AMTEC), Universiti Teknologi Malaysia,
81310 UTM Skudai, Johor Darul Ta'zim, Malaysia

[2]School of Chemical and Energy Engineering (FCEE), Faculty of Engineering Universiti
Teknologi Malaysia, 81310 UTM Skudai, Johor Darul Ta'zim, Malaysia

norhaniza@petroleum.utm.my & afauzi@utm.my

Abstract

This chapter discusses about nanofiltration (NF) membrane for water softening and desalination. The NF membrane system and thin film composite (TFC) membranes are discussed in general followed by their drawbacks. Next, recent trend of nanofillers in thin film nanocomposite (TFN) membrane is critically discussed and highlighted. The advantages and challenges of TFN membrane for water softening and desalination application are thoroughly analyzed. Lastly, the future directions of the TFN membrane for practical application are addressed.

Keywords

Thin Film Nanocomposite, Nanofiltration, Desalination, Water Softening, Membrane Fouling

Contents

1.   Introduction

Significant population growth has increased the demand of clean water which led to water shortage, specifically in water-stressed region. Different water sources such as brackish water [1], seawater [2] and ground water [3] have been explored to meet the demand of clean water. However, these types of water need to be treated in order to remove dissolved solute and hardness in the water before it is safely used for human consumption using desalination and water softening technologies, respectively. Briefly, desalination is a process of converting seawater into drinkable water by the removal of suspended and dissolved particles including water hardness from seawater using desalination technologies [4]. Desalination technologies include thermal processes such as multistage flash distillation (MSF), multi effect distillation (MED), vapor compression

(VC) and membrane processes such as sea water reverse osmosis (SWRO) and electrodialysis (ED) [5,6]. The two most popular desalination technologies are RO and MSF in which they account for 80% of world's desalination capacity [7].

Meanwhile, water softening is a process to remove water hardness. Water hardness in tap water is attributed by the presence of certain inorganic minerals such as calcium and magnesium and they are mainly accumulated within the water pipes [8]. The presence of these compounds renders the water flow that causes from clogging in piping system due to the mineral crystallization [9]. Besides that, the presence of hardness also reduces the compatibility of water with soap and detergent efficiency [3]. Therefore, it is necessary to remove hardness from water using either classical or new methods of water purification.

Conventional methods of water softening, such as lime softening and ion-exchange process have been widely employed for water softening process [10,11]. However, incomplete separation and high operational cost limit their application in water softening process. To overcome the limitations of conventional methods, attention is given to more promising methods, such as membrane technologies. Therefore, membrane technology has attracted great attention in the past recent years and grown steadily with vast applications, such as gas separation, desalination and wastewater treatment.

2. Desalination and water softening

Desalination is a process of removal salinity from saltwater or seawater. Salinity refers to concentration of dissolved salt that normally present in their ionic state, in a given volume of water. Salinity of seawater usually is around 35 g/L [12]. Since seawater consists mostly of NaCl salt, and because of NaCl dissociates into monovalent sodium (Na^+) and chloride (Cl^-) ions, studies of desalination have been focusing on the removal of NaCl [13]. As monovalent ion, Na^+ and Cl^- have relatively smaller hydrated diameters, compared to the pore diameter of the membrane, therefore NaCl possesses intrinsically higher diffusivity. This will cause separation of NaCl by membrane processes with larger pore size (refer Fig. 1) to be ineffective. Therefore, the use of NF alone for desalination would be impractical. Normally, NF is used as pre-treatment step to be coupled with reverse osmosis (RO) membrane in SWRO desalination plant [14]. Unlike other pressurized membrane processes, RO alone can be used to desalinate salt water due to its tight pore size that is able to retain small ionic species such Na^+ and Cl^-.

However, the use of other membrane processes such as microfiltration (MF), ultrafiltration (UF) or NF in hybrid with reverse osmosis (RO) membrane in SWRO desalination plant could reduce the plant's operational cost in term of energy and chemical consumption [14]. Under normal circumstances, either one of those aforementioned membrane processes is installed at the pre-treatment stage to enhance the quality of water entering the RO membrane system. Usually, MF and UF is used to remove coarse particles (Fig. 1), while NF is used to divalent salts and water hardness.

Water hardness then is referring to the presence of calcium (Ca^{2+}), magnesium (Mg^{2+}) and its companion anion, such as sulphate ions (SO_4^{2-}) in water stream [15,16]. Water softening is a recognized technology in separation of multivalent ion from various water streams. Water softening had been used in many industrial applications, such as production of drinking water, cooling water and purification of wastewater. Moreover, it has been extensively employed in softening of seawater, surface water (rivers, lakes and stream), groundwater and brackish water. Yet, in typical water hardness contain counter valent ion, such as SO_4^{2-} ions with concentration from 100 ppm in groundwater to 2500 ppm in seawater [16]. In brackish water conditions, the presence of hardness is 500 mg/L of $CaCO_3$. Thus, according to Malaysia Standard Quality, maximum acceptable limits for hard water is 500 mg/L [9]. Water softening make up of approximately 20-30% of industrial and residential waste water treatment with a specific market share of 2.5 billion dollars in 2010 for residential area and 7.3 billion in 2010 for industrial application [17,18].

Conventionally, water hardness can be eliminated by lime softening and ion exchange process. Lime softening utilizes chemical precipitation method in which lime is added to precipitate magnesium ions and calcium ions. The advantage of lime softening process is available lime source, no pollution to the natural water and able to decrease organic matter silicate as well as iron [19]. However, the major drawbacks of this process are the production of sludge and excessive use of chemicals which led to higher operating cost [10]. Furthermore, this conventional method has limited capacity as the demand of hardness removal is increased.

The ion exchange process is widely employed for water softening. The softening process utilizes salt saturated, for example, sodium chloride resin beads regenerate the resin beads [17]. In this process, capturing of calcium and magnesium occurred and sodium ions are released until saturation is reached. The saturated materials are regenerated using a very high sodium chloride solution. Ion exchange process detriments in terms of huge consumption of the salt brine used for regeneration, meanwhile calcium ions exchanged by sodium ions increase the water salinity [20].

Capacitive deionization (CDI) is another promising technology for water softening due to low energy consumption and environmental friendliness [21,22]. Basically, in CDI configuration cell, the feed solution will flow through both pair of electrode (carbon porous) and during polarization process the negative ions will be attracted to negatively surface of electrode. However, utilization of CDI technology always hindered by presence of dissolved salts in pore volume of carbon electrode resulting in energy inefficient. The available method for water softening is summarized as in Table 1.

Table 1.   Summary of available method for water softening

Method	Advantage	Disadvantage	Ref.
Lime Softening	Extensive source of lime. No pollution to natural water	Huge in volume of sludge. Excessive use of chemicals	[17]
Ion exchange	Ease of operation. Can be used in small scale	High consumption of salt brine. Increase salinity	[20]
Capacitive deionization	Environmentally friendly. Low energy consumption	Dissolved salt present in pore volume of carbon electrode	[23]
Membrane technology	Simple in operation. No chemical addition	Membrane fouling and scaling	[24]

One of the notable breakthroughs of membrane process is solving the limitation possessed by current technologies for water softening. Membrane technology can serve as potential solution of water softening process. Pressure-driven membrane can be classified into MF, UF, NF and RO as illustrated in Fig. 1.

Among these membrane processes, water softening and desalination process often utilizes either NF or RO membrane. The key feature that distinguishes RO and NF TFC membranes is that NF allows monovalent salts to pass through, while retaining multivalent ions, such as magnesium and calcium. Thus, NF is often used as pre-treatment method in SWRO desalination plant to remove water hardness. On the other hand, RO is widely employed to separate monovalent salts, such as sodium chloride (NaCl). Compared to RO, NF is favorable in water softening process mainly due to its lower operating pressure, resulting in the reduction of operation costs of separation plant. It is expected that the global growth of NF membrane to reach about $445 million with a significant increase of 157% in 2019 as compared to the year 2012, which is only $172 million as reported by BCC research [26]

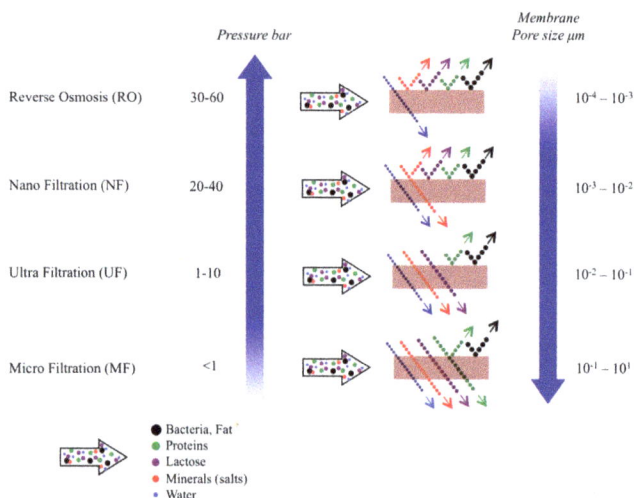

Figure 1.   *Classification of classical membrane filtration system [25].*

Most of NF and RO membranes are generally TFC. TFC membranes consist of an ultrathin layer, which is formed by interfacial polymerization (IP) over microporous substrate. In general, IP is a reaction process between two monomers; polyfunctional amine dissolved in water solution and polyfunctional acid chloride dissolved in hydrocarbon solvent. The active layer or ultra-thin layer is generally made by polyamide (PA) from the cross-linked of semi-aromatic PA. The PA layer plays a vital role in the selective barrier in water and solute transport across the membranes thus its structure considerably influences the membrane separation performances [27]. Owing to its unique structure, the properties of ultra-thin layer and porous substrate can be independently tailored to enhance the membrane performance [28]. In spite of the excellent separation performance and stable structures, the PA TFC membranes are still associated with trade-off effect between water flux and salt rejection. Many attempts have been carried out to improve the performance of TFC membrane. For example, incorporating nanofillers, such as silver nanoparticles (Ag NPs) [29], graphene oxide (GO) [30], carbon nanotube (CNT) [31], silica [32], titanate nanotube (TNT) [33] and halloysite nanotube (HNT) [34], either within PA layer or porous substrate. Upon the addition of nanofillers, the thin film membrane is normally referred to as TFN membrane

3.   Nanofiltration membranes and its drawbacks

3.1   A glance at nanofiltration process

Over the last 20 years, the NF study has shown sparks interest among the researchers in many different areas. Based on data obtained from the Scopus database, a total of 10192 papers have been published in NF membrane. As shown in Fig. 2, the main area covers by NF membrane were in chemical engineering (20%), chemistry (20%), environmental science (15%) and materials science (15%). During data collection, we noticed that the research area in NF membrane still gains interest and keeps growing steadily over the past few years. Therefore, some commercial NF membranes, such as NF270 (Dow Filtec), NF200 (Dow Filtec), N90 (Dow Filtec) and UTC20 (Toray) are available that could be used in the removal of divalent salts.

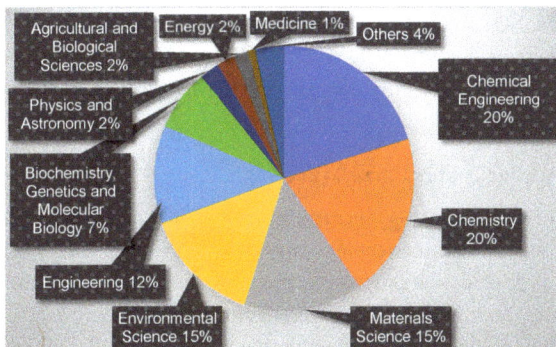

Figure 2.   Different subject areas of NF applications according to Scopus database analysis (access on 22 June 2020).

Besides, there are many review articles published, focussing on NF membranes. Hilal *et al.* (2004) published an article that was extensively focused on NF membrane and their application in water treatment [35]. Mohammad and co-workers also widely discussed the recent progress of the NF membrane, [36]. Moreover, Van der Bruggen and co-workers addressed some drawbacks of NF application and methods to overcome them [37]. Review by Lau *et al.* (2014) also discussed the characterization method for NF membrane [38]. Furthermore, the recent trend of nanomaterials in NF, RO and forward osmosis (FO) were extensively review by [24,39–41]. Despite that, another recent review has been published, focusing on chemistry and fabrication of polymeric NF membrane [42].

Most of the pressure-driven membrane separation processes (MF, UF and RO) are highly influenced by their size exclusions. However, the solute retention by most of the NF membrane is mainly controlled by membrane surface charges (Donnan effect) and size-exclusion. Based on the Donnan exclusion theory, negatively charged membranes have stronger repulsive attractions towards divalent anions (SO_4^{2-}), compared with monovalent ions (Cl^-). Moreover, SO_4^{2-} ions have larger hydrated radius, ion radius and lower diffusion coefficient encountered greater steric hindrance resistance. Due to this fact, the rejection of sulphate salts (Na_2SO_4, $MgSO_4$) was higher compared with chlorides salts ($MgCl_2$, $NaCl$). Furthermore, Mg^{2+} ions as a counter ion have greater effect on the negative electric field than Na^+. Therefore, Donnan exclusion effect between $MgSO_4$ and the membrane surface is getting weaker thus subsequently loss the retention of $MgSO_4$ compared with Na_2SO_4. According to Donnan exclusion mechanism, the rejection of the salts having the same anion primarily determined by diffusion coefficient of cations. For example, Na^+ has higher diffusivity compared with Mg^{2+} that caused the less rejection of $NaCl$ than $MgCl_2$.

3.2   Major drawbacks of NF membrane

As the growth of NF technology is rapidly increasing, it becomes more selective to remove divalent ions. It is more feasible than RO membrane in water softening and pre-treatment of water desalination process, ascribed by its intrinsic properties, such as lower operating pressure and higher water permeation. A common method to form NF membranes is through the IP of an acyl chloride monomer and an amine monomer [43]. This reaction forms the thin PA rejection layer that is often supported by a porous substrate. The loose PA layer network with small pores in TFC membrane allows higher water permeation and sustaining the higher divalent salts rejection [42,44]. However, one of the major drawbacks of TFC membrane is their performances are largely controlled by their water permeability, solutes selectivity and fouling resistances. Due to these circumstances, TFC membrane suffers from selectivity/permeability or trade-off effects in achieving the excellent performances of NF separation process. Many attempts have been made to create a TFC membrane with high water permeation without sacrificing the salts retention [45].   In past few years, various approaches have been employed to improve water flux and separation performance of TFC through the use of novel monomer [46] or incorporating inorganic nanofiller [24] either in PA layer or substrate layer. Thus, with the addition of nanofillers into the membrane, membrane properties can be improved which could increase membrane hydrophilicity or additional pathways for water transportation across the membrane structures.

Furthermore, the PA layer in TFC membrane primarily controls (mainly focus) the membrane performance in terms of permeability, selectivity and fouling resistance instead of modifying the substrate layer, which mainly plays as supporting role to the overall membrane structure [47]. With the emerging nanomaterials, layered double hydroxide, (LDH) could serve as potential fillers to be incorporated into PA layer. However, incorporation of LDH into TFC membrane is still in infancy, and fundamental understanding of this material compatibility within polymeric TFC has remained opaque.

Another major issue faced by most of the membrane-based water separation processes is extensive membrane fouling. It is a phenomenon where membrane performance declines in a long run operation. Fouling is also a major contributor that limits membrane performance, causing a decline of flux, could degrade membrane and increasing the in energy consumption [36]. Besides that, as the source of water is not only containing the minerals salts in their feeds, but also has other companion contaminant, such as protein, macromolecules and oils. Therefore, it is necessary to systematically evaluate the membrane performances with different types of foulants such as centrimonium bromide (CTAB) and bovine serum albumin (BSA).

4.   Thin film composite (TFC) membrane

Generally, commercial TFC membrane consists of three layers mainly ultra-thin top layer, a porous substrate layer and non-woven layer as demonstrated in Fig. 3. It has been a major type of NF membrane since its first introduction in 1970s. The porous substrate layer usually serves as mechanical support for thin film, whereas the properties of ultra-thin layer is important for membrane performance [48]. There were several method reported in fabrication of TFC membranes, such as photo-initiated, plasma-initiated grafting, dip-coating and IP [48–50]. Among them, IP technique is widely employed in the fabrication of TFC membrane due to better improvement in selectivity and membrane fouling [36].

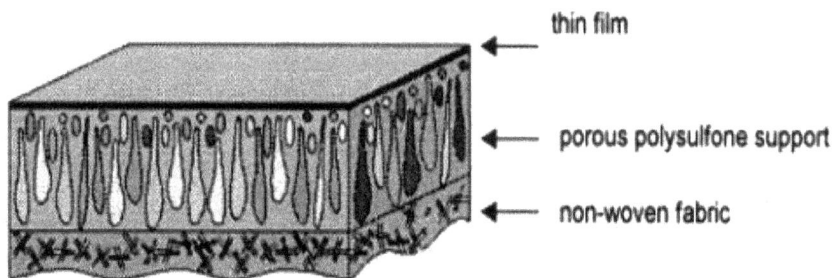

Figure 3.   Cross-section of a typical TFC membrane [51]

4.1   Interfacial polymerization (IP)

The concept of IP was first introduced by Morgan in 1965 [52]. On top of that, TFC membrane was first fabricated by Cadotte and co-workers in the 1970s and widely used in various application, such as disinfection by-products, desalinate seawater or remove heavy metals [24,40]. IP has become a practical and favourable technique for fabrication of ultra-thin layer of composite membrane like NF and RO. To date, fabrications of TFC through IP process have drawn attention in past decade significantly due to better improvement in membrane characteristics, such as membrane fouling and selectivity.

Generally, the crosslinked PA layer is formed by reacting two monomers in IP, functional amine in aqueous phase and acyl chloride in organic phase on porous substrate as depicted in Fig. 4 [28,53]. The membrane substrate such as polysulfone (PSf) and polyether sulfone (PES) is fabricated by phase inversion. The membrane is first immersed in amine aqueous solution which later transferred into acid chloride in organic solution. Alternatively, nanoparticles (NPs) may be introduced either into the aqueous amine or into the organic phase. Finally, the polymerization reaction takes place between both monomers resulting in a formation of thin film.

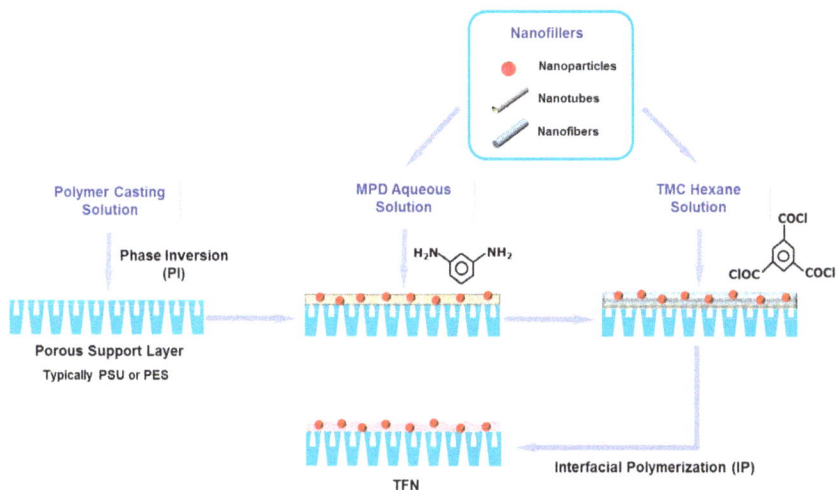

Figure 4.   Schematic diagram of IP process for TFN synthesis [24].

4.2   Parameters of NF membrane fabrication

In IP process, there were several factors that affect the formation of thin film layer on the substrate layer. Choices of monomers, type of solvents and post-treatment procedure are several parameters that influenced the formation of thin film layer. Various monomers are being used in the formation of TFC through IP, such as m-phenyldiamine (MPD), trimesoyl chloride (TMC), isophathaloyl chloride (IPC) and piperazine (PIP) as shown in Table 2. Conventionally, piperazine (PIP) and TMC established as the most favourable monomer pair in NF membrane fabrication [11,54].

Most of the fabricated NF and RO membranes are derived from cross-linked aromatic PA through IP between acyl chloride and aliphatic/aromatic diamine [28,55]. The selective thin layer or surface layer play an essential role in TFC membrane as it will determine the overall permeability, solute retention and efficiency of the membrane [36,56]. Therefore, the IP method is progressively studied to further improve their characteristics and separation performances.

The choice of solvent is another important factor in IP in which most of the organic solvents like n-hexane, cyclohexane or heptane being used. Besides that, IP reaction time is one of the main parameters in fabrication of TFC membrane. Misdan *et al.* (2015) proposed that the polymerization took place within 10-15 seconds in order to produce optimum membrane performances [57]. TFC membrane fabricated by IP is also subjected to post-treatment process, especially by using temperature. The common curing temperature is ranging from 40-120°C [58]. This post-treatment process purposely performed to stabilize thin layer by removing the residual organic solvent and crosslinking through dehydration of amine and carboxylic acids. However, too high of curing temperatures would damage the porosity of PA layer which leads to a decrease in water flux [59].

Table 2. Commonly used monomers for PA formation [46].

Monomer	Chemical structure	Molecular weight (g/mol)
Amine i) m-phenylenediamine (MPD)	m-phenylenediamine (MPD) 	108.10
ii) piperazine (PIP)	piperazine (PIP) 	86.14
Acid chloride i) trimesoyl chloride (TMC)	trimesoyl chloride (TMC) 	265.48
ii) isophathaloyl chloride (IPC)	isophthaloyl chloride (IPC) 	203.02

4.3   Other TFC membrane preparation techniques

Thin film layers could be fabricated with different techniques, such as plasma polymerization, UV grafting and spin coating. In plasma polymerization method, nitrogen-containing plasma is applied to the membrane surface to generate nitrogen functional groups that may improve hydrophilicity. In 2012, Wang and co-workers employed modification by grafting plasma 2-acrylamido-2-methylpropanesulfonic acid (AMPS) on PSf hollow fibre UF membrane with plasma method [60]. Besides that, UV grafting technique also performed by chemical bonding between active layer and substrate. Deng *et al.* (2011) used UV-grafting with methacrylatoethyl trimethyl ammonium chloride (DMC) on top of PSf substrate [61]. In spin coating method by Li and Barbari, the coating material is applied at the centre of substrate which rotates at high speed [62]. The centrifugal forces spread throughout the coating material on a substrate which led to the formation of thin film.

4.4   Advantages and challenges of TFC membrane

Fabrication of TFC membrane by IP has provided a significant advantage owing to its unique structures, such as ultra-thin selective layer and porous substrate of TFC membrane. By employing this method, both layers (PA thin layer and substrate layer) can be independently optimized and controlled to achieve desired selectively while retaining excellent mechanical strength and compression resistance. This intrinsic property allows the fabricated TFC membrane to achieve high water flux and salt rejection. The PA thin layer plays an essential role in TFC membrane as it will determine the overall permeability, solute retention and efficiency of the membrane [36,56]. Several researchers have addressed that the ultra-thin skin layer of TFC membrane is determinant in separation performance to achieve high salt rejection and improved water permeability [28,53,63].

Despite having a superior property, TFC membrane suffered from structural deterioration due to exposure of chlorine and membrane fouling. Membrane fouling is one of major hindrance in TFC membrane. It increases the operational cost and decreases membrane lifespan [64]. Therefore, chemical cleaning is necessary to remove organic fouling or biofouling on membrane surface. Chlorination is another step where it is used for disinfecting the water. When TFC exposed to chloride ion derived from water disinfectant, it will attack the amide nitrogen and ring bonded to amide nitrogen cause structural changes in PA network [65]. This process also causes the membrane degradation and lower the salt rejection [66].

Many researchers attempt to modify TFC membrane by surface modification, impregnated inorganic nanofillers in ultrathin layer or porous substrate. Addition of inorganic nanofillers is believed to improve not only membrane hydrophilicity, but also reduce the fouling tendency as well as increase the salt rejection. According to Mohammad et al. (2015) [36], incorporated NPs have gained noteworthy attention due to their ability to increase membrane permeability, mechanical properties, hydrophilicity and selectivity.

5.   Recent trends of nanofillers TFN membrane

Improvement of TFC membrane has been extensively studied in order to enhance their separation performances. In recent years, many efforts were made by researchers to improve TFC membrane, such as focusing on substrate properties, introduction of novel monomers [46] and impregnating NPs [67]. The most common NPs utilize in the fabrication of TFN membranes, are zeolites [68–70], TiO_2 [71], Ag, CNTs and silica [24]. Recently, many researchers have been focused on incorporating inorganic materials into membrane matrix. It is believed that incorporation of additives would play an important role in increasing the degree of polymerization and controlling the amine diffusion rate towards PA surface. Moreover, the inorganic nanofillers can be dispersed either in aqueous or in organic phase. NPs materials have the potential to improve membrane structure and possess the ability to control membrane fouling [72] and increase hydrophilicity [36].

Polymer nanocomposite membranes are new class membranes that contain both organic polymers and inorganic nanoscale materials which is believed to exhibit higher performance compared with standard membrane [41]. NPs have the potentiality to improve membrane structure and possess the ability to control membrane fouling [72] and increase hydrophilicity [36]. The TFN structure described in Fig. 5 is fabricated on a commercial membrane substrate. Commercial membranes are usually constructed on non-woven or fibrous which is below the porous support layer. The main purpose of this non-woven structure is to provide mechanical strength for composite membrane from deforming in operation. In the end, the efforts of embedding NPs are to enhance the performances of the membrane composite towards water softening and desalination.

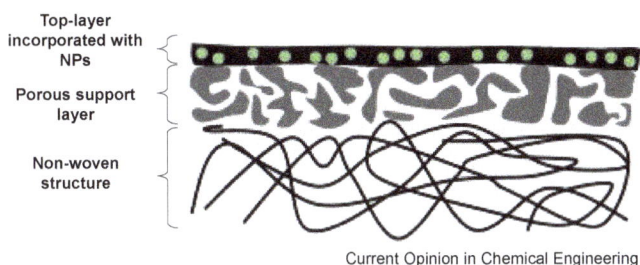

Current Opinion in Chemical Engineering

Figure 5.   *Schematic representation of TFN structure [73].*

5.1 Metal and metal oxide nanoparticles

Among the NPs that being studied, silver nanoparticles (Ag NPs) have drawn great attention owing to its antibacterial properties. A filler that has antibacterial properties can deactivate microorganisms during water filtration and reduce the bio-fouling problem as well as improve its hydrophilicity. Kim *et al.* (2012) developed TFC membrane by incorporating of Ag NPs into thin film layer and PSf substrate membrane containing CNTs [74]. The results proved that the addition of Ag NPs slightly increases the Na_2SO_4 from 94.7 to 95.6% and TFN membrane has shown an antibacterial effect. Another research reported by Ben-Sasson *et al.* (2014) where impregnated Ag NPs were used as biocides into TFC RO membrane for water purification process. Incorporating Ag NPs has demonstrated improvement of bio fouling resistance and virus removal [75].

In fabrication of TFC PA NF membrane, Ag NPs were dispersed in the organic phase [29]. TFC membrane shows improve in anti-biofouling properties with addition of Ag NPs. Moreover, immobilized Ag NPs on membrane surface also employed in order to improve TFC membrane performances. The study has shown that Ag NPs impregnated in TFN membrane exhibit increase in membrane hydrophilicity (32.2 ± 0.4°) and improved pure water flux (88.7 L/m^2.h) compared with TFC membrane (56.7 ± 2.2°) and 67.1 L/m^2.h [76].

However, the presence of Ag NPs only enhances membrane bio-fouling while ineffective for improving membrane permeability and does not have significant influence in salt rejection [74]. Moreover, the challenge possess by Ag NPs in bulk matrices is weak resistance during washing. Therefore, Ag NPs can be released as suspended NPs or silver ions in dissolved form [24]. With that, the effectiveness of Ag NPs as antimicrobial agent is diminished. Despite having great antibacterial activities, addition of Ag NPs is limited relatively to high cost thus making it not feasible to be employed [75]. Furthermore, the study investigated by Yin *et al.* (2013) [76] shown that NaCl rejection slightly decreased from 95.1% (TFC membrane) to 93.6 % with the addition of immobilized Ag NPs.

Titanium dioxide (TiO_2) has received tremendous attention due to its excellent hydrophilicity and photocatalytic ability to decompose organic compounds and antibacterial effects. Lee *et al.* (2008) incorporated TiO_2 NPs into PA layer of TFC NF membrane. Results revealed that increasing the TiO_2 loading from 0–5 wt%, the membrane flux declined slightly, whereas $MgSO_4$ rejection increased. However, further increase of TiO_2 up to 10 wt% showed significant flux increase, but extremely poor salt rejection (<5%) [77]. Pourjafar *et al.* (2012) impregnated TiO_2 into polyvinyl amide on top of PES substrate. With addition of 0.5 wt% TiO_2 NPs, the NaCl rejection was increased from 28 to 41% compared with pristine TFC membrane [78]. Another attempt was made by Rajaeian *et al.* (2013), introducing the modified TiO_2 using an amino silane coupling agent aiming to minimize particle agglomeration of TiO_2. The modified TiO_2 exhibited higher water flux compared with TFC control membrane where the pure water flux increases steadily from 11.2 to 27 L/m².h corresponding to be increasing in loading from 0.005 to 0.1 wt%. Moreover, NaCl rejection was slightly improved to 54% with small loading of modified TiO_2 [79]. Besides that, the presence of TiO_2 in membrane has improved hydrophilicity and mechanical strength of membrane [80]. However, despite excellent properties own by TiO_2 their application always hindered by particle agglomeration resulting in poor membrane performance [81].

Silica-based inorganic materials have attracted considerable interest of many membrane researchers especially in utilizing it as filler in membrane fabrication. Incorporating of small amount of silica content into PA layer can improve thermo stability of TFN membrane [82]. Moreover, silica based nanofillers possess superior properties, such as high mechanical strength, high hydrophilicity and special water transfer molecule compared with pristine TFC membrane [32]. Jadav and Singh (2009) synthesized two types of silica NPs, aiming to obtain a chemically stable PA membrane with improved performances. Introduction of silica NPs during IP process significantly improved the water flux of PA membrane from 10.2 to 40.8 L/m².h of 2000 ppm salt (NaCl) rejection under 250 psig [82].

Another reported research from Ghaemi (2017) utilized acrylate-alumoxane NPs, aiming to improve antifouling properties and selectivity of TFC membrane. Based on their reported experimental data, a significant improvement was observed in $MgSO_4$ and $MgCl_2$ rejection of 83% and 70% for $TFNC_2$ (TFN membrane prepared by 0.025 g of the NPs) compared TFC and 53% and 47% upon the addition of acrylate–alumoxane, respectively. However, $TFNC_3$ (TFN membrane prepared by 0.05 g of the NPs) exhibited unsatisfactory rejection. The low rejection of TFNC$_3$ was attributed by unsuccessful coating of nanomaterials on PES substrate. Meanwhile, enhancement in Mg^{2+} ion rejection was contributed by reduction of pore size after formation of polypyrrole (PPy) thin layer on substrate layer. Moreover, the TFC membrane possesses positively surface charge resulting in Donnan repulsion as primary separation mechanism of particular ions. Furthermore, insignificant rejection between MgSO$_4$ and MgCl$_2$ implying Mg^{2+} ions mainly rejected ions instead of SO$_4^{2-}$ ions due to positively charged surfaced membrane. Despite that, TFNC$_2$ exhibited superior results compared with other membrane due to formation of tighter and cross-linked structure with C=C bonds in acrylate-alumoxane NPs [83]. Although incorporation of acrylate alumoxane showed promising results for fabrication of TFC membrane with improved selectivity in water softening, more research still required for improvement in softening application.

Recently, mesoporous silica NPs (MSN) have significant interest because of its uniform, controllable porosity, high specific surface area, good surface hydrophilicity and low cost [40]. Another reported research by Wu *et al.* (2013) prepared modified mesoporous silica NPs (mMSN) as nanofiller into PA layer of TFN membrane. With increasing amount of mMSN NPs in PA, layer membrane has increased the water flux until it reaches to a maximum of 32.4 L/m^2.h. Moreover, the fabricated TFN membrane were able to keep a relatively high rejection of Na$_2$SO$_4$ (>80%) [84]. An attempt was made by Bao *et al.* (2013) by synthesizing monodispersed spherical mesoporous nanosilica incorporated in PA TFC RO membranes. Consequently, with addition of nanosilica, the water flux of resultant TFN RO membrane was increased from 19 to 53 L/h.m^2 compare with TFC RO while salt rejection was 96% [85]. Recently, Li *et al.* (2015) investigated the effects of silica nanospheres incorporated into poly(piperazine) amide layer aiming to improve TFC membrane separation performance. With addition silica nanospheres, the separation performances of fabricated TFN membrane was significantly increased to 95.94 ± 0.31% accompanied with permeate flux of 67.43 ± 0.85 L/m^2.h compared with TFC membrane 91.53 ± 0.21% accompanied with permeate flux of 62.53 ± 1.10 L/m^2.h [32].

Incorporating silica NPs into PA layer of TFC membrane has shown improvement in salt rejection ability and water flux as well as antifouling properties. However, despite having excellent performance, silica loading needs to be controlled because excessive amount might lead to decrease the degree of cross-linking of PA layer. This problem could arise defect in TFN membrane and affect membrane separation performance [40,86].

5.2 Carbon-based materials

Carbon-based nanomaterials, such as CNT and graphene derivative emerge as another option for modification of TFC membrane. According to Zhang *et al.* (2016), CNT has unique thermal, mechanical and electrical properties [87]. Moreover, CNT has been applying in vast application, such as gas separation membranes, membrane distillation (MD), FO and RO membrane [88]. Zarrabi *et al.* (2016) introduced amine functionalized multi-walled carbon nanotube (NH_2-MWCNT) into PA layer of TFC membrane during IP. The maximum rejection of Na_2SO_4 and NaCl salt was about 95.7% and 36.7%, respectively with the loading of 0.005wt% NH_2-MWCNT. Furthermore, improvement of the flux might come from functionalization of MWCNT with NH_2 as it increased the hydrophilicity of the membrane [89].

On the other hand, a novel thin layer of MWCNT was incorporated between skin layer of thin film and MF substrate to enhance permeation flux and salt rejection [90]. The results obtained show that modified thin film exhibited high permeation flux of 105 L/m^2h with 95% of Na_2SO_4 rejection. This improved result was due to incorporation of interlayer of MWCNT which makes the substrate more porous and defect free skin layer. Recently, Zheng *et al.* (2017) introduced sulfonated MWCNT into thin film to improve its solubility in aqueous solution. The experimented results demonstrated an increase in water flux by 79.4 ± 5.7 L/m^2.h which is 1.6 times higher than pristine TFC. Moreover, the TFN membrane able to retain higher Na_2SO_4 rejection of 96.8% upon addition of sulfonated MWCNT up to 0.02 wt% [91].

Incorporation of CNT has demonstrated improve performance upon addition in PA layer or substrate layer. Despite having excellent performances, pristine CNT has limited reaction in solvents, surfaces or polymers [92]. This can inevitably hinder the IP process and reduce the performances of membrane.

Graphene oxide (GO) is carbon-based nanomaterials with intrinsic properties, such as high surface area, great mechanical properties and superior hydrophilicity. GO has progressed rapidly into water treatment process due to its abundant oxygen functional groups, such as epoxy, hydroxyl and carboxyl as depicted in Fig. 6 [87]. Reported works by Ali *et al.,* (2016) introduced GO nanosheets into PA layer membrane in an effort to improve TFC membrane. By adding GO nanosheets, the water flux increased 39% compared with pristine TFC, which had water flux of 21.4 L/m^2.h [93]. An attempt was made by incorporating reduced graphene oxide (rGO)/TiO_2 into thin film layer in order to enhance desalination performance, antifouling and chlorine resistance properties [81]. Lai *et al.,* (2016) incorporated GO nanosheet into PSf substrate to study the separation performance of the newly developed TFN made of PSf-GO substrate. From their experimental results, we noticed that all fabricated TFN membranes exhibited high rejection of Na_2SO_4 (94%). Furthermore, TFN (0.3 wt% GO) showed that higher rejection of NaCl, $MgCl_2$ and $MgSO_4$ compared with TFC and other TFN membrane. This membrane achieved 59.5% of NaCl, 62.1% of $MgCl_2$ and 91.1% of $MgSO_4$ rejection in comparison to 31%, 67.1% and 80.4% obtained by using TFC membrane, respectively. Moreover, TFN (0.3 wt%) also exhibited highest permeability with PWP of 2.43 $L/m^2h.bar$ [94]. From the experimented results with variety of inorganic salts tested, it shows a great potential for TFN membrane in water softening application.

With the addition of hydrophilic additives, such as rGO/TiO_2, the water flux of the membrane was increased from 48.3 (TFC) to 61.4 (rGO/TiO_2) $L/m^2.h$ [95]. Besides that, Mansourpanah *et al.* (2015) also introduced novel modifier GO with polyethylene glycol (PEG) into PA layer of TFC membrane. The resulting NF membrane exhibited high Na_2SO_4 salt rejection (93%) compared with unmodified TFC membrane (63%). Besides, the enhancement of salt rejection might be attributed to the presence of the modified PEG/GO nanosheet which has –COOH and –OH functional groups that reduced the membrane pore size [50].

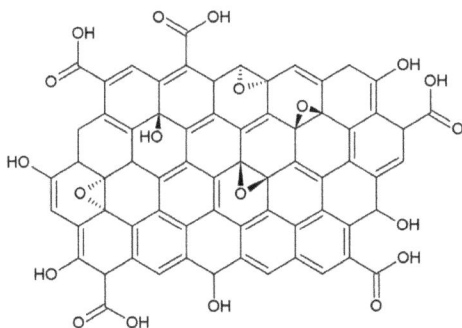

Figure 6. Schematic image of GO structure [63].

In summary, GO has demonstrated better performance upon addition in PA layer or substrate layer. Improvement of water flux, higher salt rejection and improved surface hydrophilicity are among the significant contribution of GO. Despite having a superior property, higher dispersion of GO is difficult to reach which significantly affects its modulus, strength and surface wet ability. Furthermore, increase in cost of large production of GO limits its usage in industry and the lower dispersion of GO remains as the main barrier to be overcome before its employment in another application [96].

5.3   Nanofibers materials

Nanofibers have gained a lot of attention and are widely applied in various applications, such as energy, gas separation membrane and water treatments [97]. In water treatment application, nanofibrous membranes have been either applied in NF process for removal of bivalent ions or desalination through RO and FO processes [98]. Electrospun nanofibers are produced by electrospinning method, which is a simple and versatile technique for membrane fabrication.

Another research reported by Yoon *et al.* (2009) studied polyacrylonitrile (PAN) nanofibers as support layer in comparison with PAN UF membrane in fabrication of TFC membrane. The experimental results revealed that the significant increase in permeate flux, which is 2.4 times higher than TFC membrane. The higher flux achieved might attribute to large open pore structure generated by electrospun nanofibers. Furthermore, the rejection ability of TFC was 99.4 % at 0.25% of PIP solution [99]. Besides that, Kaur *et al.* (2012) fabricated thin film nanofibrous composite (TFNC) membrane with PAN nanofibers support layer. Based on their research works, it reveals the optimum TFNC membrane with 8wt% of PAN which is able to separate $MgSO_4$ of 89% with the flux of 220 $L/m^2.h$ [100].

5.4   Metal organic framework

Metal organic framework (MOF) is a porous crystalline material that has gained a significant interest in membrane separations research [101]. Fig. 7 shows the scanning electron microscopy (SEM) image of zeolitic imidazolate framework (ZIF-8) and particle distribution of ZIF-8 within the range of 120-200 nm. In 2015, for the first time, the porous metal organic framework (MOF) was impregnated into PA layer of TFN based RO membrane [40]. Previously, MOF was widely used in polymer to form mixed matrix membrane (MMM) for gas separation [102,103]. Zeolitic imidazolate framework-8 (ZIF-8) is thermally and chemically stable MOF while possessing similar structure as zeolite and same pore configuration. The experimental results demonstrated that ZIF-8 (0.4 w/v% in TMC hexane) has increased the water flux up to 50 $L/m^2.h$, which is 162% higher than pristine TFC membrane. Meanwhile, the NaCl rejection was found to be 99% at 15.5 bar and 2000 ppm NaCl feed [104]. Despite all the excellent properties and performances of ZIF-8, extended research is required to study incorporation of MOFs into TFC membrane for desalination.

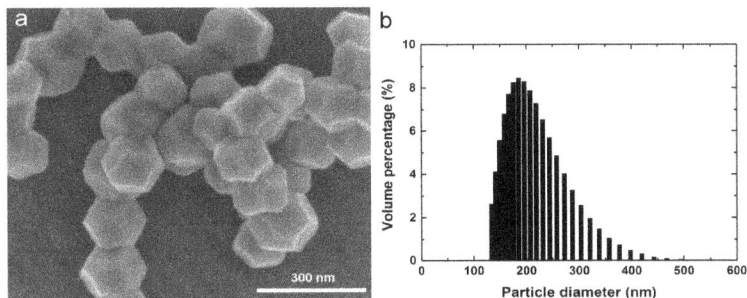

Figure 7. (a) SEM image of ZIF-8 and (b) particle distribution of ZIF-8 [104].

An overview of recent progress in inorganic fillers is demonstrated in Table 3. It can be seen that research effort has been extensively conducted by employing a wide variety of nanomaterials (e.g., silver, TiO_2, CNT, GO and MOF) for NF, RO and FO application in TFC membrane. As revealed in Table 3, it has been shown that the effects of nanofillers in TFC membrane separation, such as selectivity, antifouling properties, surface hydrophilicity and antibacterial properties are summarized. However, some materials, such as silver metal do not play role in separation performance, but it does improve antibacterial properties of fabricated membrane.

Table 3.   Summary of TFN membrane with their nanofillers for improved membrane performances.

Thin layer	Nanofiller	Membrane type	Performance	References
	TFN			
PA	Ag NPs	NF	No change in water flux and salt rejection Good antimicrobial activity at 10% of polymer in organic phase	[29]
PA	TiO_2	NF	Optimum water flux was 9.1 $L/m^2.h$ $MgSO_4$ rejection was 95%	[77]

PA	Ag NPs	NF	Surface hydrophilicity increased Salt rejection slightly increased from 94.7 to 95.6% Increased in biofouling resistance	[74]
PA	Al$_2$O$_3$ NPs	NF	Surface hydrophilicity increased Pure water flux increased No change in salt rejection	[105]
PA	Aminosilanized TiO$_2$	NF	Pure water flux and selectivity increased NaCl rejection: 54% Optimal water flux was 12.3 L/m^2.h	[79]
PA	Modified mesoporous silica	NF	Optimal water flux was 32.4 L/m^2.h Na$_2$SO$_4$ rejection was greater than 80%	[84]
PA	Ag NPs	RO	Improved bio-fouling resistance and virus removal	[75]
PA	rGO/TiO$_2$	NF	Improved water flux from 48.3 to 61.4 L/m^2.h Increased in Na$_2$SO$_4$ rejection (93.57%)	[95]
PA	Silica nanosphere	NF	Improved in membrane hydrophilicity Enhanced in water flux by 22.65 L/m^2.h Increased in MgSO$_4$ rejection (94.81%) at 0.5 Mpa	[32]
PA	PEG/GO	NF	Enhanced in Na$_2$SO$_4$ rejection (93%) Decreased in water flux from 35 to 15 L/m^2.h	[50]
PA	ZIF-8	RO	Increased in water flux about 50 L/m^2.h 98.5% of 2000 ppm NaCl rejection at 15.5 bar	[104]

PA	Mg-Al LDH	RO	Improved water flux from 36.6 to 41.7 $L/m^2.h$ Increased in NaCl rejection:99.3% Enhanced in membrane hydrophilicity	[106]
PA	NH_2-MWCNT	NF	Increased in Na_2SO_4 rejection (95.72%) Improved in antifouling ability	[89]
PA	$Al(OH)_3$	NF	Increased in water flux: 55 $L/m^2.h$ Improved $MgSO_4$ rejection: 95.6%	[107]
PPy	Acrylate-alumoxane	NF	Improved in pure water flux from 9 to 32 $L/m^2.h$ Improved salt rejection:80% Satisfying resistance against fouling	[83]
PA	Sulfonated-MWCNT	NF	Increased in water flux from 57 to 79.4 $L/m^2.h$ Improved in Na_2SO_4 rejection:96.8%	[91]

5.5   Layered double hydroxides nanofillers

Layered materials have been extensively gaining interest among researchers because of their ability to intercalate and exchange ions in layered structure. Examples of layered materials are LDH, layered perovskites, layered transition metal oxide and hydroxyl double salts [108]. From these examples, they are known to have a layered structure that consists of metal ions (either di-or trivalent) surrounded by hydroxyl ions. LDH particularly is a group of two dimensional anionic clays. The general formula of LDH structures is $[M(II)_{1-x}M(III)_x(OH)_2]^{x+}[(A^{n-})_{x/n}.yH_2O]$, where M(II) is divalent cation: Mg^{2+}, Zn^{2+}, Ca^{2+}, Cu^{2+} etc., M(III) is trivalent cation: Al^{3+}, Fe^{3+} and Cr^{3+}, whereas A^{n-} is an interlayer anion. Anions, such as Cl^- and NO^{3-} are intercalated between the layers (see Fig. 8) and classified as one member of the clay minerals family [109]. Furthermore, LDH is hydrophilic in nature due to the presence of OH^- group at its surface and its positive charge is primarily attributed from the isomorphous substitution of clay layer [110].

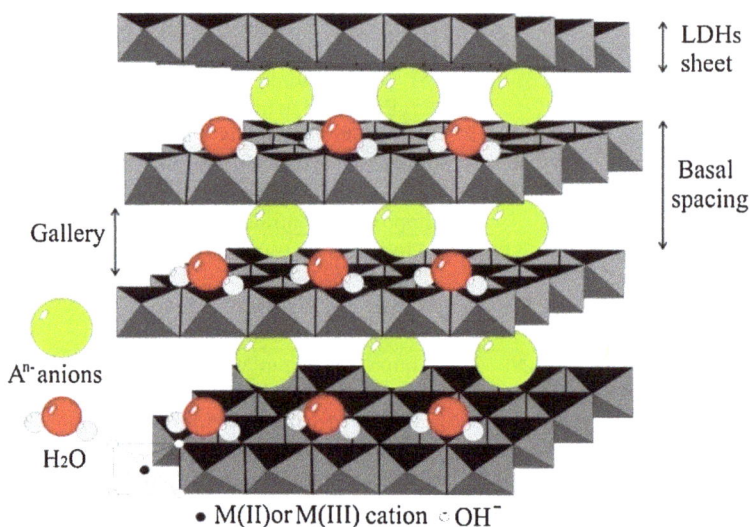

Figure 8.   Schematic diagram represented LDH structures [111].

LDH has been utilized in many applications, such as heavy metals removal, photostability and flame retardancy. With that, Mg-Al LDH is utilized for Pb removal from aqueous solution [112]. Mahjoubi and co-workers employed Zn-Al LDH as adsorbent for dye removal [113]. Furthermore, Co-Ni LDH has been utilized for energy storage devices [114]. The most common method used to synthesize LDH is co-precipitation method [115,116].

Even though LDH materials are not highly porous as zeolite or other MOF, their special structures lie in interlayer gallery which could provide pathway for water and solute transportation. These features were also supported by the hydroxyl functionality that can facilitate water molecules across the membrane. With these special features being mentioned the applicability of LDH in membrane technology could provide another new membrane design for highly permeation and water retention in their own separation process. Many LDH materials are not yet widely explored by researchers in water softening application, its implementation in different area for water treatments (adsorption of heavy metal and natural organic removals) has shown promising results and can be used in this current research to investigate its effects into TFC based membranes towards water softening by NF process.

6. Challenges of incorporating nanomaterials

The progress of TFC membrane in water purification has been extensively studied in recent years. Despite the significant interest and potential demonstrated by nanomaterials in membrane technologies, there are some obstacles in employing them. Addressing these challenges may be the key for improvement in TFN membrane for various applications in future. One of the primary drawbacks encountered by TFN membrane fabrication is the agglomeration of nanomaterials in PA layer. Limited dispersion mainly occurs due to nanoparticles agglomeration and uneven distribution in polymer matrix. The field emission scanning electron microscopy (FE-SEM) images as shown in Fig. 9(b) reveals that particle agglomeration indicated by an arrow in PA ultra-thin layer, which probably reduces the active surface area of nanoparticles and contribute to formation of defects in PA structure. The agglomeration of nanoparticles might change nanocomposite membranes properties, such as pore size, hydrophilicity and antifouling properties. This phenomenon mainly attributed to low dispersion rate of nanofillers in either organic or aqueous solutions used during IP process. Aggregation of nanoparticles may lead to non-uniform dispersion particularly in non-polar organic solvent and contribute to inhomogeneous spread out on top of substrate layer surface [24].

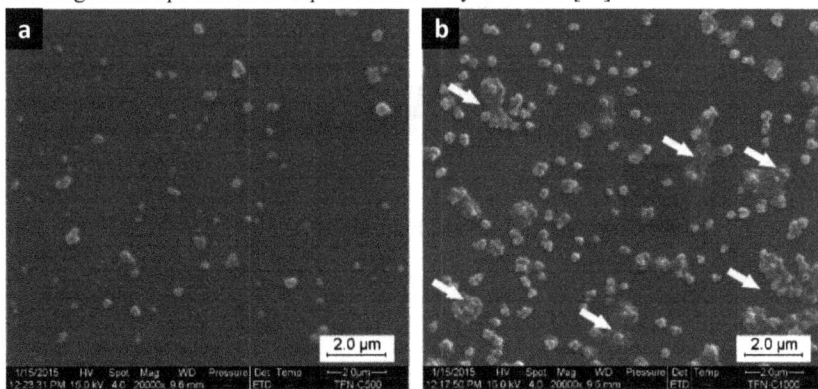

Figure 9. Zeolite NPs incorporated TFN with different NPs concentration, (a) 500 mg/L and (b) 1000 mg/L [70].

The compatibility of nanomaterials with PA matrix is crucial to determine the optimal membrane performance and stability of nanofillers in polymer matrix. Moreover, lack of interaction between nanomaterials and PA matrix would probably cause nanomaterials to be leached out during IP. Furthermore, evaluation of the potential effect in nanomaterials leaching to the environment and its environmental toxicity needs to be assessed periodically.

7.   Membrane fouling: A unique problem for water filtration

In membrane technology process, fouling is one of the main limitations that causes the declined in membrane performances. The performance of membrane separation is highly influenced by water permeability and rejection of salts. Membrane fouling can be defined as the deposition or accumulation of suspended or precipitation that occurred on the exterior of the membrane surface. Fouling can negatively impact the productivity and permeate quality in industrial application such as desalination process that can cause an increasing operating pressure. It leads to increase in energy consumption, frequent membrane cleaning and shortened membrane lifespan.

Similarly, in the context of desalination and water softening, membrane fouling is inevitable. Since seawater contains organic, inorganic and biological compound, desalination membrane would be at risk of all types of fouling as described in the next section. Meanwhile, membrane for water softening process is mainly susceptible to inorganic scaling as it only deals with inorganic divalent salt.

7.1   Common foulant types in membrane filtration

Membrane fouling can be categorised into four types which are organic fouling, inorganic scaling, biofouling and colloidal fouling as illustrated in Fig. 10. The common foulants in the water purification process including emulsified oil, natural organic matters, mineral scales, macromolecules and biofoulants [117].

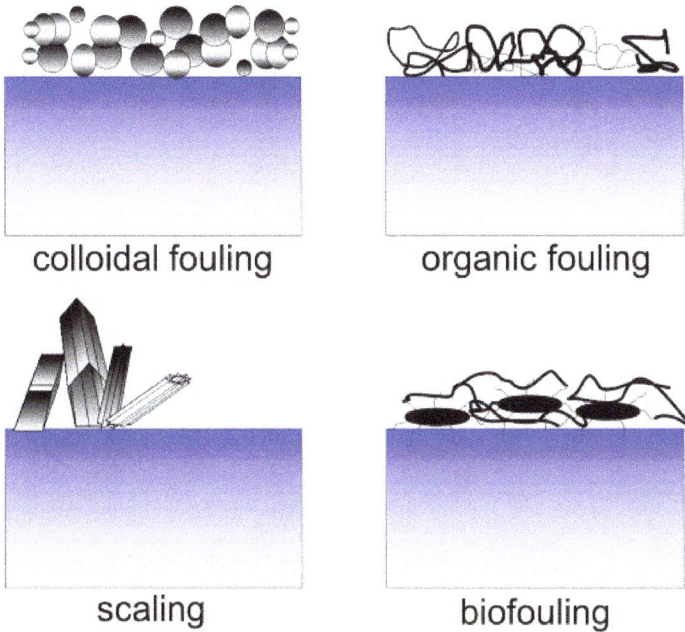

colloidal fouling   **organic fouling**

scaling   **biofouling**

Figure 10.   Formation of organic fouling on membrane [118].

As shown in Fig. 11, membrane scaling is mostly occurred due to precipitation of inorganic salts on the membrane surface and/or caused by the concentration of these salts when they reach their solubility limits. The terms scaling is referred to deposition of certain inorganic salt like $CaCO_3$, $CaSO_4$, $BaSO_4$, $MgSO_4$ etc. on the membrane surface. Moreover, high pressure membrane, which is from the concentration polarization, known as the main factor contributing to the increase in salt concentration thus leading to particle deposition [119]. Another type of fouling, organic fouling is a deposition substance in feed solution that adheres on top of the membrane surface. Natural organic matters (NOM), protein, oils, macromolecules and anti-foaming agents are some common foulants that cause the formation of organic layer on top of the membrane surface which mainly contributed from adsorption of substance. Besides that, an organic fouling is believed to be the key factor that causes the declination of water flux in membrane system specifically when involves the filtration of water containing highly natural organic matters [120].

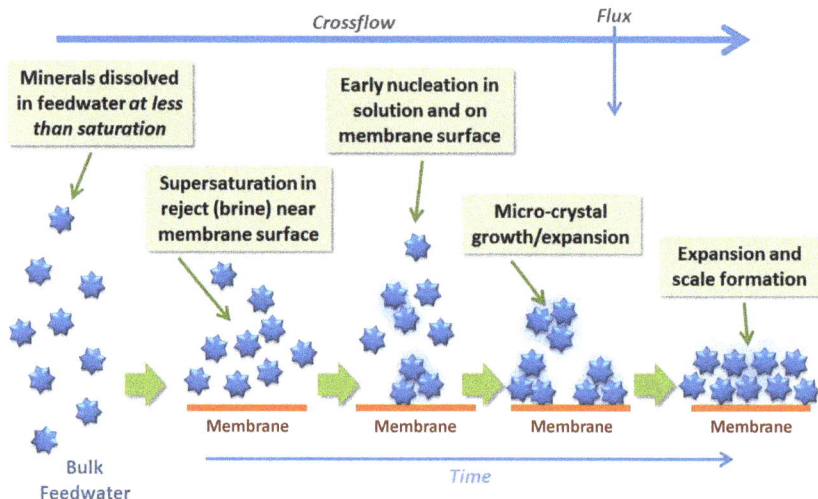

Figure 11.   Mechanism of inorganic scaling on membrane surface [121].

Membrane performances are affected by the colloidal fouling, which is caused by ubiquitous colloidal particles ranging from nanometres to few micrometres in wastewater effluent. One of the major contributors to the colloidal fouling is due to the deposition of particles, and the degree of fouling is highly influenced on the solution chemistry i.e. ionic strength, the presence of specific ions like bivalent ions or multivalent ions and pH [122]. Biofouling is another class of fouling that attributed to the growth of microbial cells with secreted biopolymers that adhere onto the membrane surface. Generally, the attached microorganisms can grow rapidly at the expenses of nutrients in feed solution thus forming biological film on top of membrane surface [123]. Consequently, the biofouling not only affected the membrane transport process, but also biodegrades the PA layer by hydrolysis. The occurrence of biofouling is closely related with the encapsulation of high molecular weight natural polymers, such as extracellular polymeric substances (EPS) [124].

7.2   How to mitigate fouling of membrane

The most obvious solution to this aforementioned problem is membrane cleaning using sodium hypochlorite solution. However, this cleaning process leads to degradation of polymer materials due to long terms exposure of harsh chemicals. One of the effective methods to mitigate fouling is by rendering the antifouling properties through the design of membranes as illustrated in Fig. 12. As of now, several novel modifications have been established to improve and customize the membrane structure in order to enhance its antifouling properties. Generally, the modifications can be developed by reducing the surface roughness, altered the surface hydrophilicity, polymer grafting and introducing the surface functionalities as can be seen in Fig. 13.

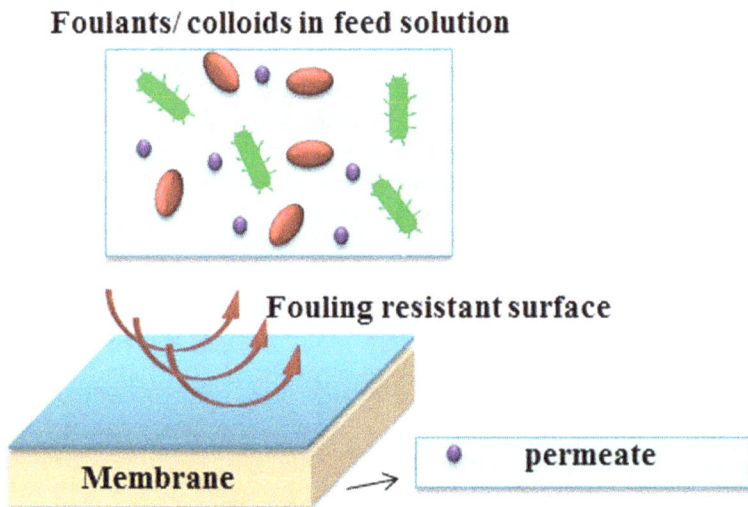

Figure 12.   Fouling resistant surface of a membrane [120].

Recently, the nanocomposite membrane has emerged as one of the advanced materials in improving the membrane fouling. Nanocomposite materials offered various advantages, such as low surface fouling, durability in harsh condition without affecting the salt rejection properties [125]. Incorporating nanofillers into thin film membrane can be designated as one of the effective methods in improving the fouling resistances in membrane process. Many nanomaterials have been widely applied in combating the fouling problems, such as carbon-based materials (CNT, graphene, GO, rGO etc.), silver NPs, copper NPs etc. Moreover, the surface modifications by polymer grafting, zwitterionic, polydopamine coating and layer by layer assembly techniques are also convenient due to their multifunctionality and simplicity.

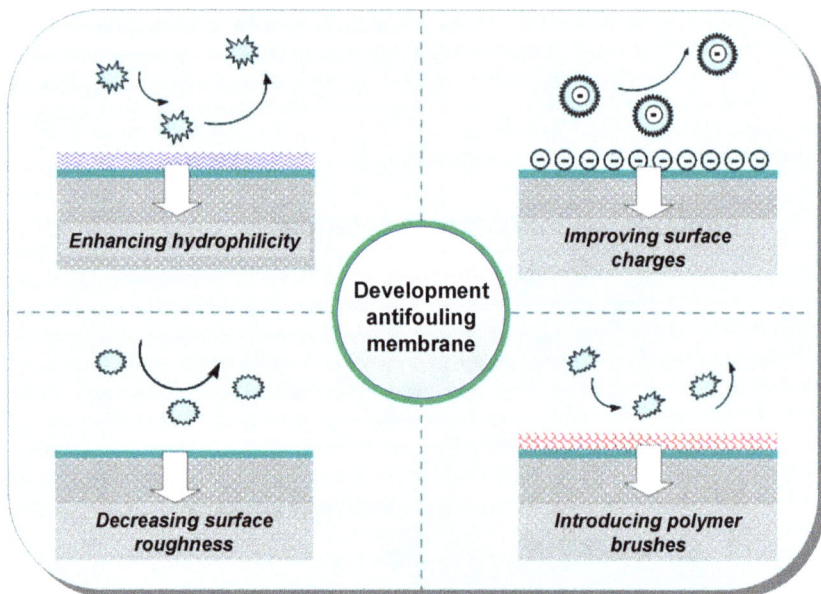

Figure 13.   Summary of surface modification for antifouling enhancement strategies in membrane process [126].

Conclusion

TFN membrane has gained a lot of attention in variety of areas or discipline and served as potential alternative in NF technology. The development and improvement of TFN membrane are an important research in not only for water softening, but also for desalination. Fabrication of TFN membrane can overcome the limitation of TFC membrane in terms of trade-off between permeability and selectivity. Practically, TFC is a good membrane to be used for desalination and treatment of water hardness due to its nanopores which results in excellent divalent salt rejection and good retention of monovalent ions. However, due to the trade-off effect, water flux is usually compromised and impacts in low water permeability.

Therefore, incorporation of nanofillers, such as silica or zeolite into PA layer is proven by exhibited enhanced water permeability. Besides that, impregnating nanofillers in PA layer has shown remarkable performance in antifouling and antibacterial properties compare with conventional TFC membrane. Moreover, with addition of nanomaterials in PA layer also improve membrane stability against chlorine resistance. It is worth to highlight that improvement of membrane by using nanomaterials can greatly affected the membrane performance.

Finally, the practical application of TFN is still in early stage for water treatment. Most of fabrication method is still in laboratory scale and very few reports in large scale production. Thus, with continuous research it is possible to evaluate the effectiveness of TFN in large scale including materials supplies and long-term stability of membranes under practical application.

References

[1]   Y. Song, T. Li, J. Zhou, Z. Li, C. Gao, Analysis of nanofiltration membrane performance during softening process of simulated brackish groundwater, Desalination. 399 (2016) 159–164. https://doi.org/0.1016/j.desal.2016.09.004

[2]   N.A. Ahmad, P.S. Goh, K.C. Wong, A.K. Zulhairun, A.F. Ismail, Enhancing desalination performance of thin film composite membrane through layer by layer assembly of oppositely charged titania nanosheet, Desalination. 476 (2020) 114167. https://doi.org/0.1016/j.desal.2019.114167

[3]   A.R. Anim-Mensah, W.B. Krantz, R. Govind, Studies on polymeric nanofiltration-based water softening and the effect of anion properties on the softening process, Eur. Polym. J. 44 (2008) 2244–2252. https://doi.org/0.1016/j.eurpolymj.2008.04.036

[4]   Z.S. Tai, M.H.A. Aziz, M.H.D. Othman, A.F. Ismail, M.A. Rahman, J. Jaafar, Chapter 8 - An Overview of Membrane Distillation, in: A.F. Ismail, M.A. Rahman, M.H.D. Othman, T. Matsuura (Eds.), Membr. Sep. Princ. Appl., Elsevier, 2019: pp. 251–281. https://doi.org/10.1016/B978-0-12-812815-2.00008-9

[5]   S. Al-Amshawee, M.Y.B.M. Yunus, A.A.M. Azoddein, D.G. Hassell, I.H. Dakhil, H.A. Hasan, Electrodialysis desalination for water and wastewater: A review, Chem. Eng. J. 380 (2020) 122231. https://doi.org/10.1016/j.cej.2019.122231

[6]   F.A. AlMarzooqi, A.A. [Al Ghaferi], I. Saadat, N. Hilal, Application of Capacitive Deionisation in water desalination: A review, Desalination. 342 (2014) 3–15. https://doi.org/10.1016/j.desal.2014.02.031

[7]   M.A. Eltawil, Z. Zhengming, L. Yuan, A review of renewable energy technologies integrated with desalination systems, Renew. Sustain. Energy Rev. 13 (2009) 2245–2262. https://doi.org/0.1016/j.rser.2009.06.011

[8]   T.N. Tuan, S. Chung, J.K. Lee, J. Lee, Improvement of water softening efficiency in capacitive deionization by ultra purification process of reduced graphene oxide, Curr. Appl. Phys. 15 (2015) 1397–1401. https://doi.org/0.1016/j.cap.2015.08.001

[9]   N.N.A. Kadir, M. Shahadat, S. Ismail, Formulation study for softening of hard water using surfactant modified bentonite adsorbent coating, Appl. Clay Sci. 137 (2017) 168–175. https://doi.org/0.1016/j.clay.2016.12.025

[10]   L. Setiawan, L. Shi, R. Wang, Dual layer composite nanofiltration hollow fiber membranes for low-pressure water softening, Polym. (United Kingdom). 55 (2014) 1367–1374. https://doi.org/0.1016/j.polymer.2013.12.032

[11]   W. Fang, L. Shi, R. Wang, Mixed polyamide-based composite nanofiltration hollow fiber membranes with improved low-pressure water softening capability, J. Memb. Sci. 468 (2014) 52–61. https://doi.org/0.1016/j.memsci.2014.05.047

[12]   N. Ghaffour, J. Bundschuh, H. Mahmoudi, M.F.A. Goosen, Renewable energy-driven desalination technologies: A comprehensive review on challenges and potential applications of integrated systems, Desalination. 356 (2015) 94–114. https://doi.org/10.1016/j.desal.2014.10.024

[13]   Y.-N. Wang, R. Wang, Chapter 1 - Reverse Osmosis Membrane Separation Technology, in: A.F. Ismail, M.A. Rahman, M.H.D. Othman, T. Matsuura (Eds.), Membr. Sep. Princ. Appl., Elsevier, 2019: pp. 1–45. https://doi.org/10.1016/B978-0-12-812815-2.00001-6

[14]   N. Ghaffour, T.M. Missimer, G.L. Amy, Technical review and evaluation of the economics of water desalination: Current and future challenges for better water supply sustainability, Desalination. 309 (2013) 197–207. https://doi.org/0.1016/j.desal.2012.10.015

[15]   A. Sharjeel, S. Anwar, A. Nasir, H. Rashid, Design, Development and Performance of Optimum Water Softener, Earth Sci. Pakistan. 3 (2019) 23–28. https://doi.org/0.26480/esp.01.2019.23.28

[16]   C. Liu, L. Shi, R. Wang, Crosslinked layer-by-layer polyelectrolyte nanofiltration hollow fiber membrane for low-pressure water softening with the presence of SO42- in feed water, J. Memb. Sci. 486 (2015) 169–176. https://doi.org/0.1016/j.memsci.2015.03.050

[17]   M.A. Arugula, K.S. Brastad, S.D. Minteer, Z. He, Enzyme catalyzed electricity-driven water softening system, Enzyme Microb. Technol. 51 (2012) 396–401. https://doi.org/0.1016/j.enzmictec.2012.08.009

[18] K.S. Brastad, Z. He, Water softening using microbial desalination cell technology, Desalination. 309 (2013) 32–37. https://doi.org/0.1016/j.desal.2012.09.015

[19] Y. Chen, R. Fan, D. An, Y. Cheng, H. Tan, Water softening by induced crystallization in fluidized bed, J. Environ. Sci. (2016) 2–9. https://doi.org/0.1016/j.jes.2016.08.014

[20] B. Van der Bruggen, H. Goossens, P.A. Everard, K. Stemgée, W. Rogge, Cost-benefit analysis of central softening for production of drinking water, J. Environ. Manage. 91 (2009) 541–549. https://doi.org/0.1016/j.jenvman.2009.09.024

[21] S.J. Seo, H. Jeon, J.K. Lee, G.Y. Kim, D. Park, H. Nojima, J. Lee, S.H. Moon, Investigation on removal of hardness ions by capacitive deionization (CDI) for water softening applications, Water Res. 44 (2010) 2267–2275. https://doi.org/0.1016/j.watres.2009.10.020

[22] T.-H. Yu, H.-Y. Shiu, M. Lee, P.-T. Chiueh, C.-H. Hou, Life cycle assessment of environmental impacts and energy demand for capacitive deionization technology, Desalination. 399 (2016) 53–60. https://doi.org/0.1016/j.desal.2016.08.007

[23] Y.J. Kim, J.H. Choi, Improvement of desalination efficiency in capacitive deionization using a carbon electrode coated with an ion-exchange polymer, Water Res. 44 (2010) 990–996. https://doi.org/0.1016/j.watres.2009.10.017

[24] J. Yin, B. Deng, Polymer-matrix nanocomposite membranes for water treatment, J. Memb. Sci. 479 (2015) 256–275. https://doi.org/0.1016/j.memsci.2014.11.019

[25] P. Wu, M. Imai, Novel Biopolymer Composite Membrane Involved with Selective Mass Transfer and Excellent Water Permeability, in: Adv. Desalin., InTech, 2012. https://doi.org/0.5772/50697

[26] S. Cumming, Global Nanofiltration Membranes Market to Reach $445.1 Million in 2019; Water and Wastewater Claim 74.6% Market Share, BCC Res. (2014)

[27] L. Lin, T.M. Weigand, M.W. Farthing, P. Jutaporn, C.T. Miller, O. Coronell, Relative importance of geometrical and intrinsic water transport properties of active layers in the water permeability of polyamide thin-film composite membranes, J. Memb. Sci. (2018) 1–36. https://doi.org/0.1016/j.memsci.2018.08.002

[28] J. Xiang, Z. Xie, M. Hoang, K. Zhang, Effect of amine salt surfactants on the performance of thin film composite poly(piperazine-amide) nanofiltration membranes, Desalination. 315 (2013) 156–163. https://doi.org/0.1016/j.desal.2012.10.038

[29] S.Y. Lee, H.J. Kim, R. Patel, S.J. Im, J.H. Kim, B.R. Min, Silver nanoparticles immobilized on thin film composite polyamide membrane : characterization ,

nanofiltration , antifouling properties, Polym. Adv. Technol. 18 (2007) 562–568. https://doi.org/0.1002/pat

[30]   M.J. Park, S. Phuntsho, T. He, G.M. Nisola, L.D. Tijing, X.M. Li, G. Chen, W.J. Chung, H.K. Shon, Graphene oxide incorporated polysulfone substrate for the fabrication of flat-sheet thin-film composite forward osmosis membranes, J. Memb. Sci. 493 (2015). https://doi.org/0.1016/j.memsci.2015.06.053

[31]   S. Morales-Torres, C.M.P. Esteves, J.L. Figueiredo, A.M.T. Silva, Thin-film composite forward osmosis membranes based on polysulfone supports blended with nanostructured carbon materials, J. Memb. Sci. 520 (2016) 326–336. https://doi.org/0.1016/j.memsci.2016.07.009

[32]   Q. Li, Y. Wang, J. Song, Y. Guan, H. Yu, X. Pan, F. Wu, M. Zhang, Influence of silica nanospheres on the separation performance of thin film composite poly(piperazine-amide) nanofiltration membranes, Appl. Surf. Sci. 324 (2015) 757–764. https://doi.org/0.1016/j.apsusc.2014.11.031

[33]   D. Emadzadeh, M. Ghanbari, W.J. Lau, M. Rahbari-Sisakht, D. Rana, T. Matsuura, B. Kruczek, A.F. Ismail, Surface modification of thin film composite membrane by nanoporous titanate nanoparticles for improving combined organic and inorganic antifouling properties, Mater. Sci. Eng. C. 75 (2017) 463–470. https://doi.org/0.1016/j.msec.2017.02.079

[34]   T. Ormanci-Acar, F. Celebi, B. Keskin, O. Mutlu-Salmanlı, M. Agtas, T. Turken, A. Tufani, D.Y. Imer, G.O. Ince, T.U. Demir, Y.Z. Menceloglu, S. Unal, I. Koyuncu, Fabrication and characterization of temperature and pH resistant thin film nanocomposite membranes embedded with halloysite nanotubes for dye rejection, Desalination. 429 (2018) 20–32. https://doi.org/0.1016/j.desal.2017.12.005

[35]   N. Hilal, H. Al-Zoubi, N.A. Darwish, A.W. Mohammad, M. Abu Arabi, A comprehensive review of nanofiltration membranes: Treatment, pretreatment, modelling, and atomic force microscopy, Desalination. 170 (2004) 281–308. https://doi.org/0.1016/j.desal.2004.01.007

[36]   A.W. Mohammad, Y.H. Teow, W.L. Ang, Y.T. Chung, D.L. Oatley-Radcliffe, N. Hilal, Nanofiltration membranes review: Recent advances and future prospects, Desalination. 356 (2015) 226–254. https://doi.org/0.1016/j.desal.2014.10.043

[37]   B. Van der Bruggen, M. Mänttäri, M. Nyström, Drawbacks of applying nanofiltration and how to avoid them: A review, Sep. Purif. Technol. 63 (2008) 251–263. https://doi.org/0.1016/j.seppur.2008.05.010

[38]   W.J. Lau, A.F. Ismail, P.S. Goh, N. Hilal, B.S. Ooi, Characterization Methods of Thin Film Composite Nanofiltration Membranes, Sep. Purif. Rev. 44 (2014) 135–156. https://doi.org/0.1080/15422119.2014.882355

[39]   W.J. Lau, S. Gray, T. Matsuura, D. Emadzadeh, J. Paul Chen, A.F. Ismail, A review on polyamide thin film nanocomposite (TFN) membranes: History, applications, challenges and approaches, Water Res. 80 (2015) 306–324. https://doi.org/0.1016/j.watres.2015.04.037

[40]   D. Li, Y. Yan, H. Wang, Recent advances in polymer and polymer composite membranes for reverse and forward osmosis processes, Prog. Polym. Sci. 61 (2015) 104–155. https://doi.org/0.1016/j.progpolymsci.2016.03.003

[41]   S. Al Aani, C.J. Wright, M.A. Atieh, N. Hilal, Engineering nanocomposite membranes: Addressing current challenges and future opportunities, Desalination. (2016). https://doi.org/0.1016/j.desal.2016.08.001

[42]   M. Paul, S.D. Jons, Chemistry and fabrication of polymeric nanofiltration membranes: A review, Polymer (Guildf). 103 (2016) 417–456. https://doi.org/0.1016/j.polymer.2016.07.085

[43]   M.B.M.Y. Ang, Y.L. Ji, S.H. Huang, H.A. Tsai, W.S. Hung, C.C. Hu, K.R. Lee, J.Y. Lai, Incorporation of carboxylic monoamines into thin-film composite polyamide membranes to enhance nanofiltration performance, J. Memb. Sci. 539 (2017) 52–64. https://doi.org/0.1016/j.memsci.2017.05.062

[44]   W. Ye, J. Lin, R. Borrego, D. Chen, A. Sotto, P. Luis, M. Liu, S. Zhao, C.Y. Tang, B. Van der Bruggen, Advanced desalination of dye/NaCl mixtures by a loose nanofiltration membrane for digital ink-jet printing, Sep. Purif. Technol. 197 (2018) 27–35. https://doi.org/0.1016/j.seppur.2017.12.045

[45]   J. Zhu, S. Yuan, A. Uliana, J. Hou, J. Li, X. Li, M. Tian, Y. Chen, A. Volodin, B. Van der Bruggen, High-flux thin film composite membranes for nanofiltration mediated by a rapid co-deposition of polydopamine/piperazine, J. Memb. Sci. 554 (2018) 97–108. https://doi.org/0.1016/j.memsci.2018.03.004

[46]   A.F. Ismail, M. Padaki, N. Hilal, T. Matsuura, W.J. Lau, Thin film composite membrane - Recent development and future potential, Desalination. 356 (2015) 140–148. https://doi.org/0.1016/j.desal.2014.10.042

[47]   Y. Ji, W. Qian, Y. Yu, Q. An, L. Liu, Y. Zhou, C. Gao, Recent developments in nanofiltration membranes based on nanomaterials, Chinese J. Chem. Eng. 25 (2017) 1639–1652. https://doi.org/0.1016/j.cjche.2017.04.014

[48]   N.K. Saha, S. V. Joshi, Performance evaluation of thin film composite polyamide nanofiltration membrane with variation in monomer type, J. Memb. Sci. 342 (2009) 60–69. https://doi.org/0.1016/j.memsci.2009.06.025

[49]   S. Yu, Q. Zhou, S. Shuai, G. Yao, M. Ma, C. Gao, Thin-film composite nanofiltration membranes with improved acid stability prepared from naphthalene-1,3,6-trisulfonylchloride (NTSC) and trimesoyl chloride (TMC), Desalination. 315 (2013) 164–172. https://doi.org/0.1016/j.desal.2012.09.011

[50]   Y. Mansourpanah, H. Shahebrahimi, E. Kolvari, PEG-modified GO nanosheets, a desired additive to increase the rejection and antifouling characteristics of polyamide thin layer membranes, Chem. Eng. Res. Des. 104 (2015) 530–540. https://doi.org/0.1016/j.cherd.2015.09.002

[51]   S.Y. Kwak, S.H. Kim, S.S. Kim, Hybrid organic/inorganic reverse osmosis (RO) membrane for bactericidal anti-fouling. 1. Preparation and characterization of TiO2 nanoparticle self-assembled aromatic polyamide thin-film-composite (TFC) membrane, Environ. Sci. Technol. 35 (2001) 2388–2394. https://doi.org/0.1021/es0017099

[52]   W.J. Lau, A.F. Ismail, N. Misdan, M.A. Kassim, A recent progress in thin film composite membrane: A review, Desalination. 287 (2012) 190–199. https://doi.org/0.1016/j.desal.2011.04.004

[53]   M. Liu, Y. Zheng, S. Shuai, Q. Zhou, S. Yu, C. Gao, Thin-film composite membrane formed by interfacial polymerization of polyvinylamine (PVAm) and trimesoyl chloride (TMC) for nanofiltration, Desalination. 288 (2012) 98–107. https://doi.org/0.1016/j.desal.2011.12.018

[54]   M. Dalwani, N.E. Benes, G. Bargeman, D. Stamatialis, M. Wessling, Effect of pH on the performance of polyamide/polyacrylonitrile based thin film composite membranes, J. Memb. Sci. 372 (2011) 228–238. https://doi.org/0.1016/j.memsci.2011.02.012

[55]   Y.-F. Mi, Q. Zhao, Y.-L. Ji, Q.-F. An, C.-J. Gao, A novel route for surface zwitterionic functionalization of polyamide nanofiltration membranes with improved performance, J. Memb. Sci. 490 (2015) 311–320. https://doi.org/0.1016/j.memsci.2015.04.072

[56]   L. Lin, R. Lopez, G.Z. Ramon, O. Coronell, Investigating the void structure of the polyamide active layers of thin-film composite membranes, J. Memb. Sci. 497 (2016) 365–376. https://doi.org/0.1016/j.memsci.2015.09.020

[57]   N. Misdan, W.J. Lau, C.S. Ong, A.F. Ismail, T. Matsuura, Study on the thin film composite poly(piperazine-amide) nanofiltration membranes made of different polymeric substrates: Effect of operating conditions, Korean J. Chem. Eng. 32 (2015) 753–760. https://doi.org/0.1007/s11814-014-0261-6

[58]   A.K. Ghosh, B.H. Jeong, X. Huang, E.M. V Hoek, Impacts of reaction and curing conditions on polyamide composite reverse osmosis membrane properties, J. Memb. Sci. 311 (2008) 34–45. https://doi.org/0.1016/j.memsci.2007.11.038

[59]   S. Hermans, R. Bernstein, A. Volodin, I.F.J. Vankelecom, Study of synthesis parameters and active layer morphology of interfacially polymerized polyamide-polysulfone membranes, React. Funct. Polym. 86 (2015) 199–208. https://doi.org/0.1016/j.reactfunctpolym.2014.09.013

[60]   X. lei Wang, J. fu Wei, Z. Dai, K. yin Zhao, H. Zhang, Preparation and characterization of negatively charged hollow fiber nanofiltration membrane by plasma-induced graft polymerization, Desalination. 286 (2012) 138–144. https://doi.org/0.1016/j.desal.2011.11.014

[61]   H. Deng, Y. Xu, Q. Chen, X. Wei, B. Zhu, High flux positively charged nanofiltration membranes prepared by UV-initiated graft polymerization of methacrylatoethyl trimethyl ammonium chloride (DMC) onto polysulfone membranes, J. Memb. Sci. 366 (2011) 363–372. doi:https://doi.org/10.1016/j.memsci.2010.10.029

[62]   R.H. Li, T.A. Barbari, Performance of poly(vinyl alcohol) thin-gel composite ultrafiltration membranes, J. Memb. Sci. 105 (1995) 71–78. https://doi.org/0.1016/0376-7388(95)00048-H

[63]   H.R. Chae, J. Lee, C.H. Lee, I.C. Kim, P.K. Park, Graphene oxide-embedded thin-film composite reverse osmosis membrane with high flux, anti-biofouling, and chlorine resistance, J. Memb. Sci. 483 (2015) 128–135. https://doi.org/0.1016/j.memsci.2015.02.045

[64]   F. Yan, H. Chen, Y. Lü, Z. Lü, S. Yu, M. Liu, C. Gao, Improving the water permeability and antifouling property of thin-film composite polyamide nanofiltration membrane by modifying the active layer with triethanolamine, J. Memb. Sci. 513 (2016) 108–116. https://doi.org/0.1016/j.memsci.2016.04.049

[65]   J.H. Lee, J.Y. Chung, E.P. Chan, C.M. Stafford, Correlating chlorine-induced changes in mechanical properties to performance in polyamide-based thin film composite membranes, J. Memb. Sci. 433 (2013) 72–79. https://doi.org/0.1016/j.memsci.2013.01.026

[66]   D. Zhao, S. Yu, A review of recent advance in fouling mitigation of NF/RO membranes in water treatment: pretreatment, membrane modification, and chemical cleaning, Desalin. Water Treat. 55 (2014) 1–22. https://doi.org/0.1080/19443994.2014.928804

[67]   S. Zhu, S. Zhao, Z. Wang, X. Tian, M. Shi, J. Wang, S. Wang, Improved performance of polyamide thin-film composite nanofiltration membrane by using polyetersulfone/polyaniline membrane as the substrate, J. Memb. Sci. 493 (2015) 263–274. https://doi.org/0.1016/j.memsci.2015.07.013

[68]   M. Fathizadeh, A. Aroujalian, A. Raisi, Effect of added NaX nano-zeolite into polyamide as a top thin layer of membrane on water flux and salt rejection in a reverse osmosis process, J. Memb. Sci. 375 (2011) 88–95. https://doi.org/0.1016/j.memsci.2011.03.017

[69]   N. Ma, J. Wei, R. Liao, C.Y. Tang, Zeolite-polyamide thin film nanocomposite membranes: Towards enhanced performance for forward osmosis, J. Memb. Sci. 405–406 (2012) 149–157. https://doi.org/0.1016/j.memsci.2012.03.002

[70]   L.X. Dong, X.C. Huang, Z. Wang, Z. Yang, X.M. Wang, C.Y. Tang, A thin-film nanocomposite nanofiltration membrane prepared on a support with in situ embedded zeolite nanoparticles, Sep. Purif. Technol. 166 (2016) 230–239. https://doi.org/0.1016/j.seppur.2016.04.043

[71]   E. Bet-moushoul, Y. Mansourpanah, K. Farhadi, M. Tabatabaei, TiO2 nanocomposite based polymeric membranes: a review on performance improvement for various applications in chemical engineering processes, Chem. Eng. J. 283 (2015) 29–46. https://doi.org/0.1016/j.cej.2015.06.124

[72]   A.L. Ahmad, A.A. Abdulkarim, B.S. Ooi, S. Ismail, Recent development in additives modifications of polyethersulfone membrane for flux enhancement, Chem. Eng. J. 223 (2013) 246–267. https://doi.org/0.1016/j.cej.2013.02.130

[73]   S. Hermans, H. Mariën, C. Van Goethem, I.F. Vankelecom, Recent developments in thin film (nano)composite membranes for solvent resistant nanofiltration, Curr. Opin. Chem. Eng. 8 (2015) 45–54. https://doi.org/0.1016/j.coche.2015.01.009

[74]   E.S. Kim, G. Hwang, M. Gamal El-Din, Y. Liu, Development of nanosilver and multi-walled carbon nanotubes thin-film nanocomposite membrane for enhanced water treatment, J. Memb. Sci. 394–395 (2012) 37–48. https://doi.org/0.1016/j.memsci.2011.11.041

[75] M. Ben-Sasson, X. Lu, E. Bar-Zeev, K.R. Zodrow, S. Nejati, G. Qi, E.P. Giannelis, M. Elimelech, In situ formation of silver nanoparticles on thin-film composite reverse osmosis membranes for biofouling mitigation, Water Res. 62 (2014) 260–270. https://doi.org/0.1016/j.watres.2014.05.049

[76] J. Yin, Y. Yang, Z. Hu, B. Deng, Attachment of silver nanoparticles (AgNPs) onto thin-film composite (TFC) membranes through covalent bonding to reduce membrane biofouling, J. Memb. Sci. 441 (2013) 73–82. https://doi.org/0.1016/j.memsci.2013.03.060

[77] H.S. Lee, S.J. Im, J.H. Kim, H.J. Kim, J.P. Kim, B.R. Min, Polyamide thin-film nanofiltration membranes containing TiO2 nanoparticles, Desalination. 219 (2008) 48–56. https://doi.org/0.1016/j.desal.2007.06.003

[78] S. Pourjafar, A. Rahimpour, M. Jahanshahi, Synthesis and characterization of PVA/PES thin film composite nanofiltration membrane modified with TiO 2 nanoparticles for better performance and surface properties, J. Ind. Eng. Chem. 18 (2012) 1398–1405. https://doi.org/0.1016/j.jiec.2012.01.041

[79] B. Rajaeian, A. Rahimpour, M.O. Tade, S. Liu, Fabrication and characterization of polyamide thin film nanocomposite (TFN) nanofiltration membrane impregnated with TiO2 nanoparticles, Desalination. 313 (2013) 176–188. https://doi.org/0.1016/j.desal.2012.12.012

[80] L.Y. Ng, A.W. Mohammad, C.P. Leo, N. Hilal, Polymeric membranes incorporated with metal/metal oxide nanoparticles: A comprehensive review, Desalination. 308 (2013) 15–33. https://doi.org/0.1016/j.desal.2010.11.033

[81] M. Safarpour, A. Khataee, V. Vatanpour, Thin film nanocomposite reverse osmosis membrane modified by reduced graphene oxide/TiO2 with improved desalination performance, J. Memb. Sci. 489 (2015) 43–54. doi:https://doi.org/10.1016/j.memsci.2015.04.010

[82] G.L. Jadav, P.S. Singh, Synthesis of novel silica-polyamide nanocomposite membrane with enhanced properties, 328 (2009) 257–267. https://doi.org/0.1016/j.memsci.2008.12.014

[83] N. Ghaemi, Novel antifouling nano-enhanced thin-film composite membrane containing cross-linkable acrylate-alumoxane nanoparticles for water softening, J. Colloid Interface Sci. 485 (2017) 81–90. https://doi.org/0.1016/j.jcis.2016.09.035

[84]   H. Wu, B. Tang, P. Wu, Optimizing polyamide thin film composite membrane covalently bonded with modified mesoporous silica nanoparticles, J. Memb. Sci. 428 (2013) 341–348. https://doi.org/0.1016/j.memsci.2012.10.053

[85]   M. Bao, G. Zhu, L. Wang, M. Wang, C. Gao, Preparation of monodispersed spherical mesoporous nanosilica-polyamide thin film composite reverse osmosis membranes via interfacial polymerization, Desalination. 309 (2013) 261–266. https://doi.org/0.1016/j.desal.2012.10.028

[86]   D. Hu, Z.-L. Xu, C. Chen, Polypiperazine-amide nanofiltration membrane containing silica nanoparticles prepared by interfacial polymerization, Des. 301 (2012) 75–81. https://doi.org/0.1016/j.desal.2012.06.015

[87]   Y. Zhang, B. Wu, H. Xu, H. Liu, M. Wang, Y. He, B. Pan, Nanomaterials-enabled water and wastewater treatment, NanoImpact. 3–4 (2016) 22–39. https://doi.org/0.1016/j.impact.2016.09.004

[88]   J.N. Shen, C.C. Yu, H.M. Ruan, C.J. Gao, B. Van der Bruggen, Preparation and characterization of thin-film nanocomposite membranes embedded with poly(methyl methacrylate) hydrophobic modified multiwalled carbon nanotubes by interfacial polymerization, J. Memb. Sci. 442 (2013) 18–26. https://doi.org/0.1016/j.memsci.2013.04.018

[89]   H. Zarrabi, M.E. Yekavalangi, V. Vatanpour, A. Shockravi, M. Safarpour, Improvement in desalination performance of thin film nanocomposite nanofiltration membrane using amine-functionalized multiwalled carbon nanotube, Desalination. 394 (2016) 83–90. https://doi.org/0.1016/j.desal.2016.05.002

[90]   M.B. Wu, Y. Lv, H.C. Yang, L.F. Liu, X. Zhang, Z.K. Xu, Thin film composite membranes combining carbon nanotube intermediate layer and microfiltration support for high nanofiltration performances, J. Memb. Sci. 515 (2016) 238–244. https://doi.org/0.1016/j.memsci.2016.05.056

[91]   J. Zheng, M. Li, K. Yu, J. Hu, X. Zhang, L. Wang, Sulfonated multiwall carbon nanotubes assisted thin-film nanocomposite membrane with enhanced water flux and anti-fouling property, J. Memb. Sci. 524 (2017) 344–353. https://doi.org/0.1016/j.memsci.2016.11.032

[92]   S. Roy, S.A. Ntim, S. Mitra, K.K. Sirkar, Facile fabrication of superior nanofiltration membranes from interfacially polymerized CNT-polymer composites, J. Memb. Sci. 375 (2011) 81–87. https://doi.org/0.1016/j.memsci.2011.03.012

[93] M.E.A. Ali, L. Wang, X. Wang, X. Feng, Thin film composite membranes embedded with graphene oxide for water desalination, Desalination. 386 (2016) 67–76. https://doi.org/0.1016/j.desal.2016.02.034

[94] G.S. Lai, W.J. Lau, P.S. Goh, A.F. Ismail, N. Yusof, Y.H. Tan, Graphene oxide incorporated thin film nanocomposite nanofiltration membrane for enhanced salt removal performance, Desalination. 387 (2016) 14–24. https://doi.org/0.1016/j.desal.2016.03.007

[95] M. Safarpour, V. Vatanpour, A. Khataee, M. Esmaeili, Development of a novel high flux and fouling-resistant thin film composite nanofiltration membrane by embedding reduced graphene oxide/TiO2, Sep. Purif. Technol. 154 (2015) 96–107. https://doi.org/10.1016/j.seppur.2015.09.039

[96] A. Ammar, A.M. Al-Enizi, M.A. AlMaadeed, A. Karim, Influence of graphene oxide on mechanical, morphological, barrier, and electrical properties of polymer membranes, Arab. J. Chem. 9 (2016) 274–286. https://doi.org/0.1016/j.arabjc.2015.07.006

[97] X. Wang, B.S. Hsiao, Electrospun nanofiber membranes, Curr. Opin. Chem. Eng. 12 (2016) 62–81. https://doi.org/0.1016/j.coche.2016.03.001

[98] S. Subramanian, R. Seeram, New directions in nano fi ltration applications — Are nano fi bers the right materials as membranes in desalination ?, 308 (2013) 198–208. https://doi.org/0.1016/j.desal.2012.08.014

[99] K. Yoon, B.S. Hsiao, B. Chu, High flux nanofiltration membranes based on interfacially polymerized polyamide barrier layer on polyacrylonitrile nanofibrous scaffolds, J. Memb. Sci. 326 (2009) 484–492. https://doi.org/0.1016/j.memsci.2008.10.023

[100] S. Kaur, S. Sundarrajan, D. Rana, T. Matsuura, S. Ramakrishna, Influence of electrospun fiber size on the separation efficiency of thin film nanofiltration composite membrane, J. Memb. Sci. 392–393 (2012) 101–111. https://doi.org/0.1016/j.memsci.2011.12.005

[101] Y. Li, L.H. Wee, J.A. Martens, I.F.J. Vankelecom, Interfacial synthesis of ZIF-8 membranes with improved nanofiltration performance, J. Memb. Sci. (2016). https://doi.org/10.1016/j.memsci.2016.09.065

[102] S. Sorribas, P. Gorgojo, C. Téllez, J. Coronas, A.G. Livingston, High flux thin film nanocomposite membranes based on metal-organic frameworks for organic solvent

nanofiltration, J. Am. Chem. Soc. 135 (2013) 15201–15208.
https://doi.org/0.1021/ja407665w

[103] Y. Lin, Metal organic framework membranes for separation applications, Curr. Opin. Chem. Eng. 8 (2015) 21–28. https://doi.org/0.1016/j.coche.2015.01.006

[104] J. Duan, Y. Pan, F. Pacheco, E. Litwiller, Z. Lai, I. Pinnau, High-performance polyamide thin-film-nanocomposite reverse osmosis membranes containing hydrophobic zeolitic imidazolate framework-8, J. Memb. Sci. 476 (2015) 303–310. https://doi.org/0.1016/j.memsci.2014.11.038

[105] T.A. Saleh, V.K. Gupta, Synthesis and characterization of alumina nano-particles polyamide membrane with enhanced flux rejection performance, Sep. Purif. Technol. 89 (2012) 245–251. https://doi.org/0.1016/j.seppur.2012.01.039

[106] H. Dong, L. Wu, L. Zhang, H. Chen, C. Gao, Clay nanosheets as charged filler materials for high-performance and fouling-resistant thin film nanocomposite membranes, J. Memb. Sci. 494 (2015) 92–103. https://doi.org/0.1016/j.memsci.2015.07.049

[107] H. Li, W. Shi, Y. Zhang, Q. Du, X. Qin, Y. Su, Improved performance of poly(piperazine amide) composite nanofiltration membranes by adding aluminum hydroxide nanospheres, Sep. Purif. Technol. 166 (2016) 240–251. https://doi.org/0.1016/j.seppur.2016.04.024

[108] S.B. Tyagi, A. Kharkwal, Nitu, M. Kharkwal, R. Sharma, Synthesis and Characterization of Layered Double Hydroxides Containing Optically Active Transition Metal Ion, Solid State Sci. 63 (2017) 93–102. https://doi.org/0.1016/j.solidstatesciences.2016.11.012

[109] M.H. Tajuddin, N. Yusof, W. Norharyati, W. Salleh, Incorporation of layered double nanomaterials in thin film nanocomposite nanofiltration membrane for magnesium sulphate removal, E3S Web Conf. 02003 (2018) 0–7. https://doi.org/10.1051/e3sconf/20183402003

[110] Z. Meng, Y. Zhang, Q. Zhang, X. Chen, L. Liu, Novel synthesis of layered double hydroxides (LDHs) from zinc hydroxide, Appl. Surf. Sci. 396 (2017) 799–803. https://doi.org/0.1016/j.apsusc.2016.11.032

[111] M.A. Djebbi, M. Braiek, P. Namour, A. Ben Haj Amara, N. Jaffrezic-Renault, Layered double hydroxide materials coated carbon electrode: New challenge to future electrochemical power devices, Appl. Surf. Sci. 386 (2016) 352–363. https://doi.org/0.1016/j.apsusc.2016.06.032

[112] D. Zhao, G. Sheng, J. Hu, C. Chen, X. Wang, The adsorption of Pb(II) on Mg2Al layered double hydroxide, Chem. Eng. J. 171 (2011) 167–174. https://doi.org/0.1016/j.cej.2011.03.082

[113] F.Z. Mahjoubi, A. Khalidi, M. Abdennouri, N. Barka, Zn–Al layered double hydroxides intercalated with carbonate, nitrate, chloride and sulfate ions: Synthesis, characterization and dyes removal properties, J. Taibah Univ. Sci. 11 (2015) 90–100. https://doi.org/0.1016/j.jtusci.2015.10.007

[114] X. Meng, M. Feng, H. Zhang, Z. Ma, C. Zhang, Solvothermal synthesis of cobalt/nickel layered double hydroxides for energy storage devices, J. Alloys Compd. 695 (2016) 3522–3529. https://doi.org/0.1016/j.jallcom.2016.11.419

[115] S.S.L. Sobhana, D.R. Bogati, M. Reza, J. Gustafsson, P. Fardim, Cellulose biotemplates for layered double hydroxides networks, Microporous Mesoporous Mater. 225 (2016) 66–73. https://doi.org/0.1016/j.micromeso.2015.12.009

[116] F.L. Theiss, G.A. Ayoko, R.L. Frost, Synthesis of layered double hydroxides containing Mg2+, Zn2+, Ca2+ and Al3+ layer cations by co-precipitation methods - A review, Appl. Surf. Sci. 383 (2016) 200–213. https://doi.org/0.1016/j.apsusc.2016.04.150

[117] Y.R. Chang, Y.J. Lee, D.J. Lee, Membrane fouling during water or wastewater treatments: Current research updated, J. Taiwan Inst. Chem. Eng. 94 (2019) 88–96. https://doi.org/0.1016/j.jtice.2017.12.019

[118] A. Tiraferri, Membrane-based water treatment to increase water supply, (2014) 1–13. http://www.colloid.ch/index.php?name=membranes

[119] P.S. Goh, W.J. Lau, M.H.D. Othman, A.F. Ismail, Membrane fouling in desalination and its mitigation strategies, Desalination. 425 (2018) 130–155. https://doi.org/0.1016/j.desal.2017.10.018

[120] J. Ayyavoo, T.P.N. Nguyen, B.M. Jun, I.C. Kim, Y.N. Kwon, Protection of polymeric membranes with antifouling surfacing via surface modifications, Colloids Surfaces A Physicochem. Eng. Asp. 506 (2016) 190–201. https://doi.org/0.1016/j.colsurfa.2016.06.026

[121] C. Piyadasa, H.F. Ridgway, T.R. Yeager, M.B. Stewart, C. Pelekani, S.R. Gray, J.D. Orbell, The application of electromagnetic fields to the control of the scaling and biofouling of reverse osmosis membranes - A review, Desalination. 418 (2017) 19–34. doi:https://doi.org/10.1016/j.desal.2017.05.017

[122] J. Kim, M. Jun, M. Park, H. Kyong, S. Kim, J. Ha, Influence of colloidal fouling on pressure retarded osmosis, Desalination. 389 (2016) 207–214

[123] A. Mollahosseini, A. Rahimpour, A new concept in polymeric thin-film composite nanofiltration membranes with antibacterial properties, Biofouling. 29 (2013) 537–548. https://doi.org/0.1080/08927014.2013.777953

[124] C.S. Ong, P.S. Goh, W.J. Lau, N. Misdan, A.F. Ismail, Nanomaterials for biofouling and scaling mitigation of thin film composite membrane: A review, Desalination. 393 (2016) 2–15. https://doi.org/0.1016/j.desal.2016.01.007

[125] J.H. Jhaveri, Z.V.P. Murthy, A comprehensive review on anti-fouling nanocomposite membranes for pressure driven membrane separation processes, Desalination. 379 (2016) 137–154. https://doi.org/0.1016/j.desal.2015.11.009

[126] G. Kang, Y. Cao, Development of antifouling reverse osmosis membranes for water treatment: A review, Water Res. 46 (2012) 584–600. doi:https://doi.org/10.1016/j.watres.2011.11.041

Chapter 5

Polymeric Reverse Osmosis and Forward Osmosis Membranes for Water Desalination

Wei Lun Ang[1,2]*, Pui Vun Chai[3]

[1]Department of Chemical and Process Engineering, Faculty of Engineering and Built Environment, Universiti Kebangsaan Malaysia, 43600 UKM Bangi, Selangor, Malaysia

[2]Centre for Sustainable Process Technology, Faculty of Engineering and Built Environment, Universiti Kebangsaan Malaysia, 43600 UKM Bangi, Selangor, Malaysia

[3] Department of Chemical and Petroleum Engineering, Faculty of Engineering, Technology and Built Environment, UCSI University, 56000, Cheras, Kuala Lumpur,Malaysia

*wl_ang@ukm.edu.my

Abstract

Polymeric reverse osmosis (RO) and forward osmosis (FO) membranes have been predominantly used in membrane applications for water desalination. The membrane science has advanced in the past decades and various efforts have been employed to improve the membrane characteristics for enhanced water flux and impurities rejection capability. In this chapter, the progress of RO and FO membranes has been discussed in three sections: synthesis methods of RO and FO membranes, modification works done on the membranes and the formulation used for the synthesis of RO and FO membranes, with particular interest given to the incorporation of nanoparticles in the synthesis of thin film composite membrane.

Keywords

Reverse Osmosis, Forward Osmosis, Desalination, Polymeric Membrane, Membrane Synthesis and Modification, Thin Film Composite Membrane

Contents

1.   Introduction

Desalination has been recognized as one of the most promising technologies to minimize the stress arising from water shortage and water scarcity. The advancement of membrane science has contributed to the widespread acceptance of desalination globally due to its advantageous features over conventional thermal-based desalination processes, such as high productivity (recovery rate), high energy efficiency (lower specific energy consumption), small physical footprint (lower space requirement due to the process and plant compactness), operational simplicity, and ease of process automation [1–3]. Furthermore, the carbon dioxide (CO_2) emission from seawater reverse osmosis plant at 1.7-2.8 $kgCO_2/m^3$ is much lower compared to thermal-based desalination processes (multi-stage flash and multi-effect distillation) at 7-25 $kgCO_2/m^3$ [4]. Currently, membrane-based seawater desalination is dominated by seawater reverse osmosis process, where it acquires a share of 65% global production [5]. Reverse osmosis (RO), being the membrane that leads to the success of membrane-based desalination, utilizes external pressure to force water permeates through the membrane while leaving the salt and other molecules behind, as shown in Fig. 1(a). Naturally, osmosis process is a process resulted from the movement of the solvent molecules (usually water) across a membrane from low to high solute concentration (usually salt) until osmotic equilibrium state for both sides is achieved. At this state, the system is said to possess a minimum pressure or osmotic pressure that maintains the equilibrium of the system. On the other hand, RO process is a process that reverses the direction of the solvent flow by the assistance of external pressure that is higher than the osmotic pressure of the system [3].

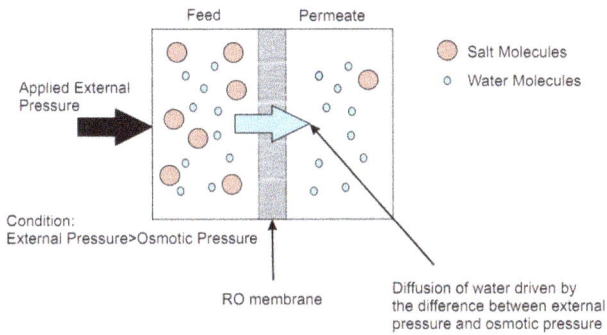

(a)

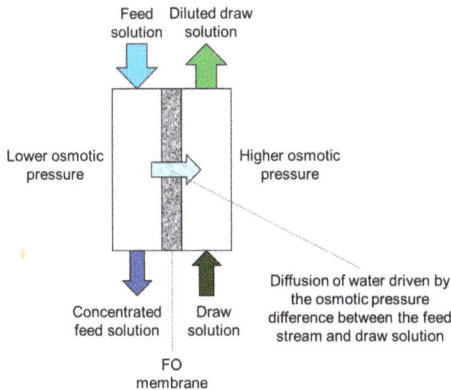

(b)

Figure 1.Working mechanism of (a) RO (adapted from [6]) and (b) FO (adapted from [7])

Another interesting membrane process that recently gained tremendous attention from the desalination researchers is forward osmosis (FO). Unlike the typical pressure-driven RO process, FO is a non-pressure driven process that utilizes the osmotic pressure difference between two streams (known as feed and draw solutions as shown in Fig. 1(b)) to allow the permeation of water from feed stream to draw solution of higher osmotic pressure [7]. The feed stream will be concentrated while the draw solution will be diluted. The latter can then be regenerated by extracting the water through additional separation processes [8]. This concentration-dilution concept of FO can be applied to desalinate water albeit indirectly (the need to have additional separation processes to extract the clean water). On another note, innovative design has been undertaken to utilize the concentration-dilution nature of FO membrane to improve the desalination process, where FO has been employed as pre-treatment to dilute the seawater (draw) with reclaimed wastewater (feed) before the seawater is sent to RO desalination process [9,10]. This offers the potential for the RO desalination process to be operated at lower operating pressure due to the reduced osmotic pressure of diluted seawater.

The successful application of RO and FO for water desalination is mainly attributed to the advancement of the membrane itself. However, membrane-based desalination still has room for improvements, such as the potential to lower the energy consumption and unit cost (to be as competitive as treated water from surface water sources) as well as more resistant to fouling (consistent performance for a long period of time with minimal downtime for cleaning and membrane replacement) [11]. To realize these aims, the RO/FO membrane with the desired characteristics such as excellent water permeability with good rejection rate for salt and foulants, and great resistance to fouling or membrane degradation should be synthesized. The minor difference between RO and FO membranes is that the latter is expected to have lower structural parameter (thinner membrane) without compromising the mechanical strength. The membrane science has come a long way with researches on the synthesis/modification approaches and material formulation has been actively conducted by the membrane researchers. These topics will be further elaborated in the following section.

2.   Development of RO and FO polymeric membranes

Among the various types of membranes, polymeric membrane acquires the largest share of membrane research and market due to its low material cost, good mechanical strength and chemical stability [11]. Similarly, almost all existing commercial RO and FO are polymeric membranes. The first RO membrane was developed by Loeb-Sourirajan where cellulose acetate (CA) polymer was used to synthesize the membrane [12]. This CA RO membrane was capable of producing potable water from sodium chloride (NaCl) brine at appreciable water flux and reasonable pressure. The invention of Loeb-Sourirajan membrane fabrication procedure in the 1960s has substantially boosted the application of membranes in the later day [1]. Although the CA RO membrane possesses desalination capability, its relatively weak chemical, thermal and physical resistance has resulted in various membrane modification and synthesis efforts to overcome these weaknesses [13]. Later, Cadotte developed thin film composite (TFC) RO membrane via interfacial polymerization (IP). The typical TFC membrane consists of active skin layer (usually polyamide (PA) layer), support layer (usually polymeric polysulfone (PSf) material) and substrate (non-woven polyester (PET) fabric), as shown in Fig. 2. The emerging of TFC RO membrane in the 1970s is one of the most remarkable advancements in the history of RO membrane [14]. In comparison to CA RO membrane, the TFC RO membrane could perform at lower pressure without compromising the salt rejection and has better physicochemical properties. Following on, biological membrane (biomimetic membrane) has been synthesized for desalination application and the performance of the membrane in term of water permeability is reportedly higher than commercially available RO membrane. However, the commercialization of the biomimetic membrane for desalination market is yet to be seen [15]. Overall, the seek for improvement in RO membrane technology continues to grow and attention has been shifted to resolve the few issues currently faced by membrane science, such as the trade-off between salt rejection and water permeability, membrane scaling, physicochemical properties and fouling.

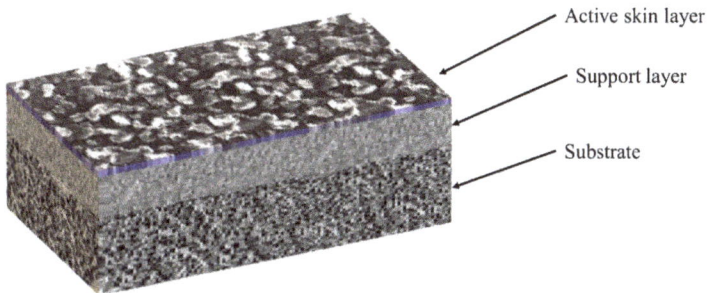

Figure 2.   *Typical structure of TFC membrane*

RO membrane has been employed for FO application prior to the available of membrane specifically synthesized for FO process. However, the water permeability was unsatisfactorily low due to the thick support layer of RO membrane (to withstand the high operating pressure) that made the diffusion of water and draw solute difficult and increased the mass transfer resistance [16]. The first commercial FO membrane was developed by Hydration Technology Innovations (HTI) where cellulose triacetate (CTA) was casted on the PET mesh support [16]. This CTA FO membrane displays some attractive features over the TFC RO membrane (only for FO application) such as greater water flux (thinner support layer), lower fouling propensity (attributed to its hydrophilic nature), and stronger chlorine resistance [17–19]. Although the cellulosic FO membrane possesses the criteria for FO application, its relatively low salt rejection (especially reverse solute flux where the draw solutes migrate to the feed side) has resulted in the shifting of interest to the development of TFC FO membrane.

The research and development of TFC FO membrane started in the year 2010 and followed by the available of few commercial TFC FO membranes [20–22]. In comparison with the TFC RO membrane, the TFC FO membrane is thinner and has higher porosity. Today, TFC FO membranes have become one of the most competitive membranes for FO application owing to the high versatility of tuning the characteristics of the membranes. For instance, adjustment of synthesis approach, modification of the membrane surface, and the formulation of the polymeric membrane are the few popular methods to improve the performance of TFC FO membrane, especially in terms of water permeation and reverse solute flux.

Table 1.   *Desirable features of RO and FO membrane for enhanced performance (summarized from [16,23–26])*

RO membranes	FO membranes
• High permeate flux	• High permeate flux
• Trade-off between permeability and salt rejection	• Compatible with the selected draw solution
• Withstand the chlorine attack and antifouling (scaling, biofouling, inorganic fouling, colloidal fouling)	• Withstand the mechanical stresses generated during operation
• High salt rejection	• High salt rejection
• Low energy consumption	• Low reverse solute flux
• Excellent chemical, thermal and mechanical stability	• Chemically stable
• Great antifouling and antibiofouling capability	• Great antifouling and antibiofouling capability
• Low concentration polarization issue	• Low concentration polarization (internal and external) issues
• Hydrophilic and porous support layer	• Thin and porous support layer for water passage

Table 1 summarizes the desirable features of RO and FO membranes which drive the effort to improve membrane characteristics. A wide range of membranes for both RO and FO have been synthesized and sold at market, with membrane in flat sheet and hollow fibre being the dominant configuration. RO is normally sold in flat sheet and spiral wound modules whereas FO membranes normally appear in flat sheet and hollow fibre configuration [3].

3.   Synthesis and modification of RO and FO membranes

One of the most popular strategies to improve the membrane properties and performance is through alteration of synthesis conditions and conduction of modification on the RO and FO membranes. This section will discuss the synthesis and modification works done on the RO and FO membranes separately.

3.1   Synthesis

Currently, TFC RO and FO membranes share the largest market in commercial RO and FO membranes, respectively. Hence, the synthesis and associated synthesis conditions of TFC will be the focus and discussed in this section. The discussion includes the works done on support and active layers of TFC membranes.

3.1.1 RO membranes

a.   Phase inversion

Most of the polymeric RO and FO membranes for desalination possess the separation thin film layer casted onto a support layer. The support layer should possess good biological, chemical, mechanical, and thermal stabilities and desirable structure and morphology [27]. The support layer is normally fabricated via phase inversion approach. The key components in preparation of the support layer are solvent, polymer, non-solvent and suitable non-woven fabrics [28]. Proper control of the formulation and synthesis conditions will lead to the synthesis of membrane with desired properties [29]. Generally, there are four major approaches based on the phase inversion concept, namely immersion precipitation (nonsolvent induced phase separation, NIPS), thermally induced phase separation (TIPS), vapour induced phase separation (VIPS) and evaporation induced phase separation (EIPS), as shown in Fig. 3. Basically, in phase inversion, the polymer solution (polymer and solvent) is first prepared followed by casting the solution on the surface of the glass sheet using film applicator. The precipitation/solidification process varied according to the selected solidification technique. Among these four approaches, immersion precipitation is the frequently used method owing to its ease of fabrication and stability of the method. For example, in the early of 1960s, Loeb and Sourirajan developed a RO membrane via phase inversion method using NIPS technique [30]. In this technique, diffusion process happened whereby solvent diffused from the casting solution to the non-solvent bath while the non-solvent diffuses into the casting solution, which caused rapid precipitation.

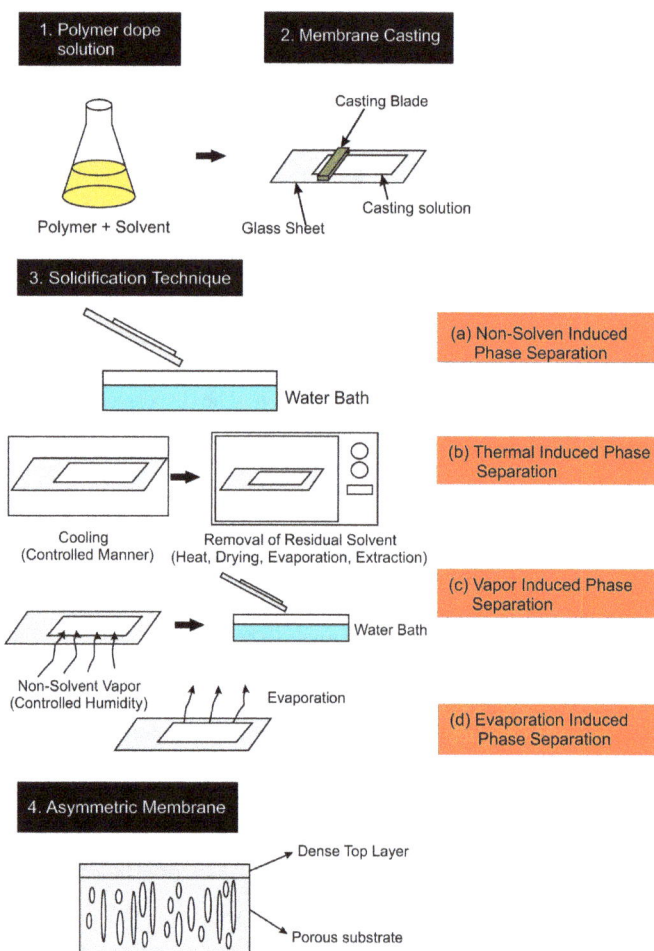

Figure 3.   Membrane Preparation via Phase Inversion Approaches (adapted from [31])

Apart from NIPS, TIPS technique is also used to prepare the support layer of RO membrane. For instance, Sabir et al. (2016) synthesized the support layer for TFC RO membrane via TIPS technique by mixing polymer cellulose acetate/polyethylene glycol (CA/ PEG3000) and silica nanoparticles (SNPs) prior to solution casting. At this point, the casted solution was quenched at lower temperature (using water ice bath) followed by heat treatment using oven to remove the remaining residual solvent. The reported hydrophilicity was measured via water content analysis technique and the modified membrane improved from 64.2% (without SNPs) to 92.1% (with 0.4wt% of SNPs). The increase in the hydrophilicity of the membrane was due to the driving force induced by the hydrophilic PEG3000 that allowed more sorption of water into the polymer matrix. This improvement resulted into the enhancement of the permeation flux for about 36.78 % (from 1.74 LMH to 2.38 LMH). Besides that, the modified membrane also showed improved salt rejection from 72% to 95% due to the presence of mesoporous SNPs that disrupted the polymer chains by forming microporous defect [32].

On the other hand, Abdelsamad et al. (2018) prepared a TFC RO membrane for desalination application via VIPS technique where the porous support layer was prepared by mixing polyethersulfone (PES) polymer, N-methyl-2-pyrrolidone (NMP) solvent, triethylenglycol and polyvinylpyrrolidone (PVP). Immediately after the casting, the glass sheet was moved to a controlled humidity air chamber to allow VIPS process to happen followed by immersion into water to remove the unwanted solvent [33]. However, there is not much work reported on the preparation of asymmetric RO membrane via phase inversion (without TFC film as an active layer).

b.   Electrospinning

Electrospinning is another emerging membrane fabrication technique that attracts the attention of researchers. This method allows the formation of interconnected and uniform pore size that is favourable in producing highly porous membrane structures [34]. It is usually applied to form a nanofibrous porous support layer prior to the non-porous active layer (PA layer) to obtain RO membrane with appreciable performance in salt rejection and water permeability [35]. The advantages of using this nanofibrous material as a support layer for RO membrane are its high surface area to volume ratio, low start-up cost, variety of polymers and materials selection and potential of fiber functionalization [36]. For instance, Wang et al. (2017) developed two thin film nanofibrous composite RO membranes by scaffolding polyacrylonitrile (PAN) nanofibrous on top of PET non-woven mat (PAN/PET) and cellulose nanofibers (CN) on top of a PAN/PET non-woven mat (CN/PAN/PET) prior to the PA thin film formation. Upon addition of the CN, the membrane showed better salt rejection (difference by around 1.0 to 1.5 %) while the flux showed slight improvement of around 0.2 to 0.5% as compared to PA/PAN/PET RO membrane. The reason for these improvements was due to the uniform thickness of membrane surface (smoother surface) for PA/CN/PAN/PET RO membrane as compared to the ridge valley layer (rougher surface) of PA/PAN/PET RO membrane [37].

c.   Interfacial polymerization

The previously discussed phase inversion method usually relies on altering the polymer concentration (>20 wt%) to form an asymmetry RO membrane and this often limits the flux improvement. In the later days of 1970s, the invention of TFC membrane via IP technique has marked another breakthrough for RO membrane. Generally, the preparation of TFC membrane usually started by soaking the support layer with the aqueous solution of m-phenylenediamine (MPD) monomer and organic solution of trimesoyl chloride (TMC) to form a dense PA layer on top of the membrane surface. This method could easily alter both flux and salt rejection by tailoring the synthesis parameters, conditions, materials (monomer and organic solvents), additives and the thin film thickness of the active layer [28,38]. For instance, the used of sulfonated diamine monomer namely 3,3'-(ethane-1,2-diylbis(azanediyl))bis(2,6-dimethylbenzesulfonic acid) instead of MPD monomer for TFC RO membrane synthesis has increased the pure water permeability by around 10% (from approximately 1.7 LMH to 1.92 LMH) as compare to MPD/TMC TFC membrane due to the created fraction of free volumes that allowed lower water transport resistance [39]. Other commonly used monomers and additives to prepare TFC membrane were also reported [40,41].

d.   Layer by layer

Apart from the commonly seen phase inversion, IP and electrospinning techniques used to prepare asymmetric RO membrane, layer by layer (LbL) is another attractive method that can be used to synthesize RO membrane owing to the possibility of tuning the thickness of the layers and various choice of materials selection. For instance, Kovacs et al. (2015) modified a PES UF membrane by spraying oppositely charged poly(allylamine hydrochloride) (PAH), poly(diallyldimethyl-ammonium) (PDAC), poly(acrylic acid) (PAA) and laponite (LAP) clay platelets to produce RO membrane. The study showed that by spraying tetralayers on the surface of the membrane (PAH/PAA/PAH/LAP) at pH 5.0, the membrane permeability and salt rejection were around 4.5×10^{-13} m^2/Pa.s and 89% (NaCl 10000 ppm), respectively. This performance was comparable with commercial Koch TFC-HR RO membrane (7.6×10^{-15} m^2/Pa.s, 98% NaCl 10000 ppm salt rejection). The reason for the appreciable salt rejection was due to the presence of stacked clay layers and polyelectrolytes that increased the tortuosity of the solute diffusion path, and thus impacted the transport of the solute diffusion across the membrane. The reason for the achieved permeability was probably due to the presence of preferential water flow channels [42].

Apart from the conventional immersion or dipping the membrane substrate with oppositely charged polyelectrolytes layers, another interesting method to deposit the polyelectrolytes layers on the membrane surface is through spin assisted LbL coating. This technique is similar to the conventional self-assembly LbL that involves sequential deposition of polyelectrolytes on the surface of the membrane followed by drying and rinsing prior to the deposition of another oppositely charged polyelectrolytes. The major difference is the polyelectrolytes that are spun cast on a spinning substrate or surface. The spinning effect helps to boost the drying and rinsing rates that are favourable in forming layer with better uniformities. For example, Fadhillah et al. (2012) used spin assisted LbL assembly technique to assemble the poly (allyl amine hydrochloride) (PAH) and PAA on the surface of the PSf membrane. The experimental runs showed that by depositing 35 layers of PAH/PAA on PSf membrane, the modified membrane achieved **88%** of NaCl (15000 ppm) salt rejection. This could be attributed to the high number of bilayers (higher Donnan potential) that favoured the rejection of ions [43].

3.1.2 FO membranes

a. Phase inversion

The morphology of the support layer such as thickness, tortuosity, and porosity has a significant impact on the characteristics of FO membrane. Factors such as the polymer composition, the nature of solvent and non-solvent, the coagulation bath, and the interaction mechanism between polymer and casting layer can affect the morphology of the fabricated membrane [11,44–46]. By properly varying these factors during the synthesis process, the thermodynamic (affinity with solvent, non-solvent, polymer, and solvent-non solvent solubility) and kinetic (viscosity of the casting solution) aspects can be controlled to obtain the desired sub-layer morphology [46,47]. For instance, the use of dimethylformamide (DMF) solvent instead of NMP for PSf sub-layer synthesis has increased the water permeability coefficient from 1.73 to 3.14 LMH due to the formation of bigger pore size and a higher porosity (2% vs 1%) of membrane support layer [48]. However, the DMF-prepared membrane recorded a slightly higher solute permeability coefficient at 0.6 LMH as compared to NMP-prepared membrane at 0.5 LMH.

Liang et al. (2017) innovatively tailored the polyvinylidene fluoride (PVDF) substrate of a TFC membrane by using bidirectional freezing synthesis process to obtain support layer with vertically oriented pores [49]. The newly developed FO support layer possessed a porosity of 74% and pore size of about 1.8 μm. SEM imaging showed that pores have an open shape in the bottom and top sides, as well as a spongy structure in the mid layer.

This structure has an excellent low tortuosity that can be proven by much higher water permeability coefficient (4.7 LMH) than the typical phase inversion method (0.63 LMH).

Co-casting is another technique that can be used to produce favourable structured support to reduce the internal concentration polarization (ICP) effect. Basically, co-casting is the fabrication of hollow fibre membrane by simultaneously casting two layers of polymer solutions using a double-blade [50–53]. Ng et al. (2015) demonstrated that co-casting with two polymer solutions of different concentrations was able to produce permeable substrate, yet still retained the surface morphology for the fabrication of active layer [50,51]. In another reported study, PEI sacrificial layer was co-casted beneath the top PSf layer. The resultant PSf substrate possessed an open bottom structure that significantly reduced the ICP effect especially at a high DS concentration [52].

The support for the casting and formulation of the polymer during phase inversion plays a role in affecting the structure of the membrane. For instance, membrane casted on a glass plate and introduced to an intermediate immersion in NMP/water bath before dipping into tap water reduced the structural parameter to 51 µm, an indicator of lower ICP in FO process [54]. Replacing the glass support with Teflon resulted in the formation of relatively dense top layer supported by a fully porous bottom. The formation of this structure was attributed to unfavourable hydrophilic (polymer)-hydrophobic (substrate) interaction. On the other hand, the inclusion of acetic acid in dioxane/acetone solvent promoted the formation of porosity of sublayer due to the complexation between acetic acid and dioxane and the nature of acetic acid as pore forming agent [55]. The inclusion of acetic acid has increased the water flux from 4 LMH to 20 LMH.

b.   Electrospinning

Compared to phase inversion, electrospinning is another synthesis technique that has a better control on the tortuosity structure of the membrane [56,57]. Synthesis conditions such as environment conditions, solution viscosity, flow rate of the solution, and the applied voltage can be fine-tuned to obtain the membrane with desirable structural characteristics [34,58]. For example, Song et al. (2014) adjusted the temperature to control the thickness of nanofibers within 50-150 nm [59]. Consequently, the nanofibers formed a support layer with special porous structure, scaffold-like, and with tight pores amongst single nanofiber. This support structure contributed to high water permeation and salt diffusion resistance, resulting in satisfying results in FO process. In another study, Tian et al. (2015) synthesized a polyethyleneimine (PEI) nanofibers support with fine (top) and coarse (bottom) structure [57]. The fine and thin nanofibers at the top are suitable for forming a rejection layer while the coarse bottom structure provides the necessary mechanical support and promotes mass transfer within the substrate.

c.   *Interfacial Polymerization*

The rejection layer of most TFC FO membranes are synthesized via IP between monomers of MPD and TMC on a microporous support as illustrated in Fig. 4. Various synthesis parameters such as composition of the monomers, reaction time, additives in the MPD/TMC solutions, and the method for removing excess MPD can be varied to fine-tune the resulting PA rejection layer [46,60]. Majority of the IP studies focus on the composition of the monomers and additives in the MPD/TMC solutions and hence will be discussed in the following section.

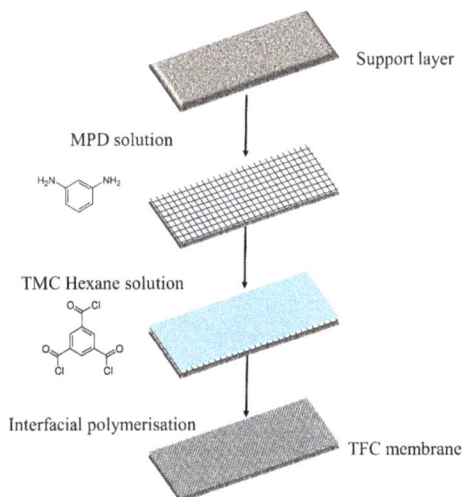

Figure 4.   *Interfacial polymerisation for the synthesis of TFC membrane (adapted from [41])*

d.   *Layer by layer*

LbL assembly on a microporous support layer is another method to fabricate FO membranes [61]. This LbL method involves the deposition of polyelectrolyte layer on the support layer through electrostatic attraction or hydrophobic interaction [61–63]. For instance, Cui et al. (2013) synthesized a multilayer membranes for FO processes by LbL assembly of positively charged PAH and negatively charged blend of PAA and poly(sodium 4-styrenesulfonate) (PSS) on top of hyperbranched PEI modified Torlon substrates and then crosslinked by glutaraldehyde [64]. The developed 3-bilayer membrane recorded water flux of 28 LMH and reverse salt flux of 1.97 gMH using 0.5 M magnesium chloride ($MgCl_2$) as the draw solution. Similar polyelectrolytes formulation has also been deposited on PES support to obtain hollow fibre FO membrane [62]. This indicated the flexibility of LbL method in synthesizing FO membrane of different forms.

Double-skinned LbL polyelectrolyte FO membrane as shown in Fig. 5 has been synthesized to enhance its antifouling property as compared to the commonly adopted single-skinned membrane [65]. Polyanion and polycation were coated LbL alternatively for multiple times on the top and bottom surfaces of PAN substrate to grant the FO membrane a double-skinned structure. The double-skinned FO membrane enjoyed lower fouling propensity since foulant clogging within the porous support has been hindered by the second active layer.

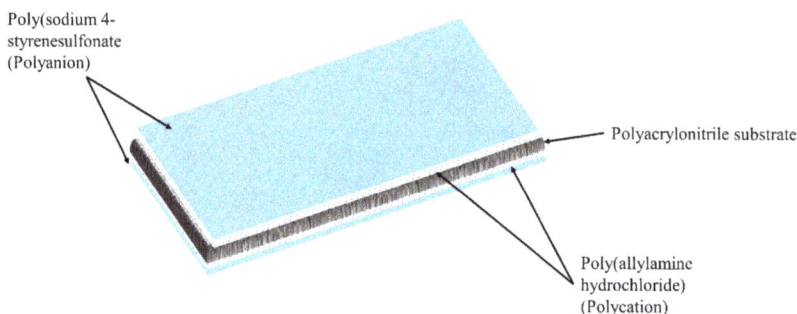

Poly(sodium 4-styrenesulfonate) (Polyanion)

Polyacrylonitrile substrate

Poly(allylamine hydrochloride) (Polycation)

Figure 5.   Double-skinned FO membrane fabricated via LbL method (adapted from [65])

Chitosan and polyacrylic acid were used as polycation and polyanion respectively for the LbL synthesis of FO membrane [66]. The membrane achieved a high water flux of 25.5 LMH and low reverse salt flux of 0.051 gMH when tested with 2 M NaCl as the draw solution. Experimental data indicated that the membrane also attained high water flux (22.25 LMH) when tested with synthetic wastewater. It was postulated that the dense nature of the LbL surface coating on substrate prevented salt leakage and led to low reverse solute flux.

3.2 Modification

Membrane modification is mainly done chemically to improve the characteristics of the RO and FO membranes, especially to alter the hydrophilicity and antifouling properties of the membrane [24]. The modification is mostly conducted on the separation layer, albeit some works on the support layer have also been reported.

3.2.1 RO membranes

a. Layer by layer

LbL is one of the versatile techniques used to alter the performance of the membrane by alternately stacking oppositely charged polyelectrolytes, namely polycation and polyanion. Attributed to its versatility, the stacking layer can be placed on the surface of the substrate, porous layer or active layer, depending on the needs of the application. The variation of choices in polyelectrolytes, surface charges and thickness can affect the membrane performance. Therefore, properly varying these parameters could lead to the synthesis of membrane with desirable properties [67]. For example, Ishigami et al. (2012) attempted to improve the hydrophilicity of commercial RO via LbL and the modification was performed by immersing the RO membrane alternately with the PSS and PAH solutions. At optimum layer number, the antifouling property of the RO membrane was improved as the relative permeability increased from around 0.72 to 0.88 while the NaCl (500 ppm) salt rejection reported slight increment from 98.25% to 99.25%. The improved antifouling property was attributed to the increased hydrophilicity and smoother membrane surface (roughness decreased from 54.9 nm to 34.8 nm). The increase of salt rejection was because of the increased hydrodynamic resistance induced by the increased layer thickness [68].

Besides stacking polyelectrolytes, oppositely charge nanomaterials can also be stacked on the surface of the membrane. For instance, Choi et al. (2013) coated GO to the PA-TFC membrane surface via LbL approach and the modified membrane still showed better flux (increased from 12.5 LMH to 14 LMH) while maintaining approximately the same rejection (around 96%) even after 10 GO bilayers. Although the deposition of bilayers was postulated to decrease the flux due to the increase of hydrodynamic resistance, interestingly flux drop was not noticed. The reason for this behaviour was because the multilayers of GO was not a strong function of thickness whereby the impact of hydrophilicity (contact angle reduced from 70.6° to 25.9°) of the GO was stronger than the hydrodynamic resistance. Attributed to this strong hydrophilicity and smoother surface induced by the presence of GO, the modified membrane also had lower fouling propensity where the flux decline was improved for almost 50% (from 34% to 15%). The multilayer of GO also helped to shield the membrane from chlorine attack that often is a challenge faced by the PA RO membrane [69]. However, the non-uniformities and the time taken for the layers to form on the surface of the membrane remain a challenge for this synthesis technique.

In another study, Ma et al. (2016) used both spray and spin assisted LbL approaches to functionalize the TFC PA RO membrane with copper nanoparticles to improve the biofouling resistance. The membrane was modified with positively charged polyethlylenimine coated copper nanoparticles and negatively charged PAA until a desired bilayer formed on the surface of the membrane. The modified membrane showed that the presence of copper nanoparticles significantly improved the antibiofouling property by mitigating the growth of *E. coli* on the surface of the membrane in the range of 94.3%-100% as compared to unmodified membrane at 14% [70]. The three methods of applying LbL on the membrane surface are illustrated in Fig. 6.

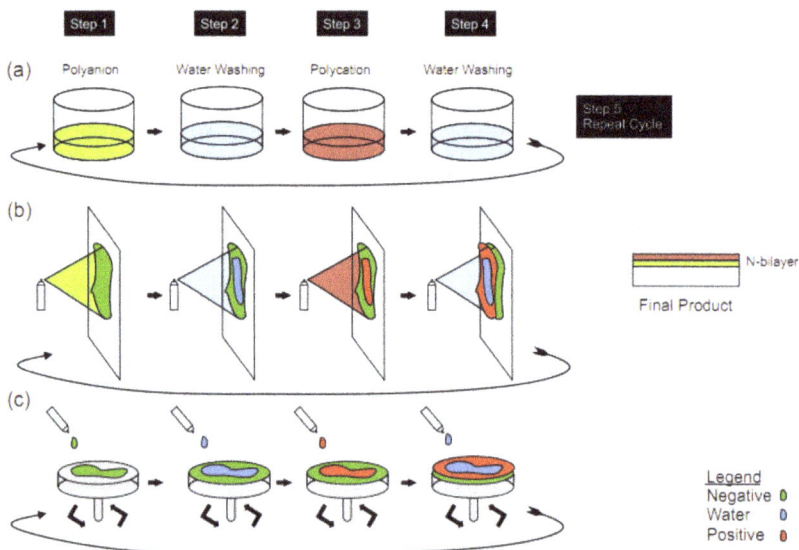

Figure 6.(a) Self-assembly (b) Spray coating (c) Spin coating (adapted from [71])

b.   *Coating*

Surface coating is a common membrane modification method used to improve the surface properties of the membrane via immersion of coating solution with or without the presence of cross-linker. For example, Zhang et al. (2016) successfully modified PA RO membrane through the coating of poly (vinyl alcohol) (PVA) polymer. Associated with the modification, the modified membrane exhibited slight improvement in both water flux and salt rejection at around 7% (from 50.5 LMH to 54 LMH) and 99.4 % to 99.5%, respectively, as compared to virgin RO membrane. The increase in water flux could be related to the hydrophilicity of PVA that overcame the hydraulic resistance caused by adsorbed PVA layer. On the other hand, the slight hydraulic resistance happened to improve the salt rejection. Additionally, the flux recovery ratio for modified membrane (60-69%) was higher than controlled RO membrane (55-60%) [72].

In another study, Shafi et al. (2015) modified a RO membrane by coating zwitterionic to improve the fouling resistance property of the membrane. Zwitterionic materials are commonly used as an antifouling coating attributed to its low protein adsorption. In this study, the zwitterionic copolymers were self-synthesized followed by deposition on a commercial Koch membrane (TFC-HR) to evaluate the improvement of fouling resistance. The modified membrane showed increased salt rejection from 96% to 98 % with a small drop in the permeate flux. The authors claimed that the reasons for the observed increase in salt rejection was most likely due to the decline in permeate flux and increased surface charge. The study also demonstrated strong resistance of modified membrane towards *E. coli* and *P. aeruginosa* cells with the number of attached cells on the surface of the membrane was reduced by 98% (from 9146 to 192 bacteria attachment) and 96.5% (from 1270 to 45 bacteria attachment) respectively attributed to the antifouling property of zwitterionic [73].

Another major challenge for the RO membrane is the vulnerability towards chlorine attack that will cause the salt rejection to deteriorate after exposure to higher chlorine containing wastewater. Associated with this, Ni et al. (2014) attempted to improve the chlorine resistance of the membrane by coating a hydrophilic copolymer poly (methylacryloxyethyldimethyl benzyl ammonium chloride-r-acryl-amide-r-2-hydroxyethyl methacrylate) followed by chemical crosslinking to covalently link the polymer chains on the commercial TFC RO membranes (LCLE and BW30). The results showed that the modified membrane can tolerate chlorine exposure 7 to 10 times better than the controlled membrane. The reason for such behaviour was because the presence of poly (methylacryloxyethyldimethyl benzyl ammonium chloride-r-acryl-amide-r-2-hydroxyethyl methacrylate) layer that protected the chlorine sensitive sites in PA film from chlorine attack [74].

Other types of polymers such as polymer sericin [75], sulfochlorinated and sulfonated polysulfone [76], PEG polymers [77], hydrophilic hyperbranched polymer, poly(amino amine) [78] and nanomaterials [79] have also been employed for the surface coating of

RO membrane. Though the modified RO membranes exhibited improved performance in terms of flux and rejection, the weak interaction between the coating layer and membrane surface has been an issue as this caused the coated layer to be easily detached from the membrane surface.

c.   Grafting

Surface modification via chemical treatment is another effective approach to alter the surface properties of the membrane to minimize the weaknesses of TFC PA RO membrane (e.g. low resistance towards fouling and chlorine). For instance, Cheng et al. (2013) successfully improved the antifouling behaviour and chlorine resistance of commercial TFC RO membrane by exposing the active layer of the membrane with hydrophilic polymer solution, N-isopropylacrylamide (NIPAm) via redox initiated graft polymerization. As a result, the modified membrane showed higher water flux by 3.5% (from 28.2 LMH to 29.2 LMH) while maintaining high NaCl salt rejection (500 mg/L) at 98.6% as compared to the unmodified membrane. The reason for improved water flux was attributed to the decrease of surface contact angle from 60.5° to 40.2° that indicated the enhancement of membrane surface hydrophilicity while the slight increase in salt rejection was mainly due to the increase of the membrane surface negative charge from -42.5 mV to -48.6 mV, which exerted stronger repulsion towards the ionic salts. Besides that, the modified membrane recorded 20.7% of flux decline as compared to controlled membrane at 45.3% over a 40-hour filtration run. The improvement of the fouling resistance was related to the enhanced hydrophilicity and membrane surface charge as well. As for the chlorine resistance, the modified membrane was able to maintain a high salt rejection of 96.5% even after exposure to high chlorine solution (5000 ppm Cl_2) as compare to controlled membrane at 78.5% salt rejection. The subsided chlorine degradation of the modified membrane was mainly due to the reduce of amine groups that were vulnerable to chlorine attack after grafting [80].

In another study, the surface modification was performed by grafting NIPAm and zinc oxide (ZnO) NPs onto the PA layer of RO membrane via gamma irradiation instead of redox initiated. The study reported on the improvement of hydrophilicity, which was indicated by the drop in the contact angle value from 74.8° to 54.8° due to the presence of active polar hydrophilic groups on the membrane surface. The salt rejection for NaCl solution of modified membrane was increased from approximately 91% to 92% while the water flux increased from around 22 LMH to 24 LMH. Such phenomenon was observed because NIPAm and ZnO improved the surface hydrophilicity and membrane surface negative charge after grafting. On the other hand, the chlorine resistance for the modified membrane was better than unmodified membrane whereby it still maintained higher than 90% salt rejection while the latter experience only 73% salt rejection at chlorine exposure of 4000 ppm. The enhancement of the chlorine resistance was related to the presence of sacrificial grafting layer [81].

Apart from hydrophilic polymer, the use of hydrophilic nanomaterial for surface grafting is also one of the emerging research areas. For instance, Huang et al. (2016) grafted azide-functionalized graphene oxide (AGO) on the surface of the commercial RO membrane to promote anti-adhesion and antifouling properties of the membrane by dipping the membrane surface with AGO dispersion solution followed by UV light exposure. The study reported a slight decrease in the water flux from 37.8 LMH to 36.3 LMH. This could be due to the increased of the hydraulic resistance after the surface grafting. The slight enhancement in terms of salt selectivity (from 94.1% to 95.3%) and significant enhancement of fouling resistance (from 70% flux decline to 40% flux decline after 7 days of filtration) were also observed. The reason for latter enhancement was due to the improved hydrophilicity, which showed by lower contact angle value (decreased from 85° to 45°) that formed a resistance towards the attachment of foulants on the membrane surface. The presence of the oxygen containing hydrophilic groups within the AGO increased the affinity between the membrane and water molecules, and thus resulted in improvement of fouling resistance. The result was further strengthened with the smoother membrane surface (roughness dropped from 44.3 nm to 29.0 nm) of the modified membrane that allowed the foulants to be easily flushed away from the surface of membrane. Additionally, the presence of GO also improved the antibacterial property to the RO membrane with the increase of seventeen folds from 5.46% *E. coli* cells coverage on the surface of the membrane (unmodified membrane) to 0.32% (modified membrane) attributed to the antimicrobial property of AGO [82].

In another study, a combination of surface coating and grafting was reported by Li et al. (2015) whereby the author coated polydopamine (PDA) on the commercial RO membrane (XLE-4040, Dow Co. Ltd) followed by the grafting of polyethyleimine (PEI). Associated with this, the hydrophilicity of the membrane was improved by 25% (reduction in contact angle value from 65.7° to 49.2 °). Although the hydrophilicity of the membrane was improved, there was still a slight drop in normalized water flux value from 1.0 to 0.8. This behaviour was due to the higher resistance towards water diffusion across the membrane as compared to the hydrophilicity effect. As a result of the layer resistance and increased electrostatic repulsion between Na^+ ions and the membrane, a minor increase (95% to 98%) in NaCl salt rejection at 1000 ppm was observed. On another note, the study reported that the changes in membrane properties in terms of hydrophilicity, roughness, and surface membrane charge resulted in lower flux decline (around 20%) as compared to the controlled membrane at 35% [83].

3.2.2 FO membranes

a.   Coating

Surface coating is a simple technique used to alter the membrane surface characteristics for better hydrophilicity, surface charge, and antifouling [84,85]. Nguyen et al. (2013) coated a commercial FO membrane surface with poly amino acid 3-(3,4-Dihydroxyphenyl)-L-alanine (L-DOPA) [86]. It was reported that L-DOPA self-polymerized in aqueous solution and attached firmly to the membrane surface [87]. This additional layer (as shown in Fig. 7) could become intensely hydrated and prevented the accumulation and adsorption of foulants on the membrane surface. The contact angle has observed a reduction from 48° to 44° and 38° for membranes coated in a period of 4 h and 12 h, respectively. Also, the membrane surface became more negatively charged after the coating process, which was useful for rejecting negatively charged foulants.

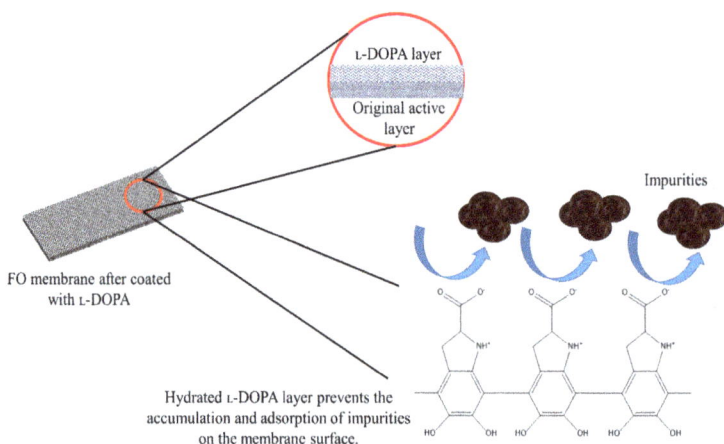

Figure 7.Schematic diagram of surface adsorption resistance for organic matter after the FO membrane coated with L-DOPA (adapted from [86])

Layered double hydroxides (LDHs) nanoparticles could also be deposited on the TFC membranes by PDA-induced coating process [88]. The pristine TFC membrane could be dip-coated into a LDHs suspension and desiccated at 30°C. The modified membrane displayed the least coverage of *E. coli.* indicating the enhanced antibiofouling capability of the membrane due to improved surface hydrophilicity by LDHs.

In general, the hydrophobicity of the support layer exacerbates the ICP phenomenon in FO membranes. Improving the wettability and hydrophilicity of the support layer would reduce the ICP issue and fouling propensity [89]. Liu et al. (2016) coated the bottom surface of a TFC FO membrane (support layer) with PDA and discovered that the contact angle recorded a reduction from 82.2° to 37.4° after a treatment time of 6 h [90]. The hydrophilicity of the bottom surface was obtained due to the phenolic hydroxyl and amine groups contributed by PDA. Though the water permeability and salt rejection did not observe obvious changes, the modified support layer had a good antifouling property.

Another popularly used coating material is PVA, owing to its ability to retain water molecules, easily available in the market, inexpensive, great chemical and physical stability, hydrophilic and antifouling properties, and remarkable film-forming characteristics [91,92]. In the study conducted by Ahn et al. (2015), CA-based membrane was coated with PVA followed by immersion in 0.01% glutaraldehyde solution for cross-linking process [93]. The hydrophilicity of the CA membrane increased substantially, as supported with the reduction of water contact angle from 73.5° to 39.3°.

Support layer can also be coated with minerals such as calcium carbonate ($CaCO_3$) to improve wettability of the membrane. The coated $CaCO_3$ can improve the hydrophilicity by connecting to water through strong ionic hydrogen bonding [94]. The modified FO membrane achieved a high water flux of 52 LMH and salt flux of 16.8 gMH using DI water and 2 M NaCl draw solution.

b.　Crosslinking

Chemical cross-linking with polyelectrolytes such as PAH, polyethyleneime, glutaldehyde, and *p*-xylylene dichloride is a more straightforward method to obtain NF-like rejection layers for FO membrane [95–98]. For instance, Setiawan et al. (2011) fabricated a positively charged PAI hollow fiber FO membranes by cross-linking with polyethleneimine under mild conditions [99]. The cross-linked FO membrane possessed tight pore size, narrow pore size distribution, and ion repulsion property from Donnan exclusion effect. Consequently, the membranes exhibited good water flux and low reverse solute flux (especially divalent salts) [100]. However, the degree of cross-linking should be optimized as excessive cross-linking could lead to the loss of mechanical strength of the membrane [101].

PEI cross-linking has also been applied to dual-layer hollow fiber poly(aminde-imide) (PAI) membrane, with promising water flux of 27.5 LMH coupled with reverse salt flux of 5.5 gMH [100]. However, the modified membrane surface possessed positive charge, which might exacerbate fouling issue due to the oppositely charge attraction towards negatively charged foulants [102]. To overcome fouling propensity, further modification with PSS polyelectrolytes has been done to alter the surface charge as well as to shrink the pores of active layer of the PEI-cross-linked membrane. It was reported that the further modified membrane maintained a stable flux of 13 LMH against feed mixture of 1000 ppm bovine serum albumin (BSA) and 2000 ppm sodium sulphate (Na_2SO_4) draw solution.

c.   *Layer by layer*

Phytic acid (PhA) is a nontoxic electrolyte that exhibits high affinity toward water molecules and strong chelation capacity with mono/divalent metal ions. This property has been utilized to incorporate antibacterial metal ions such as silver and copper for membrane surface modification [103]. Xiong et al. (2019) adopted LBL assembly to incorporate PhA and silver/copper nanoparticles on the TFC membranes [103]. As illustrated in Fig. 8, the TFC FO membrane was first modified by immersing it in an inorganic salt aqueous solution to anchor metal ions on the TFC membrane. Afterwards, the treated membrane was soaked in PhA solution and the inorganic salt solution alternatively for several cycles. The modification increased the membrane hydrophilicity with water flux enhanced by up to 57-68% and no compromising of membrane selectivity. The membranes also revealed low metal release owing to the strong attachment on the surface.

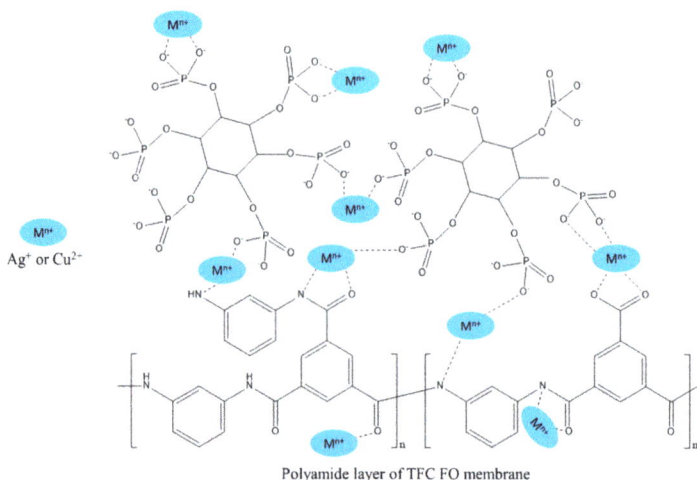

Figure 8.   *Modification of TFC membrane by LbL approach through the incorporation of PhA and metal complexes (adapted from [103])*

LBL membrane can also be synthesized by incorporating nanoparticles such as GO and oxidized carbon nanotubes (OCNTs) [104]. In a study where GO and OCNTs with five bilayers on a PES support membrane was synthesized, the contact angle of the membrane was lower than that of GO-LBL membrane due to the presence of hydroxyl groups (supported by X-ray photoelectron spectroscopy (XPS) where the former has 9.5% oxygen-containing groups while the latter only 7.6%). However, such synthesis approach that involved covalent attachment and LBL self-assembly was less appealing to industrial production due to the complexity of multiple steps. To overcome this complex synthesis approach, Hegab et al. (2015) adopted a simpler technique by immobilizing GO onto the TFC FO membrane via PDA (adhesive agent) polymer [105]. The modified membrane presented a smoother surface with improved hydrophilicity and antibiofouling capability as compared to pristine membrane.

d. Grafting

The incomplete crosslinking of the MPD and TMC monomers results in the presence of amine and carboxyl functional groups on the surface of TFC PA membrane. Upon soaked in an aqueous solution, the residual acyl chloride groups from the unreacted TMC monomers would convert to carboxyls [106]. These negatively charged moieties may serve as reactive sites for grafting modification of membrane. For instance, Li et al. (2015) grafted photosensitizer porphyrin molecules (Por) on the TFC PA surface via the reaction between the amine group of Por and the carboxyl groups on the TFC membrane to form a strong covalent bonding [84]. When exposed to sunlight radiation, the Por molecules can produce reactive oxygen species, which is highly toxic to bacteria [107,108]. This surface functionalization offered a promising method to address the biofouling issue.

In another approach, coupling reaction is used to bind functionalizing agents onto the TFC surface through other chemicals. For example, graphene oxide (GO) and carbon nanotubes (CNTs) were bound to the TFC membrane surface through N-ethyl-N'-(3-(dimethylamino)propyl)carbodiimide/N-hydroxysuccinimide (EDC/NHS) coupling reaction, where the carboxylic acids were activated with carbodiimide with the aid of NHS followed by the amine addition [109–111]. The carboxyl (–COOH) groups on GO nanosheets were first activated by EDC/NHS, and then grafted onto amine functionalized membrane through the formation of chemical bonding between the functional groups [109]. The GO-functionalized TFC FO membrane exhibited considerable antibacterial capability, which could be attributed to the physical disruption of the bacterial cell or through the release of reactive oxygen species.

Castrillon et al. (2014) functionalized the TFC FO membrane by conducting a second interfacial reaction between ethylenediamine and free carboxyl groups presence on the membrane surface [112]. The functionalized membrane reportedly possessed better antifouling (alginate) property. Similar functionalization has also been adopted by grafting superhydrophilic 3D hyperbranched polyglycerol, PEI, and polyamidoamine (PAMAM) dendrimers to the TFC surface to enhance the antifouling properties by having better rejection of foulants such as BSA, alginate, and ammonia [113–115]. These highly branched molecules offers a high density of hydrophilic functional groups on the membrane surface that could minimize the adsorption of protein and bacteria [116]. The attachment of foulants has also been prevented by the 3D star-like nanostructure of dendrimers through steric hindrance mechanism [113]. For instance, PAMAM was grafted on the TFC FO membrane through the formation of amide bonds between the unreacted carboxyl groups on the PA surface and amine groups of PAMAM dendrimers [117]. The functionalized FO membrane possessed a highly negative surface charge, which improved the ammonia rejection to 98.2% for domestic wastewater treatment.

PDA has recently emerged as one of the most popular biopolymers used in membrane surface modification to suppress biofouling due to its protonated amine groups that could potentially exert antibacterial activity through cell lysis in contact with the bacteria cell membrane [118,119]. For instance, CTA FO membrane modified with PEG-grafted PDA showed improved anti-biofouling property in a submerged FO membrane bioreactor (MBR) over 61-days testing period. It was reported that the modified membrane displayed enhanced surface hydrophilicity and lower flux decline, an indicator that it has better antifouling ability than that of the pristine CTA membrane.

Activation of polybenzimidazole (PBI) flat sheet membrane with 4-(chloromethyl) benzoic acid and subsequent surface modification with taurine, *p*-phenylene diamine (PD), ethylene diamine, and poly(acrylamide-*co*-acrylic acid)(P(Am-co-AA)) have shown to improve the hydrophilicity and surface charge of the membrane [120,121]. The water flux was increased twofold to 5.6 LMH as compared to the unmodified PBI membrane when 0.1 M NaCl solution feed and 2 M NH_4HCO_3 draw solution were used.

4.   Materials and formulation of RO and FO membranes

Apart from the synthesis conditions and modification approaches discussed previously, the formulation and materials used for the synthesis of RO and FO membranes (both support and separation layers) also play a significant role in improving the membrane characteristics. In most cases, additives especially nanomaterials and nanoparticles will be added either to the active layer or support layer of RO and FO membranes. The incorporation of additives can generally be done through several approaches, either embedded into the separation and support layers or on top of the membrane surface, as indicated in Fig. 9. The presence of additives would alter the surface properties (e.g. hydrophilicity, roughness, and charge density) and subsequently the membrane separation performance [122]. The additives must have appropriate size and internal structure, as well as compatible with the polymeric membrane to ensure the benefits associated with the additives could be fully utilized.

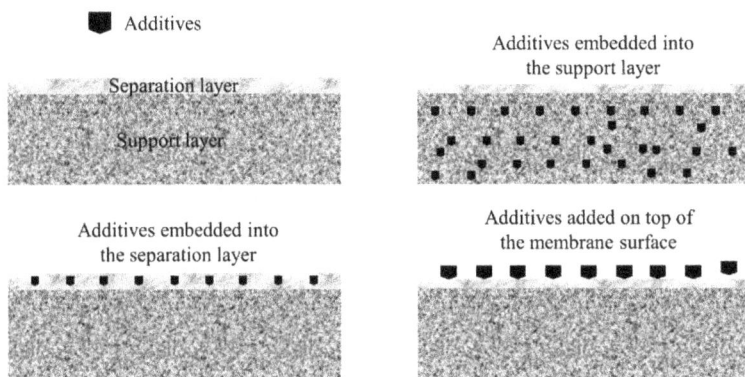

Figure 9.   *Approaches for incorporating additives into the RO and FO membranes*

4.1 RO membranes

4.1.1 Substrate

The active layer of the membrane has been considered as the key factor to the membrane performances. Associated with that, the modification of active layer is greatly researched. Although the active layer is seen to be the most important part, the formulation or synthesis conditions of the porous layer also play an important role to the performance of the membrane. Factors such as types of solvent and polymers, polymers concentration, additives and operating temperature should not be taken lightly [123].

In the literature, it was observed that the improvement of support layer water permeability and changes in the pore size can be achieved through the addition of hydrophilic additives such as PEG and PVA. For example, Fathizadeh et al. (2012) prepared a PA RO composite membrane by adding PEG additives in the PES substrate. Associated with this modification, the membrane permeability was enhanced by almost six folds from 1.2 LMH to 7.8 LMH. It was reported that the presence of the PEG enlarged and increased the macrovoid size, and thus allowing the water molecules to easily pass through the membrane with lower resistance [124]. Other studies also showed that by improving the porosity of the support layer, the water permeability could be greatly enhanced [125,126]. However, the increase of macrovoid size that caused a decrease in the thickness of the substrate might affect the mechanical strength of the RO membrane.

Although porous substrate is desirable in membrane to achieve higher flux, the challenge for porous substrate is the difficulty to withstand higher operating pressures. The compression causes the membrane macrovoid to become smaller and increase the tortuosity. When the tortuosity of the porous substrate increases, the distance of water transportation route across membrane becomes longer due to lack of connectivity between pores [127], and this possibly would lead to flux decline behaviour which is undesirable in membrane. As the high operating pressure is unavoidable especially in desalination process, alternative solutions are required to enhance the mechanical strength of the membrane. Since the introduction of nanomaterials in RO membrane in the early of 2000s, the advancement of nanotechnology in polymeric RO membrane has led to the breakthrough of this undesirable flux decline and weak mechanical strength. Not only that, the incorporation of the nanomaterials in the polymeric matrix or substrate has successfully led to improved RO membrane hydrophilicity without compromising the rejection properties of the membrane [128]. For example, Pendergast et al. (2010) observed that, there was an enhancement in flux decline behaviour between unmodified membrane (PA RO-PSf substrate membrane) and modified membrane (PA RO-PSf incorporated silica membrane & PA RO-PSf incorporated zeolite membrane). At optimal membrane performance, the modified membrane experienced lower flux decline in the range of 20% to 22% as compared to unmodified membrane at around 30%. The reason for enhanced flux decline was due to the added mechanical strength of the support structure [129].

Another interesting study of embedment of nanomaterials was reported by Yan et al. (2015) whereby the author investigated the pore formation of the membrane by assuming nanomaterials as a pore former. In this study, silicon dioxide nanomaterial (SiO_2) was embedded into the PSf supports via immersion precipitation followed by NaOH alkali treatment prior to the TFC formation on the nanocomposite support. During immersion precipitation, the pore formation happened when some of the SiO_2 embedded near the surface of the casting film diffused out to the coagulation bath. Associated with this, the total porosity of the modified membrane (optimal condition) increased by around 14 % (from 63.6 ± 2.0 to 72.3 ± 4.3). Moreover, the flux increased about 55.4% (from 35.4 LMH to 55 LMH) and the NaCl rejection achieved as high as 99.10 % as compared to

98.74% of unmodified membrane. The modified membrane also showed a membrane performance (water permeance of 3.55 LMHbar and salt rejection of 99.1%) that was comparable to some of the commercial membranes such as GE-INDUSTRIAL RO6 4040F35 (4.66 LMHbar, 99%) and DOW-BW30-400 (2.91 LMHbar, 99.5%) [130].

Carbon-based additive is another class of nanomaterials that has been adopted in membrane technology. The addition of this nanomaterial has proven its effectiveness in improving the flux, selectivity, antifouling properties and mechanical strength of the membrane. For instance, Son et al. (2014) modified the PES support layer with functionalized carbon nanotube (fCNT) prior to IP process. The presence of fCNT improved the substrate hydrophilicity level with a contact angle of 51.03° (decline from 54.76°) and increased the water permeability by 44% from 0.84 LMH/bar to 1.21 LMH/bar without sacrificing the salt rejection (maintained at around 95% for both cases). Besides, the presence of fCNT also increased the pore size (from to an average size of 10.32-14.45 nm) by connecting the disconnected pores that allowed easy water passage across the PES support. On another note, the fouling analysis showed that the flux recovery ratio increased from 74.81% to 79.87% (an increase of almost 7 %) which could be attributed to the presence of negatively charged fCNT. The negatively charged fCNT contributed to the loosely attachment of foulants on membrane surface, and thus allowing flux to be restored easily as compared to unmodified membrane [131].

Another family from carbon derivatives, GO was adopted by Chae et al. (2017) to improve the performance of the membrane. The GO was embedded into both support and active layers for RO membrane synthesis. The incorporation of GO into the support layer showed substantial improvement in terms of hydrophilicity, pore size, shifting from positive to negative surface charge property and antimicrobial property owing to the presence of hydrophilic functional groups and surface charge of GO. The support layer hydrophilicity improved as indicated by the contact angle measurement (reduced by 28.81% from 59° to 42°) whiles the average pore size of modified membrane increased from approximately 20 nm to 27 nm. Besides, there was a shifting of surface charge from 0.6 mV (unmodified, nearly neutral) to -16 mV (modified, negatively charge) which favoured the rejection of negatively charged particles. Associated with the alteration in the support layer properties, the water flux for RO membrane with the support incorporated with GO was increased by 2.6% (from 23 LMH to 29 LMH) while the salt rejection was enhanced approximately 4% (from 90.6 to 94.8%) as compared to unmodified TFC membranes. The observation of the enhanced water flux and salt rejection was due to the increased surface hydrophilicity and alteration of membrane surface charge, respectively. On top of this, the presence of GO in the support layer successfully reduced the biofilm formation with the attachment of biovolume cell was 70% lesser than unmodified TFC membrane [132].

Apart from carbon derivatives, metal organic framework (MOF) was also adopted in improving the membrane performance. For instance, Park et al. (2017) incorporated HKUST-1 [Cu$_3$(BCT$_2$)] MOF treated with sulfuric acid into PSf support layer to enhance the porosity and hydrophilicity of support layer of the TFC RO membrane. Associated with the modification, the modified PSf support layer showed improvement in water flux from 175.0 LMH to 245.8 LMH (increased by 40.46%). This was because the hydrophilicity of the support layer was enhanced, which indicated by the higher value of solid-liquid interfacial free energy (-ΔGS$_L$) 108.5 mJ/m^2 (modified support layer) as compared to 98.6 mJ/m^2 (unmodified support layer). Besides, the increase of porosity of the support layer from 40.9% to 49.4% allowed water molecules to move easily across the membrane, and thus resulting in enhancement of water flux. As a result, the TFC RO membrane with the modified support layer also experienced improved flux from 35.7 LMH to 47.6 LMH (increased by 33.33%) without compromising the salt rejection (96%-modified, 94%-unmodified) [133].

4.1.2 Active layer

The discovery of thin film nanocomposite (TFN) membrane marked the breakthrough for TFC membrane in the route of overcoming its relatively low water flux. The first attempt of combining nanomaterial with TFC RO membrane was introduced by Hoek and co-researchers. In the study, Hoek et al, incorporated zeolite nanomaterial into the hexane-TMC solution prior to membrane's modification process. The pure water permeability of the modified membrane increased almost twofold from 7.56×10^{-4} LMH (neat membrane) to 13.68×10^{-4} LMH (modified membrane) without trading off its salt (NaCl) rejection [134]. The positive impact of the TFN membrane concept led to the exploration of different types of nanomaterials, such as inorganic nanomaterials (TiO$_2$, silver, SiO$_2$, MOF, etc.), carbon-based nanomaterials (GO, CNTs, etc.) and aquaporin (protein) in the synthesis of TFC membrane [135].

For instance, by depositing titanium dioxide (TiO$_2$) on the active surface of the membrane, the resistance of the modified RO membrane towards biofouling was enhanced. It was reported that the presence of the photocatalytic bactericidal effect of TiO$_2$ inhibited the bacteria attachment on the membrane surface. Associated with the TiO$_2$ layer, the survival rate of *E.coli* cells on the RO membrane surface dropped from 60% to 0% under 4 hours of UV illumination. This subsequently minimized the flux loss from its original permeability to 15% as compared to neat RO membrane at 30% without compromising the salt rejection [136].

Carbon-based additives have shown its notable presence in improving RO membrane for water desalination processes. It has been widely employed to enhance the membrane performance in water flux, rejection, antifouling capability and stability [137]. Out of different types of carbon-based additives, graphene-based nanomaterial has attracted tremendous interest in water desalination. It is attributed to the presence of various hydrophilic functional groups, great physicochemical properties, tuneable properties and various possibilities via functionalization. For instance, Hegab et al. (2015) modified the commercial brackish water TFC membrane (BW30LE) by coating graphene oxide functionalized chitosan (GO/f-Cs) to improve the fouling resistance of the membrane. It was found out that the modified membrane showed superior performance as compared to unmodified membrane in terms of NaCl (1500 ppm) salt rejection, hydrophilicity, water flux and antifouling property. The hydrophilicity of the modified membrane was improved with the decrease of contact angle for about 70% (from 63.68° to 19.13°) due to the presence of abundance hydrophilic functional groups of GO. The membrane water flux increased from 56.14 LMH to 61.47 LMH, which could be attributed to the hydrophilicity induced by the nanomaterial and thus promoting water sorption across the membrane. Besides that, the salt rejection was 7.7% higher than the unmodified membrane due to the GO/f-Cs coated dense layer that produced compact structure on the surface of the membrane. On another note, fouling analysis showed that the modified membrane has better flux recovery ratio at 97% as compared to unmodified membrane at 86% while the flux decline has also been reduced from 18% to 6% [138]. Other improvement of GO-based TFC membrane for desalination were also reported [139,140].

Another member of carbon material, CNTs has also proved its capability to improve the membrane performance. Vatanpour et al. (2016) has successfully grafted multiwalled carbon nanotubes (MWCNTs) on a commercial PA seawater RO membrane (RE4040-SHF, South Korea) for enhanced fouling resistance, flux and rejection. The improvement was attributed to the presence of nanotubes channels for water passing and hydrophilic functional groups of MWCNTs. The channels allowed the passage of water molecules while the hydrophilic functional groups increased the affinity towards water molecule. Both mechanisms promoted the permeability of water molecules and thus increased the water flux from around 23 LMH to 25 LMH at 0.0125 wt% MWCNTs concentration while the salt rejection was slightly increased from 95.5% to 96.5% at low MWCNTs concentration (0.0125-0.05). On another note, the flux decline for modified membrane (0.025 wt% MWCNTs) was lower (around 15%) as compared to the unmodified membrane (around 19%). This proved that the modified membrane has a better fouling resistance. However, RO membrane modified with high concentration of MWCNTs was subjected to the negative impact on flux and salt rejection due to the compression of the grafting layer that eventually making it very stiff, and thus creating a resistance for water passage (increase mass transfer and decreasing the hydrophilicity effect) [141].

Silver nanomaterial displayed a remarkable antibacterial property that is useful to prevent biofouling issue in membrane process. The presence of silver nanomaterial in membrane indeed has become a great interest to membrane desalination technologies nowadays especially for the saline feed waters that contain high biological activity. Usually, the silver nanomaterials are chemically deposited on the membrane surface to suppress the formation of biofilm. For example, a commercial PA-RO membrane modified with silver nanomaterial was tested in Wukan desalination pilot plant, Penghu. It was reported that the modified membrane could reduce the flux reduction by almost 50% (from 75% to 25%), and thus prolonged the lifespan of the RO membrane performance [142]. This could be attributed to the low microbial growth since the presence of silver inhibited the growth of microbes on the membrane surface. In another study, Ben-Sasson et al, performed an in-situ functionalization of silver nanomaterials on the surface of commercial seawater RO membrane. The presence of silver nanomaterials demonstrated strong antibacterial properties towards *E. coli, P. aeruginosa and S. aureus* with the decrease of bacteria colonies (CFU) by 78%, 91%, 96% as compared to neat membrane [143]. However, the leaching stability of the nanomaterial from the membrane surface remains a major challenge for surface modification process.

Although the incorporation of nanomaterial has proven its effectiveness in tuning the membrane performance, the key challenge of using single nanomaterial (usually inorganic nanomaterial) is the tendency of forming agglomeration and aggregation that discounts the improvement brought to the membrane. To minimize this impact, Rajakumaran et al. (2019) modified the RO membrane using spherical ZnO-GO hybrid material. In this study, GO served as a platform or interface for ZnO nanomaterial dispersion and promoted better attachment with the TFC membrane due to the presence of the of hydrophilic functional groups. As a result, the modified membrane (at 0.03 wt% GO-ZnO concentration) showed almost 65% of water flux improvement (approximately from 15 LMH to 25 LMH) without compromising the salt rejection. Such improvement was attributed to the improvement in the surface hydrophilicity, as shown in the decline of contact angle value (around 20%) that facilitated the movement of water across the membrane. This decrease in contact angle measurement could be related to the presence of abundance hydrophilic functional group of the ZnO-GO hybrid material. Besides, the modified membrane also showed improved membrane stability with consistent RO membrane performance, as indicated by the enhanced chlorine stability and antifouling properties (retained more than 75% of its original flux as compared to neat membrane at 41%). The reason for this stability was due to the lower surface roughness (due to enhanced hydrophilicity) that decreased the adhesion of foulant particles on membrane surface [144].

Sulfonated materials or copolymers have also been adopted to improve the hydrophilicity of the membrane and to overcome the weak chlorine resistance of PA layer. For instance, Kim et al. (2013) modified a TFC RO membrane to TFN RO membrane by adding sulfonated poly (acrylene ether sulfone) containing amino group (aPES) material and hyper-branched aromatic PA grafted silica nanomaterial (HBP-g-silica) during the membrane synthesis. The salt rejection after chlorine exposure only dropped by 15% (from 97.7% to 82.5%) as compared to the neat membrane which dropped by 31% (from 98.0% to 67.1%). The strong chlorine resistance property was due to the formation of additional amide groups in the RO membrane from the combination of amino groups in HBP-g-silica and TMC solution. This resulted in the formation of denser membrane active layer that shielded it from chlorine attack. Furthermore, the modified membrane possessed better initial water flux (34.5 LMH) than the neat membrane (22.1 LMH) due to the presence of APES sulfonic group ($-SO_3H$) and HBP-g-silica that provided additional hydrophilicity to the membrane (drop in the contact angle value from 61.2° to 50.1°) [145].

Recently, aquaporin, (biological proteins) has emerged as one of the promising additives that can improve RO membrane performance, owing to its effective hourglass shape water transport channel and superior rejection at the same time. For example, Qi et al. (2016) synthesized an aquaporin biomimetic based RO membrane and the study revealed that the water and salt fluxes for aquaporin modified membrane improved by almost 50% (from 2.68 LMH to 4.13 LMH) and 17% (from 1.00 LMH to 1.17 LMH), respectively. The increase of flux was attributed to the improved hydrophilicity. Furthermore, the aquaporin modified membrane also recorded good operation stability by operating at 90% of its original performance even after 100 days of operation. Moreover, the energy requirement of the modified membrane was lower than unmodified membrane. The promising chemical stability (flux and rejection) and operational stability (energy and cost) indicated the great potential of aquaporin biomimetic based RO membrane (4.13 LMH water permeability, 97.2% NaCl rejection, 1.17 LMH solute permeability) in practical applications as the separation properties was comparable to the commercial BW30 RO membrane (2.83 LMH water permeability, 97.3% NaCl rejection, 0.79 LMH solute permeability) [146].

4.2 FO membranes

4.2.1 Substrate

Various materials (e.g. PES, PSf, PEI, PAN, and inorganic materials) have been used to synthesize the FO substrates, each with their own advantages and disadvantages in terms of stability, cost, and hydrophilicity [47]. Since majority of the available FO membranes in the market belong to polymeric membrane group, further discussion on the materials and formulation of FO membrane will be focusing on the polymeric membrane.

Adjusting the polymer composition is one of the most commonly used techniques to obtain porous FO support layer. For instance, decreasing the PSf concentration from 18% to 9% increased the water permeability by almost 50% (from 1.09 to 1.63 LMH/bar) [16]. Associated with this change was an elevation of FO water flux from 7.5 to 21 LMH using 1 M NaCl as draw solution and deionized water as feed solution. However, the use of lower polymer concentration might produce support layer with larger pore size, which might be too large to support the selective layer under a high hydraulic pressure [147]. While the operation of FO does not require high hydraulic pressure, sufficient mechanical strength is still required to endure the cross-flow dynamics in a large module for a long operating lifetime.

Pore forming additives such as lithium chloride (LiCl) and PEG could be utilized to modify the structure of the support layer. The incorporation of LiCl in the synthesis resulted in the collapse of the macrovoid, a reduction in pore size and porosity due to slow phase separation at high content of LiCl [148]. The reduced tortuosity has improved the water flux to 31.8 LMH. On another note, Wu et al. (2016) reported that increasing the PEG concentration from 0 to 6% increased the water permeability coefficient by 50% (1 to 1.55 LMH/bar), translating to doubled flux when the DS concentration increased from 0.5 to 1.5 M (NaCl) [149].

Instead of using single polymer, Bui et al. (2013) prepared a polymer consists of PAN and CA at different ratio for the synthesis of FO support layer [150]. The high viscous polymer solution led to the formation of nanofibers with larger diameter. This subsequently promoted the formation of bigger pore size in the support layer and improved the water permeability coefficient to 2.036 LMH as compared to commercial HTI-CTA membrane at 0.683 LMH.

Sulfonation has been adopted to promote the hydrophilicity (and thus productivity) of the FO support layer by replacing the hydrogen atoms of the membrane polymer matrix with sulfonic acid groups through electrophilic reaction [151]. For instance, when 50 wt% sulfonated polymers composed of PES and polyphenylsulfone (PPSU) was used in the synthesis of support layer, a more favourable sponge-like structure free of macrovoids was obtained [152]. The slow de-mixing in the phase inversion has produced larger pore size and porosity as compared to support layer fabricated from pure PESU. This structure helped to reduce ICP effects. Another advantage of this modification was the improvement of hydrophilicity which increased the water permeability too. However, FO membrane synthesized from sulfonated polymer is susceptible to swelling, a phenomenon that would reduce the mechanical strength of the support layer and pure water permeability [152–154].

The homogenous dispersion of nanoparticles in the polymer matrix may give rise to higher fractional free volume in mixed matrix membranes leading to high water permeability [155]. Apart from this, the aim of incorporating nanoparticles into the polymeric membranes is to promote the hydrophilicity, mechanical strength, and separation property [156]. A wide range of nanoparticles such as zeolite, silica, calcium carbonate, GO, CNTs, and TiO_2 have been explored by researchers where direct blending method is generally employed for adding the nanoparticles into the support layer polymer [24,157]. Park et al. (2015) incorporated GO nanosheets into the polymeric support layer of TFC-FO membrane [158]. The presence of abundant hydroxyl, carboxyl, and epoxy functional groups of GO increased the membrane hydrophilicity and enhanced the pure water permeation to 1.76 LMH from 0.91 LMH of pristine membrane without GO. Interestingly, the incorporation of different nanomaterials has resulted in different physicochemical properties of the support layer. For instance, the addition of PVA or carbon-TiO_2 resulted in the formation of macrovoid structure while GO and MWCNTs formed finger-like macrovoids [159].

The incorporation of titania in PSf sub-layer improved the membrane properties as in good wettability, the formation of finger-like structures, reduced tortuosity, and low ICP impacts [136]. These alterations in the support layer morphology managed to increase the water permeability to 4.27 LMH with 1.0wt% zeolite. Tang et al. (2013) demonstrated that the inclusion of 0.5% zeolite to the PSf substrate increased the substrate surface porosity and hydrophilicity as well as reduced the tortuosity [160]. Subsequently, the water permeability of the rejection layer was increased by 80% since the hydraulic resistance of the support layer has been reduced.

A more drastic formulation has been shown by Fan et al. (2019) where reduced graphene oxide and carbon nanotube have been adopted as active separation layer and substrate of hollow fibre FO membrane, respectively [161]. The GO layer was chemically reducing electrodeposited on the carbon nanotube fibre substrate. With its ultra-low friction and well-defined interlayer spacing, the FO membrane possessed high rejection rate and water permeability. Meanwhile, the carbon nanotube substrate, with tuneable porosity and good wettability could weaken the ICP phenomenon, and thus facilitated a high water flux. The improvement was quite obvious as revealed by its outstanding water flux at 22.6 LMH, which was 3 times higher than that of the commercial membrane. Besides, the reverse salt flux was only 1.6 gMH in comparison to 2.2 gMH for the commercial membrane. However, the proposed membrane might be too costly for mass production as the current price of GO and carbon nanotube are quite high. Discovery of competitive mass production approach for these materials might materialize the use of this membrane.

Tian et al. (2015) has embedded functionalized MWCNTs to strengthen the substrate strength [57]. The homogenous dispersion of the functionalized MWCNTs in the polymer solution has increased the tensile strength and stiffness. Besides, the increment of porosity from 59% to 81% indicated the formation of more interconnected pore structure. Subsequently, the water permeability coefficient was higher as compared to TFC-HTI membrane (2.6 LMH vs 1.63 LMH). However, the salt permeability coefficient was higher too (0.7 LMH vs 0.3 LMH). This phenomenon has also been observed by other researchers where the improved (higher) water flux is normally associated with higher reverse salt flux [162].

4.2.2 Active layer

The incorporation of nanoparticles in membrane synthesis has resulted in marked improvement in membrane characteristics such as high flux and rejection capability. Various nanomaterials such as zeolite, CNTs, zwitterions, dendrimers, silver, copper, GO, MOF, and aquaporin have been explored in the synthesis of FO membrane [163]. Tang et al. (2013) reported that FO membrane with 0.1 w/v% zeolite loading (in TMC/hexane) achieved approximately 50% higher flux than that of neat membrane [160]. However, the loading of nanoparticles is critical to the membrane formulation since excessive loading of nanoparticles may lead to defect formation, which might cause a high reverse solute flux [157].

Silver is one of the most extensively used biocides with reported multifold antibacterial mechanism: direction adhesion or infiltration to destroy the bacterial cells and generation of reactive oxygen species to case defects to bacteria cell structure and prevent the formation of biofilm [164,165]. Normally, silver is incorporated near the membrane surface to maximize the interactions between the nanoparticles and bacteria [166]. For example, Liu et al. (2017) utilized layer-by-layer IP approach to covalently bind the silver nanoparticles on the TFC PA surface [166]. The incorporation of silver nanoparticles resulted in stable filtration performance and excellent antibacterial properties. Another alternative biocide with lower cost is copper-based nanoparticles [167]. The copper nanoparticles could be decorated on the membrane surface by soaking the membrane in dopamine solution containing copper ions [168]. Though the membrane has lower water flux (by 28%), the antibacterial activity against *S. aureus* was increased by up to 97% with very little copper leakage into the permeate, indicating the capability of the functionalized membrane to perform long-term antibacterial activity against Gram-positive bacteria [169].

Silver nanoparticles can also be easily dip-coated onto the membrane surface through PDA layer [170]. The functionalized membrane displayed long-term silver release behaviour with the capacity to be regenerated multiple times after depletion. The thickness of the PDA layer should be controlled as the additional PDA layer would exert mass transfer resistance to water permeation. The facile regeneration of the silver nanoparticles on the PDA surface could be sustainable approach to provide long-term antibiofouling for the membrane.

GO has been recognized as the most amazing and versatile nanomaterial in this decade. It has been widely employed to improve the membrane properties be it in the form of water flux, rejection, and antifouling capability. For instance, Hegab et al. (2015) grafted GO to the TFC PA surface via poly L-Lysine (PLL) intermediary using LbL approach [171]. The modified membrane attained a higher hydrophilic and smooth surface, on top of 99% reduction in surviving bacteria. Also, the reverse salt flux was reduced too, indicating an improvement of membrane salt selectivity. Another family member of carbon-based nanoparticles is CNTs. The incorporation of 0.1 wt/v% (CNTs in MPD aqueous solution) amine-functionalized MWCNTs achieved water flux of 95.7 LMH and solute flux of 4.8 gMH, which observed an improvement of 160% higher and 30% lower than the pristine membrane [172].

The instability and dissolution of nanoparticles from the membrane are some of the hindrances that restricted the practical application of nanoparticles-functionalized membrane in water and wastewater treatment. This challenge can potentially be solved by decorating the nanoparticles onto larger support or carrier that has better compatibility with the polymer matrix (subsequently lower dissolution rate of nanoparticles) [173]. For instance, Soroush et al. (2015) demonstrated that the nanocomposite (silver nanoparticles immobilized on GO nanosheets) could be covalently bound onto the PA FO surface with enhanced hydrophilicity (contact angle below 25°) and antibacterial activity (> 96% in a static test) [174]. The GO nanosheets not only enhanced the amount of silver nanoparticles incorporated into the membrane but also stabilized the nanoparticles and decreased the ion release rate [175]. The stability was due to the covalent bonds formed between the carboxyl groups on the silver-GO nanocomposite and carboxyl groups of the TFC surface via crosslinking reaction [176].

Highly hydrophilic zwitterionic materials have also been employed as carrier to hold the biocidal silver nanoparticles. This nanocomposite possessed two functionalities, with the silver nanoparticles exert toxicity toward the bacteria while the zwitterionic materials shield the membrane surface from adsorption of organic foulants [177]. The synthesis approach first involved the grafting of zwitterionic poly(sulfobetaine methacrylate) brushes to the membrane surface via an atom transfer radical polymerization reaction. Later on, silver nanoparticles were formed in situ through chemical reduction of silver. The functionalized membrane showed superior antibacterial activity (95%) and inhibited the formation of biofilm.

Recently, MOFs have emerged as a promising additive for membrane synthesis due to their hybrid organic/inorganic structure, high surface area, highly tuneable molecular structure, and adjustable pore size [178]. It has been proposed that MOFs could house biocidal metal ions (antibacterial) for continuous gradual release of the metal ions by degradation of the framework [179]. For instance, the popularly used silver has been combined with MOFs to functionalize TFC PA membranes for FO applications [180]. The fabrication process involved only silver salt and organic ligand where both are covalently bonded to the TFC PA FO membrane at ambient temperature (without the need of additional stabilizing or reducing agent). After functionalized with Ag-MOFs, the membrane surface possessed strong antibacterial activity with more than 99% reduction of live bacteria. However, the capability of the membrane to retain its antibacterial property for long-term operation has been hindered by surface depletion of metals [177]. Apparently, regeneration of metal nanoparticles is required to preserve the antibacterial activity of the membrane surface [181].

A key challenge to FO desalination is the reverse solute flux (permeation of solute from draw to feed solutions) and forward solute flux (permeation of solute from feed to draw solutions) that would lead to several problems such as reduced osmotic driven force, salinity build-up on the feed side, feed contamination, aggravated membrane fouling, and elevated operating costs due to the need to continuously replenish draw solution [182]. To minimize the impact arises from this bidirectional solute flux, Zou et al. (2019) modified the FO membrane surface by coating zwitterion functionalized carbon nanotubes onto a commercial TFC FO membrane [183]. The zwitterion functional groups, tertiary amine group and carboxylate group, provided the necessary charge for electrostatic repulsion forces while the CNTs offered steric hindrance property. The combination of these features helped to significantly reduce reverse solute flux (55-84%) for a range draw solutes such as NaCl and $NH_4H_2PO_4$. At the same time, forward solute flux of several impurities such as Mg^{2+}, NO^{2-}-N, and NH^{4+}-N has also been reduced in the range of 30-100% as compared to unmodified FO membrane. The improved membrane characteristics have also led to consistent FO performance with enhanced fouling resistance.

Song et al. (2015) have synthesized a novel double-skinned TFN membrane for enhanced separation performance and antifouling capability in FO process [184]. CNTs were mixed with PDA which then undergone IP with trimesoylchloride to form active separation layer on both sides of PSf support layer. This resulted in the formation of two active layers sandwiching the support layer in the middle. The double-skinned FO membrane possessed better flux performance as compared to pristine TFC FO membrane or single-skinned TFC FO membrane, with approximately 25% higher water flux and 28% of lower reverse solute flux. This could be attributed to the nanocorridors formed from the CNTs provided extra passage for water transport while the dense active layer facing the draw solution prevented the salt from diffusing to feed side. In addition, the double-skinned membrane also possessed better antifouling performance where the initial water flux and normalized recovery flux were much higher compared to single-skinned TFC FO membrane.

Apart from the carbon-based and inorganic nanoparticles, aquaporin (membrane protein) has also been used to enhance the water permeability and salt rejection of FO membrane, owing to the intrinsic characteristics of the protein as a channel for water transfer and small solutes across the membrane [15]. For instance, the incorporation of aquaporin laden vesicles in the PA selective layer of FO increased the water permeability by twofold (from 3.7 to 7.6 LMH/bar) [185]. The water flux was also improved considerably from 38.5 to 49.1 LMH. The promising improvement has also been supported with the commercialization of FO membrane incorporated with aquaporin by Aquaporin A/S and its Singaporean subsidiary Aquaporin Asia Pte. Ltd. [186,187]. However, the availability of aquaporin FO membrane in the market is limited due to its complicated and difficulty in the fabrication process (high manufacturing cost). Furthermore, the long-term stability and integrity of this biomimetic FO membrane is scarce [11]. Hence, efforts are required to simplify and ease the scale-up fabrication of the membrane (reducing the cost) and more long-term operation study should be explored.

Conclusion and future prospects

In spite of the advancement of membrane science, some shortcomings are still persisting for the applications of RO and FO membranes in desalination processes. For instance, the commercially available TFC membranes are susceptible to chlorine attack and extreme pH conditions that will decompose and degrade the membrane performance. This signifies the need to seek for polymer materials that on the one hand could offer higher tolerance and resistance to chemical and mechanical conditions, but on the other hand would not sacrifice the flux. Also, membrane fouling such as deposition of organic compounds and microorganism on the membrane surface is still a big issue for long-term operation process. Fouling will shorten the membrane lifespan, which increases the cost of membrane desalination process. For RO, scaling is another challenge that leads to the degradation of membrane performance in desalination. Improper handling of scaling formation on the membrane surface will affect the membrane productivity. Moreover, additional cost is required to remove the scale, depending on the severity of the membrane surface blockage. Although the anti-scaling agents were seen to be effective in minimizing the scaling formation, search for new antiscalant and better alternative to further minimize or prevent the scale formation is still on-going.

Scaling is a lesser issue for FO process though concentration polarization is a bigger obstacle for it. Internal and external concentration polarization will weaken the osmotic pressure difference across FO membrane, which subsequently leads to lower flux performance that is infeasible for FO process as the flux of existing FO process is already low. All these phenomena indicate that fouling is an inevitable issue for membrane processes. Hence, apart from improving the membrane antifouling capability, attention should also be given to the pre-treatment process prior to the RO and FO processes such that the foulant present in feed to RO and FO can be removed to reduce membrane fouling propensity.

Though surface modifications and incorporation of additives in membrane synthesis are frequently employed to alter the membrane characteristics, some undesired side-effects have also been reported. Some modifications will result in the formation of external layer on top of membrane to increase the rejection capability but the water flux is compromised (additional resistance for water permeation). Another side-effect is the incorporation of nanoparticle additives that would induce the formation of rougher membrane surface where this characteristic is usually associated with more severe membrane fouling propensity. Thus, controlling the formulation and synthesis condition is crucial to obtain membrane with desired characteristics while suppressing the undesired side-effects.

Numerous studies have indicated that the incorporation of nanoparticles reportedly positively improving the membrane properties and performance. However, there are several challenges associated with this popular membrane modification approach. Due to its nano-size, the nanoparticle additives tend to agglomerate together and dispersed unevenly in the polymeric matrix. The agglomeration and uneven distribution of nanoparticles normally would result in the formation of defects on the membrane surface or structure, and subsequently adversely affects the membrane integrity (flux and rejection) and mechanical strength of the membrane. Another issue dealing with nanoparticles is the compatibility with the polymeric matrix. Inorganic nanoparticles do not interact or bound strongly with the organic polymeric matrix. Thus, the risk of nanoparticle leakage is apparent and the membrane will be losing its improved capability over time. This defeats the intention of utilizing the nanoparticles for enhanced membrane performance and causes financial loss as the membrane could not last long and the valuable nanoparticles are lost to environment (potentially harmful to the ecosystems and living organisms). Another challenge is related to feasibility of having large-scale production of TFC membrane incorporated with nanoparticles. Many nanoparticles used in the studies are synthesized through complicated procedures, and this will be the obstacle of commercialization of TFC membrane enhanced with nanoparticles. The nanoparticles agglomeration and leaking issues in lab-scale studies have put a doubt on the commercialization and large-scale application of RO and FO membranes enhanced with nanoparticles too. Hence, it could be seen that the road to mass production of TFC RO and FO membranes (enhanced with nanoparticles) is still quite challenging despite the positive outcomes reported in controlled-experimental conditions.

Acknowledgment

The authors sincerely appreciate and acknowledge the financial support from Universiti Kebangsaan Malaysia (UKM) for funding this work under the grant code DIP-2020-016.

References

[1]   H. Saleem, S.J. Zaidi, Nanoparticles in reverse osmosis membranes for desalination: A state of the art review. Desalination. 475 (2020) 114171. https://doi.org/10.1016/j.desal.2019.114171

[2]   A. Ali, R.A. Tufa, F. Macedonio, E. Curcio, E. Drioli, Membrane technology in renewable-energy-driven desalination. Renew. Sustain. Energy Rev. 81 (2018) 1–21. https://doi.org/10.1016/j.rser.2017.07.047

[3]   M. Qasim, M. Badrelzaman, N.N. Darwish, N.A. Darwish, N. Hilal, Reverse osmosis desalination: A state-of-the-art review. Desalination. 459 (2019) 59–104. https://doi.org/10.1016/j.desal.2019.02.008

[4]   M.W. Shahzad, M. Burhan, L. Ang, K.C. Ng, Energy-water-environment nexus underpinning future desalination sustainability. Desalination. 413 (2017) 52–64. https://doi.org/10.1016/j.desal.2017.03.009

[5]   M.A. Abdelkareem, M. El Haj Assad, E.T. Sayed, B. Soudan, Recent progress in the use of renewable energy sources to power water desalination plants. Desalination. 435 (2018) 97–113. https://doi.org/10.1016/j.desal.2017.11.018

[6]   S.S. Shenvi, A.M. Isloor, A.F. Ismail, A review on RO membrane technology : Developments and challenges. Desalination. 368 (2015) 10–26. https://doi.org/10.1016/j.desal.2014.12.042

[7]   W.L. Ang, A.W. Mohammad, D. Johnson, N. Hilal, Unlocking the application potential of forward osmosis through integrated/hybrid process. Sci. Total Environ. 706 (2020) 136047. https://doi.org/10.1016/j.scitotenv.2019.136047

[8]   D.J. Johnson, W.A. Suwaileh, A.W. Mohammed, N. Hilal, Osmotic's potential: An overview of draw solutes for forward osmosis. Desalination. 434 (2018) 100–120. https://doi.org/10.1016/j.desal.2017.09.017

[9]   F. Volpin, E. Fons, L. Chekli, J.E. Kim, A. Jang, H.K. Shon, Hybrid forward osmosis-reverse osmosis for wastewater reuse and seawater desalination: Understanding the optimal feed solution to minimise fouling. Process Saf. Environ. Prot. 117 (2018) 523–532. https://doi.org/10.1016/j.psep.2018.05.006

[10]   S.J. Im, S. Jeong, S. Jeong, A. Jang, Techno-economic evaluation of an element-scale forward osmosis-reverse osmosis hybrid process for seawater desalination. Desalination. 476 (2020) 114240. https://doi.org/10.1016/j.desal.2019.114240

[11]   D. Li, Y. Yan, H. Wang, Recent advances in polymer and polymer composite membranes for reverse and forward osmosis processes. Prog. Polym. Sci. 61 (2016) 104–155. https://doi.org/10.1016/j.progpolymsci.2016.03.003

[12]   S. Loeb, S. Sourirajan, Sea water demineralization by means of an osmotic membrane. Adv. Chem. Am. Chem. Soc. (1963) 117–132. https://doi.org/10.1021/ba-1963-0038.ch009

[13]   J.S. Lee, S.A. Heo, H.J. Jo, B.R. Min, Preparation and characteristics of cross-linked cellulose acetate ultrafiltration membranes with high chemical resistance and mechanical strength. React. Funct. Polym. 99 (2016) 114–121. https://doi.org/10.1016/j.reactfunctpolym.2015.12.014

[14]   J.E. Cadotte, R.J. Petersen, R.E. Larson, E.E. Erickson, Interfacially synthesized reverse osmosis membrane, 1981.

[15]   C.Y. Tang, Y. Zhao, R. Wang, C. Hélix-Nielsen, A.G. Fane, Desalination by biomimetic aquaporin membranes: Review of status and prospects. Desalination. 308 (2013) 34–40. https://doi.org/10.1016/j.desal.2012.07.007

[16]   A. Tiraferri, N.Y. Yip, W.A. Phillip, J.D. Schiffman, M. Elimelech, Relating performance of thin-film composite forward osmosis membranes to support layer formation and structure. J. Memb. Sci. 367 (2011) 340–352. https://doi.org/10.1016/j.memsci.2010.11.014

[17]   R.C. Ong, T.S. Chung, B.J. Helmer, J.S. De Wit, Novel cellulose esters for forward osmosis membranes. Ind. Eng. Chem. Res. 51 (2012) 16135–16145. https://doi.org/10.1021/ie302654h

[18]   J.R. McCutcheon, R.L. McGinnis, M. Elimelech, A novel ammonia-carbon dioxide forward (direct) osmosis desalination process. Desalination. 174 (2005) 1–11. https://doi.org/10.1016/j.desal.2004.11.002

[19]   C.Y. Tang, Q. She, W.C.L. Lay, R. Wang, A.G. Fane, Coupled effects of internal concentration polarization and fouling on flux behavior of forward osmosis membranes during humic acid filtration. J. Memb. Sci. 354 (2010) 123–133. https://doi.org/10.1016/j.memsci.2010.02.059

[20]   N.Y. Yip, A. Tiraferri, W.A. Phillip, J.D. Schiffman, M. Elimelech, High Performance Thin-Film Membrane. Environ. Sci. Technol. 44 (2010) 3812–3818. https://doi.org/10.1021/es1002555

[21]   K.L. Hickenbottom, J. Vanneste, M. Elimelech, T.Y. Cath, Assessing the current state of commercially available membranes and spacers for energy production with pressure retarded osmosis. Desalination. 389 (2016) 108–118. https://doi.org/10.1016/j.desal.2015.09.029

[22]   R. Wang, L. Shi, C.Y. Tang, S. Chou, C. Qiu, A.G. Fane, Characterization of novel forward osmosis hollow fiber membranes. J. Memb. Sci. 355 (2010) 158–167. https://doi.org/10.1016/j.memsci.2010.03.017

[23]   G.R. Xu, S.H. Wang, H.L. Zhao, S.B. Wu, J.M. Xu, L. Li, X.Y. Liu, Layer-by-layer (LBL) assembly technology as promising strategy for tailoring pressure-driven desalination membranes. J. Memb. Sci. 493 (2015) 428–443. https://doi.org/10.1016/j.memsci.2015.06.038

[24]   W.A. Suwaileh, D.J. Johnson, S. Sarp, N. Hilal, Advances in forward osmosis membranes: Altering the sub-layer structure via recent fabrication and chemical modification approaches. Desalination. 436 (2018) 176–201. https://doi.org/10.1016/j.desal.2018.01.035

[25]   R.H. Hailemariam, Y.C. Woo, M.M. Damtie, B.C. Kim, K.D. Park, J.S. Choi, Reverse osmosis membrane fabrication and modification technologies and future trends: A review. Adv. Colloid Interface Sci. 276 (2020) 102100. https://doi.org/10.1016/j.cis.2019.102100

[26]   M. Asadollahi, D. Bastani, S.A. Musavi, Enhancement of surface properties and performance of reverse osmosis membranes after surface modification: A review. Desalination. 420 (2017) 330–383. https://doi.org/10.1016/j.desal.2017.05.027

[27]   A.K. Ghosh, E.M.V. Hoek, Impacts of support membrane structure and chemistry on polyamide-polysulfone interfacial composite membranes. J. Memb. Sci. 336 (2009) 140–148. https://doi.org/10.1016/j.memsci.2009.03.024

[28]   L. Boor Singh, V. Kochkodan, R. Hashaikeh, N. Hilal, A review on membrane fabrication : Structure , properties and performance relationship. Desalination. 326 (2013) 77–95. https://doi.org/10.1016/j.desal.2013.06.016

[29]   G.R. Guillen, Y. Pan, M. Li, E.M.V. Hoek, Preparation and characterization of membranes formed by nonsolvent induced phase separation: A review. Ind. Eng. Chem. Res. 50 (2011) 3798–3817. https://doi.org/10.1021/ie101928r

[30]   K.P. Lee, T.C. Arnot, D. Mattia, A review of reverse osmosis membrane materials for desalination — Development to date and future potential. J. Memb. Sci. 370 (2011) 1–22. https://doi.org/10.1016/j.memsci.2010.12.036

[31]   D.M. Wang, J.Y. Lai, Recent advances in preparation and morphology control of polymeric membranes formed by nonsolvent induced phase separation. Curr. Opin. Chem. Eng. 2 (2013) 229–237. https://doi.org/10.1016/j.coche.2013.04.003

[32]   A. Sabir, M. Shafiq, A. Islam, F. Jabeen, A. Shafeeq, A. Ahmad, M.T. Zahid Butt, K.I. Jacob, T. Jamil, Conjugation of silica nanoparticles with cellulose acetate/polyethylene glycol 300 membrane for reverse osmosis using MgSO4 solution. Carbohydr. Polym. 136 (2016) 551–559. https://doi.org/10.1016/j.carbpol.2015.09.042

[33]   A.M.A. Abdelsamad, A.S.G. Khalil, M. Ulbricht, Influence of controlled functionalization of mesoporous silica nanoparticles as tailored fillers for thin-film nanocomposite membranes on desalination performance. J. Memb. Sci. 563 (2018) 149–161. https://doi.org/10.1016/j.memsci.2018.05.043

[34]   F.E. Ahmed, B.S. Lalia, R. Hashaikeh, A review on electrospinning for membrane fabrication: Challenges and applications. Desalination. 356 (2015) 15–30. https://doi.org/10.1016/j.desal.2014.09.033

[35]   M.K. Selatile, S. Ray, Recent developments in polymeric electrospun nano fibrous membranes for seawater desalination. RSC Adv. (2018) 37915–37938. https://doi.org/10.1039/C8RA07489E

[36]   S.S. Ray, S. Chen, C. Li, C. Nguyen, A comprehensive review : electrospinning technique for fabrication and surface modi fi cation. RSC Adv. 6 (2016) 85495–85514. https://doi.org/10.1039/C6RA14952A

[37]   X. Wang, H. Ma, B. Chu, B.S. Hsiao, Thin- film nano fibrous composite reverse osmosis membranes for desalination. Desalination. 420 (2017) 91–98. https://doi.org/10.1016/j.desal.2017.06.029

[38]   Z. Yang, Y. Zhou, Z. Feng, X. Rui, T. Zhang, Z. Zhang, A review on reverse osmosis and nanofiltration membranes for water purification. Polymers (Basel). 11 (2019) 1252. https://doi.org/10.3390/polym11081252

[39]   J. Zheng, Y. Yao, M. Li, L. Wang, X. Zhang, A non-MPD-type reverse osmosis membrane with enhanced permselectivity for brackish water desalination. J. Memb. Sci. 565 (2018) 104–111. https://doi.org/10.1016/j.memsci.2018.08.015

[40]   W.J. Lau, A.F. Ismail, N. Misdan, M.A. Kassim, A recent progress in thin film composite membrane: A review. Desalination. 287 (2012) 190–199. https://doi.org/10.1016/j.desal.2011.04.004

[41]   J.M. Gohil, P. Ray, A review on semi-aromatic polyamide TFC membranes prepared by interfacial polymerization: Potential for water treatment and desalination. Sep. Purif. Technol. 181 (2017) 159–182. https://doi.org/10.1016/j.seppur.2017.03.020

[42]   J.R. Kovacs, C. Liu, P.T. Hammond, Spray Layer-by-layer assembled clay composite thin films as selective layers in reverse osmosis membranes. ACS Appl. Mater. Interfaces. 7 (2015) 13375–13383. https://doi.org/10.1021/acsami.5b01879

[43]   P. Taylor, F. Fadhillah, S.M.J. Zaidi, Z. Khan, M. Khaled, T. Paula, Desalination and Water Treatment Reverse osmosis desalination membrane formed from weak polyelectrolytes by spin assisted layer by layer technique by spin assisted layer by layer technique. Desalin. Water Treat. 34 (2012) 44–49. https://doi.org/10.5004/dwt.2011.2856

[44]   I.L. Alsvik, M.B. Hägg, Pressure retarded osmosis and forward osmosis membranes: Materials and methods. Polymers (Basel). 5 (2013) 303–327. https://doi.org/10.3390/polym5010303

[45]   C. Boo, S. Lee, M. Elimelech, Z. Meng, S. Hong, Colloidal fouling in forward osmosis: Role of reverse salt diffusion. J. Memb. Sci. 390–391 (2012) 277–284. https://doi.org/10.1016/j.memsci.2011.12.001

[46]   M. Mulder, Basic Principles of Membrane Technology, 2nd ed., Kluwer Academic Publishers, The Netherlands, 1996.

[47]   Y.N. Wang, K. Goh, X. Li, L. Setiawan, R. Wang, Membranes and processes for forward osmosis-based desalination: Recent advances and future prospects. Desalination. 434 (2018) 81–99. https://doi.org/10.1016/j.desal.2017.10.028

[48]   X. Lu, L.H. Arias Chavez, S. Romero-Vargas Castrillón, J. Ma, M. Elimelech, Influence of active layer and support layer surface structures on organic fouling propensity of thin-film composite forward osmosis membranes. Environ. Sci. Technol. 49 (2015) 1436–1444. https://doi.org/10.1021/es5044062

[49] H.Q. Liang, W.S. Hung, H.H. Yu, C.C. Hu, K.R. Lee, J.Y. Lai, Z.K. Xu, Forward osmosis membranes with unprecedented water flux. J. Memb. Sci. 529 (2017) 47–54. https://doi.org/10.1016/j.memsci.2017.01.056

[50] X. Liu, H.Y. Ng, Fabrication of layered silica-polysulfone mixed matrix substrate membrane for enhancing performance of thin-film composite forward osmosis membrane. J. Memb. Sci. 481 (2015) 148–163. https://doi.org/10.1016/j.memsci.2015.02.012

[51] X. Liu, H.Y. Ng, Double-blade casting technique for optimizing substrate membrane in thin-film composite forward osmosis membrane fabrication. J. Memb. Sci. 469 (2014) 112–126. https://doi.org/10.1016/j.memsci.2014.06.037

[52] P. Xiao, L.D. Nghiem, Y. Yin, X.M. Li, M. Zhang, G. Chen, J. Song, T. He, A sacrificial-layer approach to fabricate polysulfone support for forward osmosis thin-film composite membranes with reduced internal concentration polarisation. J. Memb. Sci. 481 (2015) 106–114. https://doi.org/10.1016/j.memsci.2015.01.036

[53] G. Chen, R. Liu, H.K. Shon, Y. Wang, J. Song, X.M. Li, T. He, Open porous hydrophilic supported thin-film composite forward osmosis membrane via co-casting for treatment of high-salinity wastewater. Desalination. 405 (2017) 76–84. https://doi.org/10.1016/j.desal.2016.12.004

[54] S. Zhang, K.Y. Wang, T.S. Chung, H. Chen, Y.C. Jean, G. Amy, Well-constructed cellulose acetate membranes for forward osmosis: Minimized internal concentration polarization with an ultra-thin selective layer. J. Memb. Sci. 360 (2010) 522–535. https://doi.org/10.1016/j.memsci.2010.05.056

[55] R.C. Ong, T.S. Chung, Fabrication and positron annihilation spectroscopy (PAS) characterization of cellulose triacetate membranes for forward osmosis. J. Memb. Sci. 394–395 (2012) 230–240. https://doi.org/10.1016/j.memsci.2011.12.046

[56] N.N. Bui, M.L. Lind, E.M.V. Hoek, J.R. McCutcheon, Electrospun nanofiber supported thin film composite membranes for engineered osmosis. J. Memb. Sci. 385–386 (2011) 10–19. https://doi.org/10.1016/j.memsci.2011.08.002

[57] M. Tian, R. Wang, K. Goh, Y. Liao, A.G. Fane, Synthesis and characterization of high-performance novel thin film nanocomposite PRO membranes with tiered nanofiber support reinforced by functionalized carbon nanotubes. J. Memb. Sci. 486 (2015) 151–160. https://doi.org/10.1016/j.memsci.2015.03.054

[58] M. Obaid, M.A. Abdelkareem, S. Kook, H.Y. Kim, N. Hilal, N. Ghaffour, I.S. Kim, Breakthroughs in the fabrication of electrospun-nanofiber-supported thin film composite/nanocomposite membranes for the forward osmosis process: A review. Crit.

Rev. Environ. Sci. Technol. (2019) 1–69.
https://doi.org/10.1080/10643389.2019.1672510

[59]   E.L. Tian, H. Zhou, Y.W. Ren, Z. a. mirza, X.Z. Wang, S.W. Xiong, Novel design
of hydrophobic/hydrophilic interpenetrating network composite nanofibers for the
support layer of forward osmosis membrane. Desalination. 347 (2014) 207–214.
https://doi.org/10.1016/j.desal.2014.05.043

[60]   J. Wei, X. Liu, C. Qiu, R. Wang, C.Y. Tang, Influence of monomer concentrations
on the performance of polyamide-based thin film composite forward osmosis
membranes. J. Memb. Sci. 381 (2011) 110–117.
https://doi.org/10.1016/j.memsci.2011.07.034

[61]   Q. Saren, C.Q. Qiu, C.Y. Tang, Synthesis and characterization of novel forward
osmosis membranes based on layer-by-layer assembly. Environ. Sci. Technol. 45
(2011) 5201–5208. https://doi.org/10.1021/es200115w

[62]   C. Liu, W. Fang, S. Chou, L. Shi, A.G. Fane, R. Wang, Fabrication of layer-by-
layer assembled FO hollow fiber membranes and their performances using low
concentration draw solutions. Desalination. 308 (2013) 147–153.
https://doi.org/10.1016/j.desal.2012.07.027

[63]   C. Liu, L. Shi, R. Wang, Enhanced hollow fiber membrane performance via semi-
dynamic layer-by-layer polyelectrolyte inner surface deposition for nanofiltration and
forward osmosis applications. React. Funct. Polym. 86 (2015) 154–160.
https://doi.org/10.1016/j.reactfunctpolym.2014.07.018

[64]   Y. Cui, H. Wang, H. Wang, T.S. Chung, Micro-morphology and formation of
layer-by-layer membranes and their performance in osmotically driven processes.
Chem. Eng. Sci. 101 (2013) 13–26. https://doi.org/10.1016/j.ces.2013.06.011

[65]   S. Qi, C.Q. Qiu, Y. Zhao, C.Y. Tang, Double-skinned forward osmosis
membranes based on layer-by-layer assembly-FO performance and fouling behavior.
J. Memb. Sci. 405–406 (2012) 20–29. https://doi.org/10.1016/j.memsci.2012.02.032

[66]   P. Pardeshi, A.A. Mungray, Synthesis, characterization and application of novel
high flux FO membrane by layer-by-layer self-assembled polyelectrolyte. J. Memb.
Sci. 453 (2014) 202–211. https://doi.org/10.1016/j.memsci.2013.11.001

[67]   G. Xu, S. Wang, H. Zhao, S. Wu, J. Xu, L. Li, X. Liu, S. Wang, H. Zhao, S. Wu, J.
Xu, L. Li, X. Liu, Layer by layer (LBL) assembly technology as promising strategy for
tailoring pressure-driven desalination membranes. J. Memb. Sci. 493 (2015) 428–443.
https://doi.org/10.1016/j.memsci.2015.06.038

[68]   T. Ishigami, K. Amano, A. Fujii, Y. Ohmukai, E. Kamio, T. Maruyama, H. Matsuyama, Fouling reduction of reverse osmosis membrane by surface modification via layer-by-layer assembly. Sep. Purif. Technol. 99 (2012) 1–7. https://doi.org/10.1016/j.seppur.2012.08.002

[69]   W. Choi, J. Choi, J. Bang, J. Lee, Layer-by-layer assembly of graphene oxide nanosheets on polyamide membranes for durable reverse-osmosis applications. ACS Appl. Mater. Interfaces. 5 (2013) 12510–12519. https://doi.org/10.1021/am403790s

[70]   W. Ma, A. Soroush, T. Van Anh, G. Brennan, S. Rahaman, B. Asadishad, N. Tufenkji, Spray- and spin-assisted layer-by-layer assembly of copper nanoparticles on thin- film composite reverse osmosis membrane for biofouling mitigation. Water Res. 99 (2016) 188–199. https://doi.org/10.1016/j.watres.2016.04.042

[71]   J. Saqib, I.H. Aljundi, Membrane fouling and modification using surface treatment and layer-by-layer assembly of polyelectrolytes : State-of-the-art review. J. Water Process Eng. 11 (2016) 68–87. https://doi.org/10.1016/j.jwpe.2016.03.009

[72]   Q. Zhang, C. Zhang, J. Xu, Y. Nie, S. Li, S. Zhang, C. Peg, Effect of poly (vinyl alcohol) coating process conditions on the properties and performance of polyamide reverse osmosis membranes. Desalination. 379 (2016) 42–52. https://doi.org/10.1016/j.desal.2015.10.012

[73]   S. Hafiz zahid, Z. Khan, R. Yang, K.K. Gleason, Surface modification of reverse osmosis membranes with zwitterionic coating for improved resistance to fouling. Desalination. 362 (2015) 93–103. https://doi.org/10.1016/j.desal.2015.02.009

[74]   L. Ni, J. Meng, X. Li, Y. Zhang, Surface coating on the polyamide TFC RO membrane for chlorine resistance and antifouling performance improvement. J. Memb. Sci. 451 (2014) 205–215. https://doi.org/10.1016/j.memsci.2013.09.040

[75]   S. Yu, G. Yao, B. Dong, H. Zhu, X. Peng, J. Liu, Improving fouling resistance of thin-film composite polyamide reverse osmosis membrane by coating natural hydrophilic polymer sericin. Sep. Purif. Technol. 118 (2013) 285–293. https://doi.org/10.1016/j.seppur.2013.07.018

[76]   Y. Zhao, L. Dai, Q. Zhang, S. Zhou, S. Zhang, Chlorine-resistant sulfochlorinated and sulfonated polysulfone for reverse osmosis membranes by coating method. J. Colloid Interface Sci. 541 (2019) 434–443. https://doi.org/10.1016/j.jcis.2019.01.104

[77]   N. Misdan, A.F. Ismail, N. Hilal, Recent advances in the development of (bio) fouling resistant thin fi lm composite membranes for desalination ☆. Desalination. 380 (2015) 105–111. https://doi.org/10.1016/j.desal.2015.06.001

[78]   D. Nikolaeva, C. Langner, A. Ghanem, M. Abdel, B. Voit, J. Meier-haack, Hydrogel surface modi fi cation of reverse osmosis membranes. J. Memb. Sci. 476 (2015) 264–276. https://doi.org/10.1016/j.memsci.2014.11.051

[79]   M.M. Armendáriz-ontiveros, A.G. García, S.D.L.S. Villalobos, Biofouling performance of RO membranes coated with Iron NPs on graphene oxide. Desalination. (2018) 0–1. https://doi.org/10.1016/j.desal.2018.07.005

[80]   Q. Cheng, Y. Zheng, S. Yu, H. Zhu, X. Peng, Surface modi fi cation of a commercial thin- fi lm composite polyamide reverse osmosis membrane through graft polymerization of N-isopropylacrylamide followed by acrylic acid. J. Memb. Sci. 447 (2013) 236–245. https://doi.org/10.1016/j.memsci.2013.07.025

[81]   M.B. El-arnaouty, A.M.A. Ghaffar, M. Eid, M.E. Aboulfotouh, N.H. Taher, E. Soliman, Nano-modification of polyamide thin film composite reverse osmosis membranes by radiation grafting. J. Radiat. Res. Appl. Sci. (2018) 1–13. https://doi.org/10.1016/j.jrras.2018.01.005

[82]   X. Huang, K.L. Marsh, B.T. Mcverry, E.M. V Hoek, R.B. Kaner, Low-fouling antibacterial reverse osmosis membranes via surface grafting of graphene oxide. ACS Appl. Mater. Interfaces. 8 (2016) 14334–14338. https://doi.org/10.1021/acsami.6b05293

[83]   H. Li, L. Peng, Y. Luo, P. Yu, Enhancement in membrane performances of a commercial polyamide reverse osmosis membrane via surface coating of polydopamine followed by the grafting of polyethylenimine. RSC Adv. 5 (2015) 98566–98575. https://doi.org/10.1039/C5RA20891B

[84]   J. Li, L. Yin, G. Qiu, X. Li, Q. Liu, J. Xie, A photo-bactericidal thin film composite membrane for forward osmosis. J. Mater. Chem. A. 3 (2015) 6781–6786. https://doi.org/10.1039/C5TA00430F

[85]   D. Li, H. Wang, Recent developments in reverse osmosis desalination membranes. J. Mater. Chem. 20 (2010) 4551–4566. https://doi.org/10.1039/b924553g

[86]   A. Nguyen, S. Azari, L. Zou, Coating zwitterionic amino acid l-DOPA to increase fouling resistance of forward osmosis membrane. Desalination. 312 (2013) 82–87. https://doi.org/10.1016/j.desal.2012.11.038

[87]   S. Azari, L. Zou, Using zwitterionic amino acid l-DOPA to modify the surface of thin film composite polyamide reverse osmosis membranes to increase their fouling resistance. J. Memb. Sci. 401–402 (2012) 68–75. https://doi.org/10.1016/j.memsci.2012.01.041

[88]   P. Lu, S. Liang, T. Zhou, T. Xue, X. Mei, Q. Wang, Layered double hydroxide nanoparticle modified forward osmosis membranes via polydopamine immobilization with significantly enhanced chlorine and fouling resistance. Desalination. 421 (2017) 99–109. https://doi.org/10.1016/j.desal.2017.04.030

[89]   J.T. Arena, S.S. Manickam, K.K. Reimund, B.D. Freeman, J.R. McCutcheon, Solute and water transport in forward osmosis using polydopamine modified thin film composite membranes. Desalination. 343 (2014) 8–16. https://doi.org/10.1016/j.desal.2014.01.009

[90]   X. Liu, S.L. Ong, H.Y. Ng, Fabrication of mesh-embedded double-skinned substrate membrane and enhancement of its surface hydrophilicity to improve anti-fouling performance of resultant thin-film composite forward osmosis membrane. J. Memb. Sci. 511 (2016) 40–53. https://doi.org/10.1016/j.memsci.2016.03.015

[91]   S. Kim, R. Roque, B. Birgisson, A. Guarin, Porosity of the dominant aggregate size range to evaluate coarse aggregate structure of asphalt mixtures. J. Mater. Civ. Eng. 21 (2009) 32–39. https://doi.org/10.1061/(ASCE)0899-1561(2009)21:1(32)

[92]   F. Peng, X. Huang, A. Jawor, E.M.V. Hoek, Transport, structural, and interfacial properties of poly(vinyl alcohol)-polysulfone composite nanofiltration membranes. J. Memb. Sci. 353 (2010) 169–176. https://doi.org/10.1016/j.memsci.2010.02.044

[93]   H.R. Ahn, T.M. Tak, Y.N. Kwon, Preparation and applications of poly vinyl alcohol (PVA) modified cellulose acetate (CA) membranes for forward osmosis (FO) processes. Desalin. Water Treat. 53 (2015) 1–7. https://doi.org/10.1080/19443994.2013.834516

[94]   Q. Liu, J. Li, Z. Zhou, J. Xie, J.Y. Lee, Hydrophilic mineral coating of membrane substrate for reducing internal concentration polarization (ICP) in forward Osmosis. Sci. Rep. 6 (2016) 1–10. https://doi.org/10.1038/srep19593

[95]   C. Qiu, L. Setiawan, R. Wang, C.Y. Tang, A.G. Fane, High performance flat sheet forward osmosis membrane with an NF-like selective layer on a woven fabric embedded substrate. Desalination. 287 (2012) 266–270. https://doi.org/10.1016/j.desal.2011.06.047

[96]   L. Setiawan, R. Wang, S. Tan, L. Shi, A.G. Fane, Fabrication of poly(amide-imide)-polyethersulfone dual layer hollow fiber membranes applied in forward osmosis by combined polyelectrolyte cross-linking and depositions. Desalination. 312 (2013) 99–106. https://doi.org/10.1016/j.desal.2012.10.032

[97]   K.Y. Wang, Q. Yang, T.S. Chung, R. Rajagopalan, Enhanced forward osmosis from chemically modified polybenzimidazole (PBI) nanofiltration hollow fiber

membranes with a thin wall. Chem. Eng. Sci. 64 (2009) 1577–1584. https://doi.org/10.1016/j.ces.2008.12.032

[98] L. Setiawan, L. Shi, R. Wang, Dual layer composite nanofiltration hollow fiber membranes for low-pressure water softening. Polymer (Guildf). 55 (2014) 1367–1374. https://doi.org/10.1016/j.polymer.2013.12.032

[99] L. Setiawan, R. Wang, K. Li, A.G. Fane, Fabrication of novel poly(amide-imide) forward osmosis hollow fiber membranes with a positively charged nanofiltration-like selective layer. J. Memb. Sci. 369 (2011) 196–205. https://doi.org/10.1016/j.memsci.2010.11.067

[100] L. Setiawan, R. Wang, L. Shi, K. Li, A.G. Fane, Novel dual-layer hollow fiber membranes applied for forward osmosis process. J. Memb. Sci. 421–422 (2012) 238–246. https://doi.org/10.1016/j.memsci.2012.07.020

[101] K. Goh, L. Setiawan, L. Wei, R. Si, A.G. Fane, R. Wang, Y. Chen, Graphene oxide as effective selective barriers on a hollow fiber membrane for water treatment process. J. Memb. Sci. 474 (2015) 244–253. https://doi.org/10.1016/j.memsci.2014.09.057

[102] L. Setiawan, R. Wang, K. Li, A.G. Fane, Fabrication and characterization of forward osmosis hollow fiber membranes with antifouling NF-like selective layer. J. Memb. Sci. 394–395 (2012) 80–88. https://doi.org/10.1016/j.memsci.2011.12.026

[103] S. Xiong, S. Xu, A. Phommachanh, M. Yi, Y. Wang, Versatile Surface Modification of TFC Membrane by Layer-by-Layer Assembly of Phytic Acid-Metal Complexes for Comprehensively Enhanced FO Performance. Environ. Sci. Technol. 53 (2019) 3331–3341. https://doi.org/10.1021/acs.est.8b06628

[104] H. Kang, W. Wang, J. Shi, Z. Xu, H. Lv, X. Qian, L. Liu, M. Jing, F. Li, J. Niu, Interlamination restrictive effect of carbon nanotubes for graphene oxide forward osmosis membrane via layer by layer assembly. Appl. Surf. Sci. 465 (2019) 1103–1106. https://doi.org/10.1016/j.apsusc.2018.09.255

[105] H.M. Hegab, L. Zou, Graphene oxide-assisted membranes: Fabrication and potential applications in desalination and water purification. J. Memb. Sci. 484 (2015) 95–106. https://doi.org/10.1016/j.memsci.2015.03.011

[106] A. Tiraferri, Y. Kang, E.P. Giannelis, M. Elimelech, Highly hydrophilic thin-film composite forward osmosis membranes functionalized with surface-tailored nanoparticles. ACS Appl. Mater. Interfaces. 4 (2012) 5044–5053. https://doi.org/10.1021/am301532g

[107] C. Xing, L. Liu, H. Tang, X. Feng, Q. Yang, S. Wang, G.C. Bazan, Design guidelines for conjugated polymers with light-activated anticancer activity. Adv. Funct. Mater. 21 (2011) 4058–4067. https://doi.org/10.1002/adfm.201100840

[108] C. Ringot, V. Sol, R. Granet, P. Krausz, Porphyrin-grafted cellulose fabric: New photobactericidal material obtained by "Click-Chemistry" reaction. Mater. Lett. 63 (2009) 1889–1891. https://doi.org/10.1016/j.matlet.2009.06.009

[109] P.K.S. Mural, S. Jain, S. Kumar, G. Madras, S. Bose, Unimpeded permeation of water through biocidal graphene oxide sheets anchored on to 3D porous polyolefinic membranes. Nanoscale. 8 (2016) 8048–8057. https://doi.org/10.1039/C6NR01356B

[110] X. Zhao, R. Zhang, Y. Liu, M. He, Y. Su, C. Gao, Z. Jiang, Antifouling membrane surface construction: Chemistry plays a critical role. J. Memb. Sci. 551 (2018) 145–171. https://doi.org/10.1016/j.memsci.2018.01.039

[111] C. Wang, Q. Yan, H.B. Liu, X.H. Zhou, S.J. Xiao, Different EDC/NHS activation mechanisms between PAA and PMAA brushes and the following amidation reactions. Langmuir. 27 (2011) 12058–12068. https://doi.org/10.1021/la202267p

[112] S. Romero-Vargas Castrillón, X. Lu, D.L. Shaffer, M. Elimelech, Amine enrichment and poly(ethylene glycol) (PEG) surface modification of thin-film composite forward osmosis membranes for organic fouling control. J. Memb. Sci. 450 (2014) 331–339. https://doi.org/10.1016/j.memsci.2013.09.028

[113] L. Shen, X. Zhang, J. Zuo, Y. Wang, Performance enhancement of TFC FO membranes with polyethyleneimine modification and post-treatment. J. Memb. Sci. 534 (2017) 46–58. https://doi.org/10.1016/j.memsci.2017.04.008

[114] Z. Liu, X. An, C. Dong, S. Zheng, B. Mi, Y. Hu, Modification of thin film composite polyamide membranes with 3D hyperbranched polyglycerol for simultaneous improvement in their filtration performance and antifouling properties. J. Mater. Chem. A. 5 (2017) 23190–23197. https://doi.org/10.1039/C7TA07335F

[115] X. Bao, Q. Wu, W. Shi, W. Wang, Z. Zhu, Z. Zhang, R. Zhang, X. Zhang, B. Zhang, Y. Guo, F. Cui, Insights into simultaneous ammonia-selective and anti-fouling mechanism over forward osmosis membrane for resource recovery from domestic wastewater. J. Memb. Sci. 573 (2019) 135–144. https://doi.org/10.1016/j.memsci.2018.11.072

[116] A. Schulze, M. Went, A. Prager, Membrane functionalization with hyperbranched polymers. Materials (Basel). 9 (2016). https://doi.org/10.3390/ma9080706

[117] X. Bao, Q. Wu, W. Shi, W. Wang, H. Yu, Z. Zhu, X. Zhang, Z. Zhang, R. Zhang, F. Cui, Polyamidoamine dendrimer grafted forward osmosis membrane with superior

ammonia selectivity and robust antifouling capacity for domestic wastewater concentration. Water Res. 153 (2019) 1–10.
https://doi.org/10.1016/j.watres.2018.12.067

[118] R.R. Choudhury, J.M. Gohil, S. Mohanty, S.K. Nayak, Antifouling, fouling release and antimicrobial materials for surface modification of reverse osmosis and nanofiltration membranes. J. Mater. Chem. A. 6 (2018) 313–333.
https://doi.org/10.1039/C7TA08627J

[119] R. Zhang, Y. Liu, M. He, Y. Su, X. Zhao, M. Elimelech, Z. Jiang, Antifouling membranes for sustainable water purification: Strategies and mechanisms. Chem. Soc. Rev. 45 (2016) 5888–5924. https://doi.org/10.1039/C5CS00579E

[120] M.F. Flanagan, I.C. Escobar, Novel charged and hydrophilized polybenzimidazole (PBI) membranes for forward osmosis. J. Memb. Sci. 434 (2013) 85–92.
https://doi.org/10.1016/j.memsci.2013.01.039

[121] R. Hausman, B. Digman, I.C. Escobar, M. Coleman, T.S. Chung, Functionalization of polybenzimidizole membranes to impart negative charge and hydrophilicity. J. Memb. Sci. 363 (2010) 195–203.
https://doi.org/10.1016/j.memsci.2010.07.027

[122] B.S. Lalia, V. Kochkodan, R. Hashaikeh, N. Hilal, A review on membrane fabrication: Structure, properties and performance relationship. Desalination. 326 (2013) 77–95. https://doi.org/10.1016/j.desal.2013.06.016

[123] J. Li, M. Wei, Y. Wang, NU SC State Key Laboratory of Materials-Oriented Chemical Engineering , Jiangsu National Synergetic. Chinese J. Chem. Eng. (2017).

[124] M. Fathizadeh, A. Aroujalian, A. Raisi, Effect of lag time in interfacial polymerization on polyamide composite membrane with different hydrophilic sub layers. Desalination. 284 (2012) 32–41. https://doi.org/10.1016/j.desal.2011.08.034

[125] J. Lee, R. Wang, T.H. Bae, High-performance reverse osmosis membranes fabricated on highly porous microstructured supports. Desalination. 436 (2018) 48–55.
https://doi.org/10.1016/j.desal.2018.01.037

[126] T.H. Lee, M.Y. Lee, H.D. Lee, J.S. Roh, H.W. Kim, H.B. Park, Highly porous carbon nanotube/polysulfone nanocomposite supports for high-flux polyamide reverse osmosis membranes. J. Memb. Sci. 539 (2017) 441–450.
https://doi.org/10.1016/j.memsci.2017.06.027

[127] W. Sobieski, M. Matyka, J. Gołembiewski, S. Lipiński, The Path Tracking Method as an alternative for tortuosity determination in granular beds. Granul. Matter. 20 (2018). https://doi.org/10.1007/s10035-018-0842-x

[128] A.F. Ismail, M. Padaki, N. Hilal, T. Matsuura, W.J. Lau, Thin film composite membrane-Recent development and future potential. Desalination. 356 (2015) 140–148. https://doi.org/10.1016/j.desal.2014.10.042

[129] M.T.M. Pendergast, J.M. Nygaard, A.K. Ghosh, E.M. V Hoek, Using nanocomposite materials technology to understand and control reverse osmosis membrane compaction. Desalination. 261 (2010) 255–263. https://doi.org/10.1016/j.desal.2010.06.008

[130] W. Yan, Z. Wang, J. Wu, S. Zhao, J. Wang, Enhancing the flux of brackish water TFC RO membrane by improving support surface porosity via a secondary pore-forming method. J. Memb. Sci. 498 (2015) 227–241. https://doi.org/10.1016/j.memsci.2015.10.029

[131] M. Son, H. Choi, L. Liu, E. Celik, H. Park, H. Choi, Efficacy of carbon nanotube positioning in the polyethersulfone support layer on the performance of thin-film composite membrane for desalination. Chem. Eng. J. (2014). https://doi.org/10.1016/j.cej.2014.12.108

[132] H. Chae, C. Lee, P. Park, I. Kim, J. Kim, Synergetic effect of graphene oxide nanosheets embedded in the active and support layers on the performance of thin- film composite membranes. J. Memb. Sci. 525 (2017) 99–106. https://doi.org/10.1016/j.memsci.2016.10.034

[133] H.M. Park, K.Y. Jee, Y.T. Lee, Preparation and characterization of a thin-film composite reverse osmosis membrane using a polysulfone membrane including metal-organic frameworks. J. Memb. Sci. 541 (2017) 510–518. https://doi.org/10.1016/j.memsci.2017.07.034

[134] B. Jeong, E.M. V Hoek, Y. Yan, A. Subramani, X. Huang, G. Hurwitz, A.K. Ghosh, A. Jawor, Interfacial polymerization of thin film nanocomposites : A new concept for reverse osmosis membranes. J. Memb. Sci. 294 (2007) 1–7. https://doi.org/10.1016/j.memsci.2007.02.025

[135] D.L. Zhao, S. Japip, Y. Zhang, M. Weber, C. Maletzko, T.S. Chung, Emerging thin-film nanocomposite (TFN) membranes for reverse osmosis: A review. Water Res. 173 (2020) 115557. https://doi.org/10.1016/j.watres.2020.115557

[136] S. Ho, S. Kwak, B. Sohn, T. Hyun, Design of TiO2 nanoparticle self-assembled aromatic polyamide thin-film-composite (TFC) membrane as an approach to solve biofouling problem. J. Memb. Sci. 211 (2003) 157–165. https://doi.org/10.1016/S0376-7388(02)00418-0

[137] K.A. Mahmoud, B. Mansoor, A. Mansour, M. Khraisheh, Functional graphene nanosheets : The next generation membranes for water desalination. Desalination. 365 (2014) 208–225. https://doi.org/10.1016/j.desal.2014.10.022

[138] H.M. Hegab, Y. Wimalasiri, M. Ginic-markovic, L. Zou, Improving the fouling resistance of brackish water membranes via surface modification with graphene oxide functionalized chitosan. Desalination. 365 (2015) 99–107. https://doi.org/10.1016/j.desal.2015.02.029

[139] J. Yin, G. Zhu, B. Deng, Graphene oxide (GO) enhanced polyamide (PA) thin-film nanocomposite (TFN) membrane for water purification. Desalination. 379 (2016) 93–101. https://doi.org/10.1016/j.desal.2015.11.001

[140] L. He, L.F. Dumée, C. Feng, L. Velleman, R. Reis, F. She, W. Gao, L. Kong, Promoted water transport across graphene oxide – poly (amide) thin film composite membranes and their antibacterial activity. Desalination. 365 (2015) 126–135. https://doi.org/10.1016/j.desal.2015.02.032

[141] V. Vatanpour, N. Zoqi, Surface modification of commercial seawater reverse osmosis membranes by grafting of hydrophilic monomer blended with carboxylated multiwalled carbon nanotubes. Appl. Surf. Sci. 396 (2016) 1478–1489. https://doi.org/10.1016/j.apsusc.2016.11.195

[142] H. Yang, J.C. Lin, C. Huang, Application of nanosilver surface modification to RO membrane and spacer for mitigating biofouling in seawater desalination. Water Res. 43 (2009) 3777–3786. https://doi.org/10.1016/j.watres.2009.06.002

[143] M. Ben-sasson, X. Lu, E. Bar-zeev, K.R. Zodrow, S. Nejati, G. Qi, E.P. Giannelis, M. Elimelech, In situ formation of silver nanoparticles on thin-film composite reverse osmosis membranes for biofouling mitigation. Water Res. 62 (2014) 260–270. https://doi.org/10.1016/j.watres.2014.05.049

[144] R. Rajakumaran, V. Boddu, M. Kumar, M.S. Shalaby, H. Abdallah, Effect of ZnO morphology on GO-ZnO modified polyamide reverse osmosis membranes for desalination. Desalination. 467 (2019) 245–256. https://doi.org/10.1016/j.desal.2019.06.018

[145] S.G. Kim, J.H. Chun, B. Chun, S.H. Kim, Preparation , characterization and performance of poly (aylene ether sulfone)/ modified silica nanocomposite reverse osmosis membrane for seawater desalination ☆. Desalination. 325 (2013) 76–83. https://doi.org/10.1016/j.desal.2013.06.017

[146] S. Qi, R. Wang, G. Krishna, M. Chaitra, J. Torres, X. Hu, A. Gordon, Aquaporin-based biomimetic reverse osmosis membranes : Stability and long term performance. J. Memb. Sci. 508 (2016) 94–103. https://doi.org/10.1016/j.memsci.2016.02.013

[147] L. Huang, J.R. McCutcheon, Impact of support layer pore size on performance of thin film composite membranes for forward osmosis, 483 (2015) 25-33. https://doi.org/10.1016/j.memsci.2015.01.025

[148] E. Fontananova, J.C. Jansen, A. Cristiano, E. Curcio, E. Drioli, Effect of additives in the casting solution on the formation of PVDF membranes. Desalination. 192 (2006) 190–197. https://doi.org/10.1016/j.desal.2005.09.021

[149] Y. Wu, H. Zhu, L. Feng, L. Zhang, Effects of polyethylene glycol on the structure and filtration performance of thin-film PA-Psf composite forward osmosis membranes. Sep. Sci. Technol. 51 (2016) 862–873. https://doi.org/10.1080/01496395.2015.1119846

[150] N.N. Bui, J.R. McCutcheon, Hydrophilic nanofibers as new supports for thin film composite membranes for engineered osmosis. Environ. Sci. Technol. 47 (2013) 1761–1769. https://doi.org/10.1021/es304215g

[151] B. Van der Bruggen, Chemical modification of polyethersulfone nanofiltration membranes: A review. J. Appl. Polym. Sci. 114 (2009) 630–642. https://doi.org/10.1002/app.30578

[152] N. Widjojo, T.S. Chung, M. Weber, C. Maletzko, V. Warzelhan, The role of sulphonated polymer and macrovoid-free structure in the support layer for thin-film composite (TFC) forward osmosis (FO) membranes. J. Memb. Sci. 383 (2011) 214–223. https://doi.org/10.1016/j.memsci.2011.08.041

[153] X. Zhang, J. Tian, Z. Ren, W. Shi, Z. Zhang, Y. Xu, S. Gao, F. Cui, High performance thin-film composite (TFC) forward osmosis (FO) membrane fabricated on novel hydrophilic disulfonated poly(arylene ether sulfone) multiblock copolymer/polysulfone substrate. J. Memb. Sci. 520 (2016) 529–539. https://doi.org/10.1016/j.memsci.2016.08.005

[154] D. Möckel, E. Staude, M.D. Guiver, Static protein adsorption, ultrafiltration behavior and cleanability of hydrophilized polysulfone membranes. J. Memb. Sci. 158 (1999) 63–75. https://doi.org/10.1016/S0376-7388(99)00028-9

[155] T.S. Chung, L.Y. Jiang, Y. Li, S. Kulprathipanja, Mixed matrix membranes (MMMs) comprising organic polymers with dispersed inorganic fillers for gas separation. Prog. Polym. Sci. 32 (2007) 483–507. https://doi.org/10.1016/j.progpolymsci.2007.01.008

[156] L.Y. Ng, A.W. Mohammad, C.P. Leo, N. Hilal, Polymeric membranes incorporated with metal/metal oxide nanoparticles: A comprehensive review. Desalination. 308 (2013) 15–33. https://doi.org/10.1016/j.desal.2010.11.033

[157] W.J. Lau, S. Gray, T. Matsuura, D. Emadzadeh, J. Paul Chen, A.F. Ismail, A review on polyamide thin film nanocomposite (TFN) membranes: History, applications, challenges and approaches. Water Res. 80 (2015) 306–324. https://doi.org/10.1016/j.watres.2015.04.037

[158] M.J. Park, S. Phuntsho, T. He, G.M. Nisola, L.D. Tijing, X.M. Li, G. Chen, W.J. Chung, H.K. Shon, Graphene oxide incorporated polysulfone substrate for the fabrication of flat-sheet thin-film composite forward osmosis membranes. J. Memb. Sci. 493 (2015) 496–507. https://doi.org/10.1016/j.memsci.2015.06.053

[159] S. Morales-Torres, C.M.P. Esteves, J.L. Figueiredo, A.M.T. Silva, Thin-film composite forward osmosis membranes based on polysulfone supports blended with nanostructured carbon materials. J. Memb. Sci. 520 (2016) 326–336. https://doi.org/10.1016/j.memsci.2016.07.009

[160] N. Ma, J. Wei, S. Qi, Y. Zhao, Y. Gao, C.Y. Tang, Nanocomposite substrates for controlling internal concentration polarization in forward osmosis membranes. J. Memb. Sci. 441 (2013) 54–62. https://doi.org/10.1016/j.memsci.2013.04.004

[161] X. Fan, Y. Liu, X. Quan, A novel reduced graphene oxide/carbon nanotube hollow fiber membrane with high forward osmosis performance. Desalination. (2019) 117–124. https://doi.org/10.1016/j.desal.2018.07.020

[162] H. gyu Choi, M. Son, H. Choi, Integrating seawater desalination and wastewater reclamation forward osmosis process using thin-film composite mixed matrix membrane with functionalized carbon nanotube blended polyethersulfone support layer. Chemosphere. 185 (2017) 1181–1188. https://doi.org/10.1016/j.chemosphere.2017.06.136

[163] M.D. Firouzjaei, S.F. Seyedpour, S.A. Aktij, M. Giagnorio, N. Bazrafshan, A. Mollahosseini, F. Samadi, S. Ahmadalipour, F.D. Firouzjaei, M.R. Esfahani, A. Tiraferri, M. Elliott, M. Sangermano, A. Abdelrasoul, J.R. McCutcheon, M. Sadrzadeh, A.R. Esfahani, A. Rahimpour, Recent advances in functionalized polymer membranes for biofouling control and mitigation in forward osmosis. J. Memb. Sci. 596 (2020). https://doi.org/10.1016/j.memsci.2019.117604

[164] D.Y. Koseoglu-Imer, B. Kose, M. Altinbas, I. Koyuncu, The production of polysulfone (PS) membrane with silver nanoparticles (AgNP): Physical properties,

filtration performances, and biofouling resistances of membranes. J. Memb. Sci. 428 (2013) 620–628. https://doi.org/10.1016/j.memsci.2012.10.046

[165] K. Chamakura, R. Perez-Ballestero, Z. Luo, S. Bashir, J. Liu, Comparison of bactericidal activities of silver nanoparticles with common chemical disinfectants. Colloids Surfaces B Biointerfaces. 84 (2011) 88–96. https://doi.org/10.1016/j.colsurfb.2010.12.020

[166] Z. Liu, L. Qi, X. An, C. Liu, Y. Hu, Surface engineering of thin film composite polyamide membranes with silver nanoparticles through layer-by-layer interfacial polymerization for antibacterial properties. ACS Appl. Mater. Interfaces. 9 (2017) 40987–40997. https://doi.org/10.1021/acsami.7b12314

[167] R. Hausman, T. Gullinkala, I.C. Escobar, Development of copper-charged polypropylene feedspacers for biofouling control. J. Memb. Sci. 358 (2010) 114–121. https://doi.org/10.1016/j.memsci.2010.04.033

[168] Z. Liu, Y. Hu, C. Liu, Z. Zhou, Surface-independent one-pot chelation of copper ions onto filtration membranes to provide antibacterial properties. Chem. Commun. 52 (2016) 12245–12248. https://doi.org/10.1039/C6CC06015C

[169] A.C. Abreu, R.R. Tavares, A. Borges, F. Mergulhão, M. Simões, Current and emergent strategies for disinfection of hospital environments. J. Antimicrob. Chemother. 68 (2013) 2718–2732. https://doi.org/10.1093/jac/dkt281

[170] Z. Liu, Y. Hu, Sustainable antibiofouling properties of thin film composite forward osmosis membrane with rechargeable silver nanoparticles loading. ACS Appl. Mater. Interfaces. 8 (2016) 21666–21673. https://doi.org/10.1021/acsami.6b06727

[171] H.M. Hegab, A. ElMekawy, T.G. Barclay, A. Michelmore, L. Zou, C.P. Saint, M. Ginic-Markovic, Fine-tuning the surface of forward osmosis membranes via grafting Graphene oxide: Performance patterns and biofouling propensity. ACS Appl. Mater. Interfaces. 7 (2015) 18004–18016. https://doi.org/10.1021/acsami.5b04818

[172] M. Amini, M. Jahanshahi, A. Rahimpour, Synthesis of novel thin film nanocomposite (TFN) forward osmosis membranes using functionalized multi-walled carbon nanotubes. J. Memb. Sci. 435 (2013) 233–241. https://doi.org/10.1016/j.memsci.2013.01.041

[173] Q. Liu, G.R. Xu, Graphene oxide (GO) as functional material in tailoring polyamide thin film composite (PA-TFC) reverse osmosis (RO) membranes. Desalination. 394 (2016) 162–175. https://doi.org/10.1016/j.desal.2016.05.017

[174] A. Soroush, W. Ma, Y. Silvino, M.S. Rahaman, Surface modification of thin film composite forward osmosis membrane by silver-decorated graphene-oxide nanosheets. Environ. Sci. Nano. 2 (2015) 395–405. https://doi.org/10.1039/C5EN00086F

[175] A. Soroush, W. Ma, M. Cyr, M.S. Rahaman, B. Asadishad, N. Tufenkji, In situ silver decoration on graphene oxide-treated thin film composite forward osmosis membranes: Biocidal properties and regeneration potential. Environ. Sci. Technol. Lett. 3 (2016) 13–18. https://doi.org/10.1021/acs.estlett.5b00304

[176] A.F. Faria, C. Liu, M. Xie, F. Perreault, L.D. Nghiem, J. Ma, M. Elimelech, Thin-film composite forward osmosis membranes functionalized with graphene oxide–silver nanocomposites for biofouling control. J. Memb. Sci. 525 (2017) 146–156. https://doi.org/10.1016/j.memsci.2016.10.040

[177] C. Liu, A.F. Faria, J. Ma, M. Elimelech, Mitigation of biofilm development on thin-film composite membranes functionalized with zwitterionic polymers and silver nanoparticles. Environ. Sci. Technol. 51 (2017) 182–191. https://doi.org/10.1021/acs.est.6b03795

[178] S. Khoshhal, A.A. Ghoreyshi, M. Jahanshahi, M. Mohammadi, Study of the temperature and solvent content effects on the structure of Cu-BTC metal organic framework for hydrogen storage. RSC Adv. 5 (2015) 24758–24768. https://doi.org/10.1039/C5RA01890K

[179] M. Berchel, T. Le Gall, C. Denis, S. Le Hir, F. Quentel, C. Elléouet, T. Montier, J.M. Rueff, J.Y. Salaün, J.P. Haelters, G.B. Hix, P. Lehn, P.A. Jaffrès, A silver-based metal-organic framework material as a "reservoir" of bactericidal metal ions. New J. Chem. 35 (2011) 1000–1003. https://doi.org/10.1039/c1nj20202b

[180] S.F. Seyedpour, A. Rahimpour, G. Najafpour, Facile in-situ assembly of silver-based MOFs to surface functionalization of TFC membrane: A novel approach toward long-lasting biofouling mitigation. J. Memb. Sci. 573 (2019) 257–269. https://doi.org/10.1016/j.memsci.2018.12.016

[181] M. Ben-Sasson, K.R. Zodrow, Q. Genggeng, Y. Kang, E.P. Giannelis, M. Elimelech, Surface functionalization of thin-film composite membranes with copper nanoparticles for antimicrobial surface properties. Environ. Sci. Technol. 48 (2014) 384–393. https://doi.org/10.1021/es404232s

[182] Y. Lu, Z. He, Mitigation of salinity buildup and recovery of wasted salts in a hybrid osmotic membrane bioreactor-electrodialysis system. Environ. Sci. Technol. 49 (2015) 10529–10535. https://doi.org/10.1021/acs.est.5b01243

[183] S. Zou, E.D. Smith, S. Lin, S.M. Martin, Z. He, Mitigation of bidirectional solute flux in forward osmosis via membrane surface coating of zwitterion functionalized carbon nanotubes. Environ. Int. 131 (2019) 104970. https://doi.org/10.1016/j.envint.2019.104970

[184] X. Song, L. Wang, C.Y. Tang, Z. Wang, C. Gao, Fabrication of carbon nanotubes incorporated double-skinned thin film nanocomposite membranes for enhanced separation performance and antifouling capability in forward osmosis process. Desalination. 369 (2015) 1–9. https://doi.org/10.1016/j.desal.2015.04.020

[185] M. Yasukawa, S. Mishima, M. Shibuya, D. Saeki, T. Takahashi, T. Miyoshi, H. Matsuyama, Preparation of a forward osmosis membrane using a highly porous polyketone microfiltration membrane as a novel support. J. Memb. Sci. 487 (2015) 51–59. https://doi.org/10.1016/j.memsci.2015.03.043

[186] H.T. Madsen, N. Bajraktari, C. Hélix-Nielsen, B. Van der Bruggen, E.G. Søgaard, Use of biomimetic forward osmosis membrane for trace organics removal. J. Memb. Sci. 476 (2015) 469–474. https://doi.org/10.1016/j.memsci.2014.11.055

[187] C. Tang, Z. Wang, I. Petrinić, A.G. Fane, C. Hélix-Nielsen, Biomimetic aquaporin membranes coming of age. Desalination. 368 (2015) 89–105. https://doi.org/10.1016/j.desal.2015.04.026

Polymeric Membranes for Water Purification and Gas Separation
Materials Research Foundations **113** (2021) 171-202

Materials Research Forum LLC
https://doi.org/10.21741/9781644901632-6

Chapter 6

Polymeric Membranes for O₂/N₂ Separation

Ragib Shakil[1#], Yeasin Arafat Tarek[1#], Mahamudul Hasan Rumon[1#], Chanchal Kumar Roy[1], Al-Nakib Chowdhury[1*], Rasel Das[2,3*]

[1]Department of Chemistry, Bangladesh University of Engineering and Technology, Dhaka-1000, Bangladesh

[2]Department of Biochemistry and Biotechnology, University of Science and Technology Chittagong, Foy's lake, Chittagong-4202, Bangladesh

[3]Department of Chemistry, Stony Brook University, Stony Brook, NY 11794, USA
[#]These authors are equally contributed

* nakib@chem.buet.ac.bd, raseldas@daad-alumni.de

Abstract

Over the last few decades, polymeric membranes-based O₂/N₂ separation techniques have been progressed from a laboratory curiosity to a commercial reality. These membranes show various advantages *i.e.*, low energy consumption and cost, compared to the other conventional methods for O₂ and N₂ separation. These benefits generate a great deal of interest in industry and academia to accelerate the commercial feasibility of polymeric membranes for O₂/N₂ separation. For this, various materials have been developed in order to enhance both O₂ permeability and O₂/N₂ selectivity of the polymeric membranes. In this chapter, the recent development of various polymeric membranes, including a different polymer matrix and polymer inorganic composite for O₂/N₂ separation, is discussed.

Keywords

Membrane Technology, Polymeric Membrane, O₂/N₂ Separation, Gas Separation, Selectivity and Permeability

Contents

1.   Introduction

In the modern industrialization era, the demand for high-quality oxygen (O_2) and nitrogen (N_2) in many industrial and medical applications is constantly uprising. The technological advancement for the separation of O_2/N_2 thus plays an essential part. Many methods are still under development for highly efficient technology bring-up and reducing the production cost for O_2/N_2 purification. Membrane-based technology has shown a significant impact on gas separation over the traditional approaches *i.e.,* cryogenic distillation, solvent absorption, and solid adsorption owing to their relatively lower cost, chemical, and energy consumption, and compact design [1, 2]. Except for O_2/N_2 gas pair, the efficacy of membrane technologies has also been shown for several other gas pairs [3]. Among the membrane materials, carbon-, inorganic-, polymer-, polymer-inorganic composite-based are noteworthy [4].

However, membrane-based gas transportation is nothing new, researchers are trying to develop this process for hundreds of years [5]. Usually, there are two types of membrane available such as porous membrane and non-porous or tight membrane. Porous membrane separation includes Knudsen-diffusion, and molecular sieving, whereas the non-porous membrane follows solution diffusion mechanism [6]. In between Knudsen-diffusion and molecular sieving, the pore size allowed for Knudsen flow is relatively higher which limits the separation factor of Knudsen-diffusion [7]. But the molecular sieving shows more efficacies in gas separation due to its smaller pore size, which only allows passing the smaller molecule. Alternatively, diffusion is the only mode of transport in the solution-diffusion model. The gas molecules to be transported must be dissolved in the membrane first. The solution-diffusion model takes the general approach of assuming that the chemical potentials of the O_2/N_2 and permeate fluids are in equilibrium with the neighboring membrane surfaces, allowing sufficient chemical potential expressions. The mechanisms of porous and non-porous membrane technology for gas separation are shown in Fig. 1.

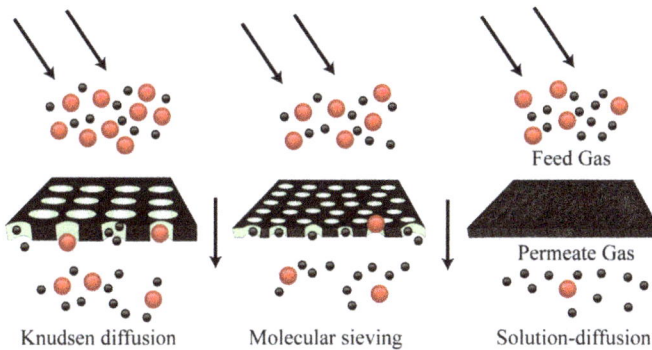

Figure 1. Different mechanisms of porous and non-porous membrane technology for gas separation.

Polymer-based membrane materials have drawn great attention because of their lightweight, low cost, and easy fabrication process [8–10]. Rather, the efficacy of polymeric membranes is quite high because it usually follows the solution-diffusion mechanism for gas separation which is controlled by the permeability and selectivity. The gas permeability can be calculated by using Eq. 1. According to the mass transfer mechanism, the diffusion coefficient depends indispensably on the size of the feed gasses (N_2, O_2, CO_2, CH_4, etc.) where a gas with a smaller molecular size has a higher diffusion coefficient [11]. This confirms the purity and selectivity of the polymeric membranes. The conventional polymer membranes are rigid, fragile, and glassy, which attribute free volume in the polymeric matrix [12]. Whereas, the flexible rubbery polymeric membranes have a higher free volume which depicts a higher diffusion coefficient [13]. The solubility is the ratio shown in Eq. 2 of polymer concentration and adjacent polymeric gas pressure.

$$P_g = S \times D \tag{1}$$

$$S = \frac{c}{p} \tag{2}$$

where P_g is gas permeability, S is gas solubility, D is the effective diffusion coefficient, c is the concentration of gas in polymer, and p is the pressure of the gas.

However, at low temperatures and pressure, the polymeric membranes should achieve limited permeability and selectivity, which makes these membranes limited in applications. To overcome the permeability and selectivity issues, polymer surface modification including functionalization and mixed polymer matrix are being adopted nowadays [14–17]. The surface functionalization of a polymer membrane improves the surface properties. Whereas, a combination of an inorganic, carbonaceous, metal-organic framework (MOF), etc. with polymer membranes generates a new type of polymeric membrane termed as mixed polymer matrix membrane (MPMM). When the pore diameters of the additive materials *i.e.;* MOFs/zeolites are higher than the size of one gas molecule but smaller than others, the permeability will be improved during the separation process [18, 19]. This leads to an economical advancement in large-scale separation for medical and industrial applications. This MPMM was firstly introduced in the early '80s and still booming to the next generation gas separation membrane technology [20].

The motivation of the chapter is to provide an overview based on various requirements of the polymeric membrane's technology being used through the present day to enhance the selectivity of O_2/N_2 separation. Different types of physical and chemical modifications of membranes for better performance are discussed. An attempt to describe the current strategies in the developments of various polymeric membranes and their suitability for O_2/N_2 enrichment are also considered and well corroborated in this chapter.

2.   Types of polymeric membranes

Generally, polyamide (PA), polydimethylsiloxane (PDMS), polyaniline (PANI), polysulfone (PSf), polyimide (PI), polyurethane (PU), polypropylene (PP) etc. are the basic components of the polymeric membranes used in O_2/N_2 gas separation. Among them, PSf shows its superiority over others because of its higher selectivity and permeability with significant mechanical properties. But the PSf possess moderate chemical and thermal stabilities [21]. To overcome this, nowadays polyimide has become popular for its excellent thermal and chemical stability with generous mechanical properties [22]. Polyurethane-based membranes have microstructure but its hard segments of urea or urethane in polymer network limit the applications in gas separation [23]. Efforts are continuously putting on the improvement of overcoming those limitations and enhancing the gas separation efficiency. In addition, detailed discussion on various polymer compositions and their membrane fabrication techniques can be found in the following sections.

2.1   Asymmetric membranes

Asymmetric membranes are composed of a dense and thin top-layer supported by a porous substrate-layer in the same polymer matrix as shown in Fig. 2. The thin layer termed as skin layer or active layer that directly affects the separation process, whereas the thicker porous membrane called as sub-layer or bottom layer that ensures mechanical support to the top layer. The top layer fabrication processes and their optimization play a governing role in the gas separation and may influence the gas permeability and overall selectivity. However, to confirm the significant permeability, the top layer must be very thin, because gas permeability has an inversely proportional relation with the layer thickness for asymmetric membrane [25]. Based on the thickness, the top layer is two types *i.e.;* ultrathin and hyperthin. The ultrathin layer will have a thickness of 1000–5000 Å, where the thickness of the hyperthin layer is less than 1000 Å [26]. Moreover, it should be confirmed that no defects are existing in the top layer because even a tiny defect could be a sensitive issue for the gas separation [24]. Contamination may occur at the permeate gas, which is unwanted, can pass through a separation membrane having defects in the thin layer. The defect-free layers follow the solution diffusion mechanism, whereas the defective layer's transport gas via the Knudsen mechanism during gas separation as discussed in the introduction [24]. However, the bottom layer attributes no significant effects on the gas separation but the structural design e.g. having large pore size to minimize excess mass transfer resistance [27].

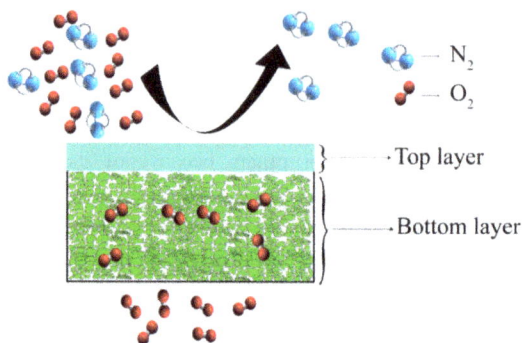

Figure 2. Gas separation mechanism in a typical asymmetric membrane.

Interfacial polymerization is commonly used for preparing an asymmetric membrane [28]. The morphology of the membrane is directed by the solidification and the phase separation of the polymer solution [29]. Usually, the phase separation is conducted by dry-wet phase, wet phase, or dry phase processes [30]. Nonetheless, there are several factors including the rheological properties, solvent ratio, the concentration of polymer solution, and the solidification time, etc. affect the gas separation performance [29]. Table 1 summarizes some recent progress in the asymmetric membrane for O_2/N_2 separation.

Table 1. Asymmetric polymeric membrane for O_2/N_2 separation

Membrane Compositions	O_2/N_2 Selectivity	O_2 Permeability GPU	Top layer thickness (Å)	References
PI	6.3	65	-	[31]
[a]PC	5.3	36	-	[31]
PI	6.6	12	730	[32]
[b]PES	5.8	9.3	470	[33]
PSf	7.3	6.9	-	[34]
PES	6.2	12	-	[35]
PES	5.7	15.8	-	[35]
PES	6.8	14.3	600	[36]
[c]PLA	1	0.3	-	[37]
PSf	4.5	3.4	-	[38]

[a]Polycarbonate (PC), [b]Polyethersulfone (PES), and [c]poly (lactic acid) (PLA).

2.2   Copolymer and blended polymeric membrane

There is a certain difference between copolymer and blended polymeric membrane. The copolymer membrane is the result of the co-existence of multiple monomers [39]. On the contrary, polymer blend membranes contain homopolymers, which generally appear after the synthesis [40]. The copolymerization including (a) homopolymer, (b) alternating copolymer, (c) random copolymer, (d) block copolymer, (e) graft copolymer, etc. Of them, block copolymers have shown excellent structural formation for better gas separation [41]. The resulting membrane yields high free volume sites, high thermal decomposition temperature, high glass transition temperature, and high selectivity towards the O_2/N_2 separation [41]. However, some very common examples of block copolymer-based membranes are as follows: tetra-methyl-hexafluoro-naphthalene polysulfone (TMHF-NPSF), hexafluoro-naphthalene polysulfone (HF-NPSF), tetramethyl- naphthalene polysulfone (TM-NPSF), etc. as shown in Fig. 3. Among those block copolymers-based membranes that follow the O_2/N_2 gas separation permeability order: TM-NPSF < HF-NPSF < TMHF-NPSF [24]. This order is closely related to the size of the free volume membranes cavity. As the size of the groups present in the NPSF moiety increased, the cavity size also improved. As a result, increasing of gas flux is expected with the available free space in the moiety [12]. Moreover, this cavity size could affect the selectivity and the permeability of the membrane performance in the O_2/N_2 gas separation. However, the cavity size is inversely related to selectivity and directly proportional with gas permeability [42]. Some exceptions are also observed. Such as in the case of TMHF-NPSF membrane, high permeability with high selectivity of O_2/N_2 separation [42].

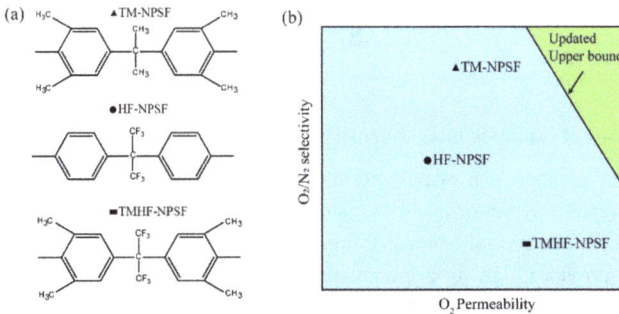

Figure 3. Effect of side groups on the selectivity and permeability of O_2/N_2 separation. (a) Structures of TM-NPSF, HF-NPSF, and TMHF-NPSF. (b) Relation of O_2 permeability with O_2/N_2 selectivity.

Polymer blending is a membrane fabrication technique where multiple polymers are combined together. As a result, different physical properties arise in the fabricated polymer blend membrane from different constituents. One of the major concerns during the fabrication process is the miscibility of different polymers which have direct effects on gas selectivity and permeability as revealed in Table 2. The homogeneity attributes to the quality of the fabricated polymer membrane. For example, PLA homopolymers blended from L96D4:L88D12 with the ratio of 8:2 show much better O_2/N_2 selectivity than alone L96D4 homopolymers [43]. The higher crystallinity of (L96D4) homopolymer in the PLA (L96D4:L88D12) polymer blend is the reason behind the enhanced permeability. It is mentionable that the permeability is proportional to the crystalline structure of the homopolymers [44].

Table 2. Composition of polymer blends in O_2/N_2 separation.

Membrane Compositions	O_2/N_2 Selectivity	O_2 Permeability (barrer)	References
[a]SBR	5.8	602	[45]
[b]SBR/NR	2.9	1457	[45]
PLA ($L_{96}D_4/L_{88}D_{12}$)	6.2	0.34	[43]
[c]PVP/ [d]EC	5.9	3.10	[46]
[e]6FDA-TAB	5.9	15.2	[47]
6FDA/PMDA-TAB	12.5	0.453	[47]

[a]Styrene-Butadiene Rubber (SBR), [b]Natural Rubber (NR), [c]Poly(4-Vinyl-Pyridine) (PVP), [d]Ethylcellulose (EC), and [e]4,4'-Hexafluoroisopropylidene Diphthalic Anhydride-1,2,4,5-Tetraaminobenzene (6FDA-TAB).

2.3　Polymer of intrinsic microporosity

To further the advancement of polymeric membranes, researchers introduce a new type of polymer called polymer of intrinsic microporosity (PIM) [48]. The amorphous PIM, as shown in Fig. 4, has special twisted or bending features with porous and rigid polymeric network that provides high surface area and large free volume. As a result, it shows high O_2 permeability and O_2/N_2 selectivity. For example, PIM ladder polymers have introduced two side groups of trifluoromethyl and phenylsulfone which provide high selectivity for O_2/N_2, as well as the O_2 permeability, remain high due to the extensive surface area/free volume [49].

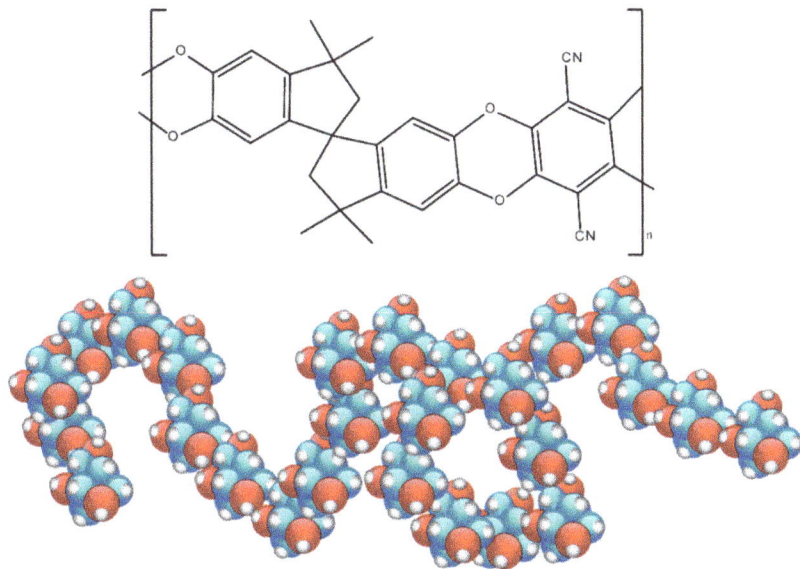

Figure 4. Structure of an amorphous and intrinsic microporous polymers.

Besides this, another type of microporous polymer can be fabricated by thermal rearrangement (TR), and it is termed as TR polymers [50]. During this fabrication process, a lot of cavities are found which are responsible for the microporous structure of the polymer matrix. The rigid microporous structure possesses improved surface area with high free cavity volume [51]. Nevertheless, the hardness control of the TR polymer is still a challenge to the researcher. During the thermal conversion at high temperature, the structure of these TR polymers become brittle [52]. Researchers are still trying to fabricate some higher degree of TR polymers with high tensile and elongation strength to enhance the gas separation performance. In Table 3 some of the PIM and TR-based microporous polymers are listed out.

Table 3. O_2/N_2 selectivity and O_2 permeability in PIM and TR-based microporous polymeric membrane.

Membrane Compositions	O_2/N_2 Selectivity	O_2 Permeability (barrer)	Ref.
PIM-1	4	370	[53]
PIM-7	4.6	190	[53]
PIM-1	3.3	1610	[53]
PIM-1	3.3	785	[54]
PIM-1	3.2	1133	[55]
[a]TFMPSPIM 1	4.6	156	[55]
TFMPSPIM 2	4.1	307	[55]
TFMPSPIM 3	3.5	560	[55]
TFMPSPIM 4	3.3	736	[55]
PIM-PI-1	3.1	151	[56]
PIM-PI-3	3.6	85	[56]
PIM-PI-8	3.4	545	[56]
[b]TR-DAR-350	7.1	5.1	[57]
TR-DAR-350	538	4.1	[57]

[a]Trifluoromethyl phenylsulfonyl polymer intrinsic membrane (TFMPSPIM), [b]Thermal rearrangement-diaminoresorcinol (TR-DAR)

3. Drawbacks of conventional polymeric membranes

Fabricating a polymeric membrane with high permeability and selectivity is the most demanded topic in O_2/N_2 gas separation. Most of the conventional polymeric membranes show a trade-off between permeability and selectivity, (describe in section 2.2). This can be depicted in the Robenson trade-off diagram [58]. According to this trade-off, most of the conventional polymeric membranes occupy around the prior upper bound. This may result from the lower surface area, less free volume size, uncontrolled morphology, etc. However, to overcome these challenges, new approaches like surface modification of polymer surface or mixing of inorganic/other materials with a polymer to produce mixed polymeric membrane have been adopted.

4. Surface modification of polymeric membrane

Modification of polymeric membrane surface has taken great attention to improve the polymer/membrane surface properties which could enhance O_2/N_2 selectivity and O_2 permeability. Chemical modification of polymeric membrane results in a compact and tight polymeric structure, thus enhance the selectivity and permeability [59, 60].

Crosslinking modification and layer coating are the most common modification processes for polymeric membranes for O_2/N_2 separation.

4.1   Crosslinking modification

Generally, cross-linking results in compact packing density and can hinder the mobility of polymer chains [61]. It often enhances the membrane performance and stability in O_2/N_2 separation process. There are several techniques introduced for cross-linking modification *i.e.,* thermal modification, chemical modification, ion beam irradiation, plasma treatment, and UV irradiation [44-47]. In the UV irradiation technique, the O_2/N_2 selectivity increases but there is a certain decrease in O_2 permeability [62, 63]. As the mobility of polymer chains is restricted in the polymer matrix that decreases in the gas permeability. To overcome these issues, a new technique is introduced where reactive oxygen is combined with UV radiation [64-66]. This process is termed as UV-ozone treatment shown in Fig. 5. However, long exposure to UV may cause a decrease in both O_2/N_2 selectivity and O_2 permeability [67].

Usually, chemical modification involves wet or chemical reactions, often used to modify the polymer membrane structure for better O_2/N_2 gas separation [65]. A general approach for chemical modification is dipping or soaking the polymeric membrane into a reagent solution [14]. The soaking/dipping time is very important to control the degree of cross-linking [65]. Generally, the chemical modification includes oxidation, grafting, etching, plasma polymerization, etc.

Heat treatment is another approach in which the polymeric membrane is condensed/compressed/shrieked resulting in a higher packing density of the polymer chains. Due to this higher packing density, substance resistance increases. Thus, the permeability of the large molecule *i.e.* N_2 decreases and the O_2/N_2 selectivity of the membrane increases after this treatment [68]. The heat applied in this process plays a vital role where a higher temperature results in a decrease in both selectivity and permeability. This method is often utilized as a post-treatment procedure for O_2/N_2 separation.

To date, the ion beam techniques have shown their extensive applicability for the modification of the membrane surface to enhance the O_2/N_2 gas separation performance. As ion beam irradiation-induced the polymer chain functionalization, various properties of the membrane can be tailored by adopting this technique [69]. The increase in free volume and the rigidity due to the cross-linking of the polymer chains attribute a high O_2/N_2 selectivity and O_2 permeability of the polymeric membrane. However, a non-selective resistance layer is formed when high irradiation is applied and decreases both in selectivity and permeability [70]. Generally, ion beam techniques include ion irradiation,

ion implantation, and focus ion beam where the ion irradiation and the ion implantation involve *ex-situ* and alternatively, the focus ion beam involve *in-situ* modification [71]. They are all pure physical processes as none of them introduces any foreign species into the targeted polymeric membrane. In addition, a great advantage of ion beam techniques over other techniques is their controllability and repeatability.

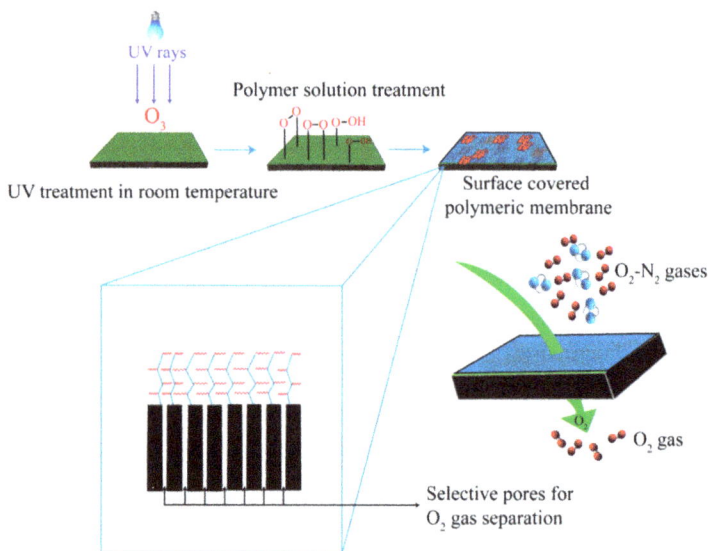

Figure 5. A visualization of UV-Ozone treatment on the surface modification of the polymeric membrane.

The plasma treatment gives the most satisfactory techniques compared to the other polymer modification techniques. The advantages are the multilayer film formation, controlled surface modification, eco-friendly, deposition of highly cross-linked polymer, etc. The fundamental objective of plasma treatment is to modify the surface with specific functional groups like ($-NH_2$, $-OH$, $-CHO$, $-COOH$) groups are introduced for pharmaceutical and medical purposes [72].

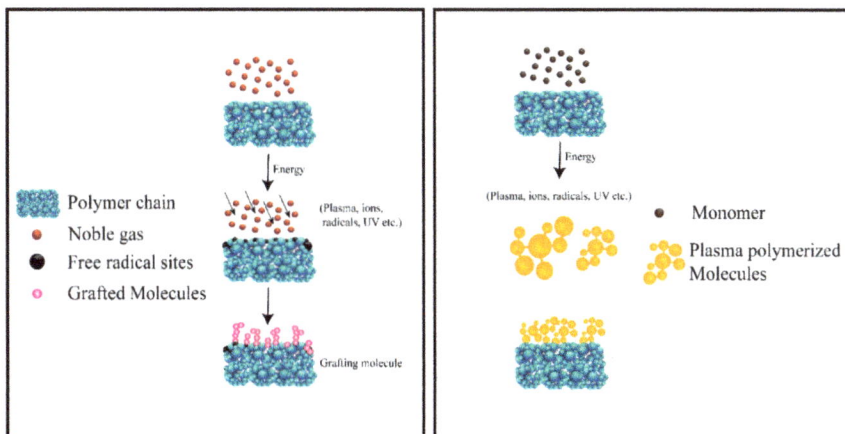

Figure 6. Surface modification of polymeric membrane by plasma treatment and plasma polymerization mechanism.

These surface groups attribute higher gas selectivity. Generally, for O_2/N_2 separation, various types of gas like N_2-containing gases, florin/chlorine-containing gases, inert gases, etc. are used in plasma treatment for generating free radicals under plasma conditions to initiate the modification process [73]. The free radicals are produced from the inelastic collision between the electron in the plasma and the polymer surface [74]. The free radicals play a crucial factor as the degree of modification highly depends on the density of the free radicals. For a complete modification, a high density of free radical is required where a low density free radical results in incomplete modification. In between them the NH_3 and N_2 plasma treatment enhance both the O_2/N_2 selectivity and O_2 permeability [74]. Whereas, the florin-containing gases/fluorination and the chlorine-containing gases/chlorination may result in an increase in selectivity but a decrease in permeability [75]. The higher selectivity attributes from the renovation of the physical structure of the polymeric membrane and the low permeability results from the decrease in segmental mobility and cross-linking formation of the polymer chain. Modification with other plasma-like ethylenediamine plasma can also attribute a higher selectivity and permeability [24]. The treatment time and plasma power play important roles in plasma technique. Longer treatment time does not have any effect on the gas selectivity, meanwhile, a high-power plasma accompanies microcracking in the surface layer which results in a decrease in selectivity [76]. A schematic illustration of plasma polymerization mechanism and plasma-induced modification of polymer membrane is shown in Fig. 6.

4.2   Coating

Surface coating is an advanced technology for the modification of the membrane that is used vastly in large-scale industrial applications. In this technique, the membrane surface is coated with another layer which even has the possibility to be penetrated into the pores of the membrane. Though this technique suppresses the cost and power utility, the major challenge is to modify the polymeric surface with a stable coating layer [77]. This is due to the weak interaction between the coated layer and the surface of the membrane. Some common methods like cross-linking and sulfonation are introduced to address the coating layer on the polymeric surface. The coating is generally carried out via solution coating and plasma polymerization.

The layer-by-layer technique has shown to be a successful approach in membrane surface modification for O_2/N_2 gas separation process. As an example, PANI has great O_2 selectivity owing to its dense polymeric structure and electrical conductivity [78]. On the other hand, pure PANI films have weak mechanical characteristics, and thus PANI membranes have lower permeability. When PANI is coated with another polymeric membrane-like porous polyvinylidene difluoride, both the mechanical properties and permeability are enhanced [79].

On the other hand, plasma polymerization coating that is also termed as vapor deposition coating provides ultrathin layers with largescale pinhole. In presence of a plasma discharge exposure, organic gas molecules undergo repeated fragmentation and provide molecular radicals. As a result, polymerization occurred via the combination of these radical molecules in the very next stage [80]. The discharge power and monomer flow rate are the major issues in controlling the morphology and the deposition rate. However, as discussed in section 4.1, the presence of halide shows better performance in the O_2/N_2 separation. For example, if fluorine present in the solvent, the efficiency of oxygen sorption improved. Fluorine content might be enhanced, if the percentage of hexafluorthene increased in the mixed gases. This leads to high performance in O_2/N_2 gas separation via improving both selectivity and permeability. Some O_2/N_2 selectivity and permeability of thin plasma polymeric membranes are listed in Table 4.

Table 4. Thin plasma polymers-based membranes for O_2/N_2 separation.

Membrane Compositions	Ratio of monomer gas mixtures	O_2/N_2 Selectivity	O_2 Permeability (barrer)	Ref.
PSf	-	5.1	4.5	[81]
MP Milipore VSWP 4700	CH_4/[a]HFP (1:3 mole ratio)	2.8	65	[82]
PSf	[b]4-Vpy	4.7	1.3	[81]
Aluminium oxide and sinter glass	H_2/C_2F_6 (1:1: volume ratio)	3.4	20	[83]
NR	4-Vpy	5.8	135	[81]

[a]Hexafluoropropene (HFP), [b]Vinylpiridine (Vpy).

5. Mixed polymer membrane

Mixed matrix membrane (MMM) is a scalable membrane technology to make it more applicable by dealing with the conventional polymeric membrane drawbacks. MMM is a heterogeneous polymer composite in which a polymer matrix contains some other filler materials. The combination of the polymer matrix and the filler materials provide significant improvements, especially for the glassy polymer. Especially, the particle surface has poor adhesion interaction including hydrogen bonds, and London forces, etc. within the polymer network. That might have improved its cavity-free surface area through the formation of voids around the filler materials. As a result, high permeability is expected, but selectivity reduces. To overcome this, the researchers introduce several surface modification techniques including surface functionalization methods, and modification in fabrication technique, etc. In the surface functionalization process, various coupling agents are used to improve the binding side among the inorganic fillers and the polymer network [64, 65]. Long-chain alkyl amine is a very well-known coupling agent. It is reported that when filler particles are functionalized with this coupling agent, the polymer membrane enhances significant permeability by controlling higher selectivity [84].

It is also possible to enhance the significant separation performance by small changes in the fabrication process, either using priming or annealing techniques [66, 67]. In the priming process, the inorganic filler is firstly being enveloped with a tiny amount of polymeric solution. In the very next step, the rest of the polymers are dispersed in the solution. This two steps process provides a defect-free MMM by improving the filler and polymer network interactions [24]. Zeolite is able to get focus on MMM membrane fabrications for their suitable size and cavity shapes. These help to reduce the gas sieving discrimination. Among other zeolites, zeolite 4A (pore size of 3.8 Å) is more suitable for the O_2/N_2 separation through the precise selectivity of the O_2 and N_2 gas depending on the size and shape. Though alone zeolite is able to improve the selectivity but permeably suffers in the O_2/N_2 separation. To overcome this, sometimes zeolite is used as a filler material with functionalized polymer matrix [85, 86], and hence the O_2/N_2 permeability with maintaining the selectivity can be achieved.

On the other hand, MOF shows even better O_2/N_2 separation in comparison with zeolite and other fillers. In a typical MOF structure, organic moiety connected with metals or metal oxides generates microporous cavities with multiple dimensions [19, 88]. When MOF present in the MMM, the structural improvement provides sufficient surface area to the membrane for O_2/N_2 gas separation. For example, in presence of MOF in the MMM matrix, enhanced significant O_2/N_2 separation performance [89]. Particularly, ZIF-8 and PSf composite membrane exhibits large O_2 permeability and O_2/N_2 selectivity compare to pure PSf alone. The tiny cavity of ZIF-8 helps to enhance the resulting O_2/N_2 separation with improved the ZIF-8-PSf membrane performance with increasing adverse effect including the total defective surface area in membrane surface. In Table 5, the O_2/N_2 selectivity and permeability of some mixed polymers are listed out.

Table 5. O_2/N_2 selectivity and permeability of some mixed polymers-based membrane

Membrane Compositions	Filler	Inorganic filler loading (wt%)	O_2/N_2 Selectivity	O_2 Permeability (barrer)	Ref.
PSf	Silicalite-1	16	4.3	2	[89]
PSf	Zeolite 4A	60	4	23.30	[86]
PSf	Silica	0.1	6.4	15.83	[90]
PSf	SWNTs	10	5.4	1.23	[84]
PSf	ZIF-8	16	8.3	2.6	[89]
[a]PMMM	Zeolite 4A	20	5.01	2.04	[24]
PI (Matrimid)	[d]CMS	36	7.9	3	[87]
[b]PTMSP	Zeolite	5	3.03	885	[91]
[c]PVAc	[e]CuTPA	15	6.79	0.624	[92]

[a]Poly (Methy-l-Methacrylate) (PMMA), [b]Polytrimethylsilylpropyne, (PTMSP), [c]Poly (Vinyl Acetate) (PVAc), [d]Carbon Molecular Sieve (CMS), and [e]Copper and Terephthalic Acid (CuTPA).

6. Polymer magnetic membranes

The most realistic strategy to obtain "smart" membranes is to combine polymeric membranes with inorganic magnetic nanoparticles that typically show attractive surface properties when activated with static or alternating magnetic fields of different frequencies. The magnetic properties of O_2 and N_2 are different where O_2 is paramagnetic also having relatively high magnetic susceptibility and N_2 is diamagnetic having negative and low magnetic susceptibility. When magnetic nanoparticles are dispersed in one side or in both sides of the polymeric membrane, the O_2/N_2 shows different response in the externally applied magnetic field as the magnetic properties of O_2/N_2 are different. Because of the paramagnetic behavior and higher magnetic susceptibility of O_2, the magnetic membrane shows higher affinity towards the O_2 molecules [93]. Thus, the O_2 molecules can pass through the magnetic membrane. This is the basic mechanism of the magnetic polymeric membrane separation process. To synthesis polymeric magnetic membrane, the membranes that can effectively change the pore size in the ultrafiltration, either by static magnetic fields or by high-frequency magnetic fields and thermo-responsive hydrogels are used [94]. The polymer magnetic membrane can be synthesized in two ways: (i) dispersing the magnetic nanoparticles (as magnetic filler nanomaterial) in the polymeric matrix, and (ii) magnetic nanoparticle coating on the polymeric membrane (surface modification) [95-98]. The preparation method highly affects the membrane morphology, thus O_2/N_2 separation performance of the polymeric membrane. Compared with magnetic nanoparticle dispersion on the polymeric membrane, the magnetic nanoparticle coating or surface modification shows higher O_2 permeability as well as the O_2 mass transfer coefficient. Generally, ethyl cellulose (EC), linear polyimide (LPI), hyperbranched polyimide (HBPI) and poly 2,6-dimethyl-1,4-phenylene oxide, and magnetic nanoparticles like ferroferric oxide (Fe_2O_3 and Fe_3O_4) or a permanent magnet like neodymium, praseodymium, ferrite, etc. are used to fabricate a smart polymer magnetic membrane [70–73]. The permeability of O_2 highly depends on the magnetic field induction. The O_2 permeability increases with the increase in the magnetic field. HBPI polymer-based membrane shows better magnetic field induction and thus, better O_2 permeability than that from LPI and EC [99, 100]. However, the O_2/N_2 separation not only depends on the external magnetic field but also on the loading of magnetic nanoparticles. In between the magnetic nanoparticles mentioned above, ferroferric oxide attributes higher O_2 permeability and O_2/N_2 selectivity owing to the increased magnetic field of Fe_2O_3 and Fe_3O_4 nanoparticles [24]. However, the O_2/N_2 selectivity is not that high. Hence, further research/investigation is needed to control the surface morphology. A schematic illustration of the polymeric magnetic membrane is shown in Fig. 7.

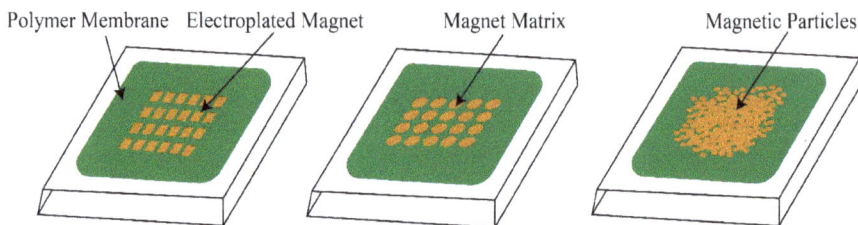

Figure 7. Illustration of the polymeric magnetic membrane.

7. Factors affecting O_2/N_2 separation

Like any other fabrication process in polymer science, there are some crucial operating conditions exist whose control the O_2/N_2 separation performance. First of all, the gas feeding pressure, secondly the temperature issue, thirdly the composition of gases, and lastly feed flow rate through membrane all of which affect the permeability and selectivity of O_2/N_2 gas during the separation process. With all the modifications related to the membrane described above in sections 4 and 5, the proper optimization of those conditions is also a must.

7.1 Pressure

The actual influence of pressure on the O_2/N_2 gas separation varies from membrane to membrane. Based on the composition of the membrane, the driving force of O_2/N_2 gas is controlled by the pressure at the feed side, atmospheric pressure applied for permeate gas, and the vacuum nature of the permeate gas [101]. Usually, it has been seen that the increase in the feed gas pressure with the decreased permeate pressure, the productivity and recovery enhanced. But this often leads to decrease the membrane surface area may be due to collapsing of some pores. Some studies have shown that the separation factors of O_2/N_2 eventually decrease upon increasing feed pressure [102]. Alternatively, some study shows that the feed pressure can control the permeability to a certain point. Above that certain point, the increase in the feed pressure will decrease the permeability of O_2/N_2 gas in the membrane [103]. A more interesting issue has also been reported that the O_2/N_2 permeability is completely independent of feed gas pressure as well as the selectivity of the O_2/N_2 is also constant [104]. So, the above-stated conditions regarding pressure are valid for different sorts of polymeric membranes.

7.2 Temperature

The temperature has a definite effect on the O_2/N_2 selectivity and permeability. For a nonporous polymer membrane, the gas permeability (P_g) is related to the solubility and diffusivity coefficient as shown in preceding Eq. 1. Again, this solubility and diffusivity are the functions of temperature. As a result, the permeability is prominent and selectivity is controlled by the temperature during the gas separation process. The permeability of O_2 usually enhances upon the raise of the temperature which is ultimately good for the O_2/N_2 separation process [105]. Nevertheless, increasing the temperature may hinder selectivity. It is due to high temperature often causes a lower O_2/N_2 solubility ratio, which results in lower diffusivity and ultimately lower selectivity [106]. The lower selectivity is always unwanted for efficient gas separation. For a high-quality O_2/N_2 separation, a higher degree of selectivity is always required. All of these evidences clear that the operative temperature should be optimized for balancing both the permeability and selectivity for a better-quality gas separation.

7.3 Feed flow rate and flow configuration

Most of the commercially available membranes for O_2/N_2 gas separation are composed of asymmetric hollow fibers. The hollow fibers have two different sides *viz;* bore side and inner side for flowing O_2/N_2 gas molecules during the separation process. Usually, most of the feed gas molecules pass through the bore side or internal side and it results in better separation performance than the shell-side or the outer side of the fibers. The pressure of the feed gas is really important to maintain the gas separation process efficiently in long term processes. The increase in the feed gas pressure can result in the expansion of the bore but compress the shell-side of the fiber [107]. If the bore size as showed in the Fig. 8 increases, that may enhance the permeability of gases but can reduce the selectivity [108]. As well as, the compression of the shell-side can also negatively affect the overall performance of the gas separation. However, the expansion in the bore side may occur due to the stacked gas molecules, and this effect is termed as concentration polarization [109]. Concentration polarization is destructive for the longevity and the separation performance of the membrane. The solution is related to the proper optimization feed flow. Usually, at low atmospheric pressure and high feed flow rate, the enrichment of O_2 is the highest. On the contrary, for better N_2 enrichment, high atmospheric pressure and low feed flow rate are required.

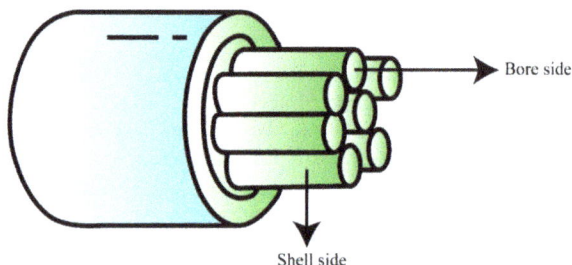

Figure 8. Illustration of the hollow fiber in a polymeric membrane.

7.4   Feed composition

As feed composition directly impacts the solubility of gas molecules and their sorption-diffusion process during the O_2/N_2 gas separation, this is also considered as a vital factor to be considered. In a multiple component-based feed gas, the separation process is hindered by the coupling effects of multiple components [110]. Besides, if there are any condensable species being present in the feed gas, then the membrane could be plasticized. Therefore, the longevity issue is mostly dependent on the composition of the feed gas. In the case of only O_2/N_2, if the feed pressure increases, the permeability of O_2 increases but reduces in the case of N_2. Thus, the optimization of feed gas composition and other controlling parameters are simultaneously important for controlling the performance of O_2/N_2 gas separation.

Conclusion

Over the past few years, the polymeric membrane technology improved enormously for O_2/N_2 gas separation. Researchers are now trying to optimize these classical membranes' performance using various nanomaterials. In addition, fabrication of new membranes using some novel polymers is an ongoing process for gas separation. Besides synthetic polymers, we encourage to use biopolymer such as nanocellulose, chitin nanofiber etc. for such application because of their biodegradability and cost-effectivity. Overall, the efficiency of the polymeric membranes highly depends on the polymeric membrane materials, manufacturing process as well as membrane operating conditions. In order to boost up the permeability and selectivity of gas molecules, the mixed polymeric membranes, polymer magnetic membranes, cross-linked and selective layer membranes are still need to be improved. Though the fabrication of polymeric membrane with great molecular selectivity is very challenging, this technology is consequently anticipated to

Polymeric Membranes for Water Purification and Gas Separation   Materials Research Forum LLC
Materials Research Foundations **113** (2021) 171-202   https://doi.org/10.21741/9781644901632-6

be capable of competing with the existing separation techniques, to massively separate O_2/N_2 gas to meet industrial and medicinal demands.

Acknowledgments

Supports from the University Grants Commission (UGC) of Bangladesh and Bangladesh University of Engineering and Technology (BUET) are highly appreciated.

References

[1] T. A. Saleh, and V. K. Gupta, An Overview of Membrane Science and Technology, Nano. Pol. Mem., Elsevier (2016) 1–23. https://doi.org/10.1016/b978-0-12-804703-300001-2

[2] Z. Y. Yeo, T. L. Chew, P. W. Zhu, A. R. Mohamed, and S. P. Chai, Conventional processes and membrane technology for carbon dioxide removal from natural gas: A review, J. Nat. Gas Chem., Elsevier 21(3) (2012) 282–298. https://doi.org/10.1016/S1003-9953(11)60366-6

[3] X. Y. Chen, H. Vinh-Thang, A. A. Ramirez, D. Rodrigue, and S. Kaliaguine, Membrane gas separation technologies for biogas upgrading, RSC Adv. 5(31) (2015). https://doi.org/10.1039/c5ra00666j

[4] A. F. Ismail, K. C. Khulbe, and T. Matsuura, Gas separation membranes: Polymeric and inorganic, Springer 10 (2015) 978-3. https://doi.org/ 10.1007/978-3-319-01095-3

[5] L. Li, G. Ma, Z. Pan, N. Zhang, and Z. Zhang, Research progress in gas separation using hollow fiber membrane contactors, Membranes (Basel) 10(12) (2020) 1–20. https://doi.org/10.3390/membranes10120380

[6] E. Nagy, Membrane Gas Separation, Basic Equations of Mass Transport Through a Membrane Layer, Elsevier (2019) 457–481. https://doi.org/10.1016/B978-0-12-813722-2.00018-2

[7] J. Gilron, and A. Soffer, Knudsen diffusion in microporous carbon membranes with molecular sieving character, J. Memb. Sci., 209(2) (2002) 339–352. https://doi.org/10.1016/S0376-7388(02)00074-1

[8] C. Z. Liang, T. S. Chung, and J. Y. Lai, A review of polymeric composite membranes for gas separation and energy production, Prog. Pol. Sci., Elsevier 97 (2019) 101-141 https://doi.org/10.1016/j.progpolymsci.2019.06.001

[9] D. F. Sanders, Z. P. Smith, R. Guo, L. M. Robeson, J. E. McGrath, D. R. Pail, and B. D. Freeman, Energy-efficient polymeric gas separation membranes for a sustainable future: A review, Polymer, Elsevier 54(18) (2013) 4729–4761. https://doi.org/10.1016/j.polymer.2013.05.075

[10] N. Sazali, A review of the application of carbon-based membranes to hydrogen separation, J. Mat. Sci., Springer (2020) 11052–11070. https://doi.org/10.1007/s10853-020-04829-7

[11] B. D. Freeman, Basis of permeability/selectivity tradeoff relations in polymeric gas separation membranes, Macromolecules, 32(2) (1999) 375–380. https://doi.org/10.1021/ma9814548

[12] E. Kianfar, and V. Cao, Polymeric membranes on base of PolyMethyl methacrylate for air separation: A review, J. Mat. Res. Tech., Elsevier 10 (2021) 1437–1461. https://doi.org/10.1016/j.jmrt.2020.12.061

[13] B. D. Freeman, and I. Pinnau, Polymeric Materials for Gas Separations, (1999). https://doi.org/10.1021/bk-1999-0733.ch001

[14] K. C. Khulbe, C. Feng, and T. Matsuura, The art of surface modification of synthetic polymeric membranes, J. Appl. Polym. Sci., 115(2) (2010) 855–895. https://doi.org/10.1002/app.31108

[15] A. B. Gil'man, Low-Temperature Plasma Treatment as an Effective Method for Surface Modification of Polymeric Materials, High Energy Chem., 37(1) (2003) 17–23. https://doi.org/10.1023/A:1021957425359

[16] Ş. B. Tantekin-Ersolmaz, Ç. Atalay-Oral, M. Tatlier, A. Erdem-Şenatalar, B. Schoeman, and J. Sterte, Effect of zeolite particle size on the performance of polymer-zeolite mixed matrix membranes, J. Memb. Sci., 175(2), (2000) 285–288. https://doi.org/10.1016/S0376-7388(00)00423-3

[17] J. Dechnik, J. Gascon, C. J. Doonan, C. Janiak, and C. J. Sumby, Mixed-Matrix Membranes, Angewandte Chemie - International Edition, 56(32) (2017) 9292–9310. https://doi.org/10.1002/anie.201701109

[18] J. Caro, Are MOF membranes better in gas separation than those made of zeolites?, Curr. Opi. Chem. Eng., Elsevier 1(1) (2011) 77–83. https://doi.org/10.1016/j.coche.2011.08.007

[19] H. B. Tanh Jeazet, C. Staudt, and C. Janiak, Metal-organic frameworks in mixed-

matrix membranes for gas separation, Dalt. Trans., RSC 41(46) (2012) 14003–14027. https://doi.org/10.1039/c2dt31550e

[20] T. S. Chung, L. Y. Jiang, Y. Li, and S. Kulprathipanja, Mixed matrix membranes (MMMs) comprising organic polymers with dispersed inorganic fillers for gas separation, Prog. Pol. Sci. 32(4) (2007) 483–507. https://doi.org/10.1016/j.progpolymsci.2007.01.008

[21] N. Y. Abu-Thabit, S. A. Ali, S. M. J. Zaidi, and K. Mezghani, Novel sulfonated poly(ether ether ketone)/phosphonated polysulfone polymer blends for proton conducting membranes, J. Mater. Res. 27(15) (2012) 1958–1968. https://doi.org/10.1557/jmr.2012.145

[22] A. Ram, Description of Major Plastics: Structure, Properties and Utilization, Fun. Pol. Eng., Springer (1997) 48–215. https://doi.org/10.1007/978-1-4899-1822-2_6

[23] O. Mehmood et al., Optimization analysis of polyurethane based mixed matrix gas separation membranes by incorporation of gamma-cyclodextrin metal organic frame work, Chem. Pap. 74(10) (2020) 3527–3543. https://doi.org/10.1007/s11696-020-01179-1

[24] N. F. Himma, A. K. Wardani, N. Prasetya, P. T. P. Aryanti, and I. G. Wenten, Recent progress and challenges in membrane-based O2/N2 separation, Rev. Chem. Eng. 35(5) (2019) 591–625. https://doi.org/10.1515/revce-2017-0094

[25] N. Abdullah, M. A. Rahman, M. H. D. Othman, J. Jaafar, and A. F. Ismail, Membranes and Membrane Processes: Fundamentals, in Current Trends and Future Developments on (Bio-) Membranes: Photocatalytic Membranes and Photocatalytic Membrane Reactors, Elsevier 2018 45–70. https://doi.org/10.1016/B978-0-12-813549-5.00002-5

[26] A. F. Ismail, and P. Y. Lai, Effects of phase inversion and rheological factors on formation of defect-free and ultrathin-skinned asymmetric polysulfone membranes for gas separation, Sep. Purif. Tech. 33(2) (2003) 127–143. https://doi.org/10.1016/S1383-5866(02)00201-0

[27] B. W. Rowe, B. D. Freeman, and D. R. Paul, Chapter 3. Physical Aging of Membranes for Gas Separations (2011) 58–83. https://doi.org/10.1039/9781849733472-00058

[28] P. Banerjee, R. Das, P. Das, and A. Mukhopadhyay, Membrane technology, Carbon Nano., Springer (2018) 127–150

[29] F. Mostafapoor et al., Interface analysis of compatibilized polymer blends, in Compatibilization of Polymer Blends: Micro and Nano Scale Phase Morphologies, Interphase Characterization, and Properties, Elsevier (2019) 349–371. https://doi.org/10.1016/B978-0-12-816006-0.00012-8

[30] J. M. Gohil, and R. R. Choudhury, Introduction to Nanostructured and Nano-enhanced Polymeric Membranes: Preparation, Function, and Application for Water Purification, Nano. Mat. Wat. Pur., Elsevier (2018) 25–57. https://doi.org/10.1016/B978-0-12-813926-4.00038-0

[31] I. Pinnau, and W. J. Koros, Structures and gas separation properties of asymmetric polysulfone membranes made by dry, wet, and dry/wet phase inversion, J. Appl. Polym. Sci. 43(8) (1991) 1491–1502. https://doi.org/10.1002/app.1991.070430811

[32] D. T. Clausi, and W. J. Koros, Formation of defect-free polyimide hollow fiber membranes for gas separations, J. Memb. Sci. 167(1) (2000) 79–89. https://doi.org/10.1016/S0376-7388(99)00276-8

[33] T. S. Chung, S. K. Teoh, and X. Hu, Formation of ultrathin high-performance polyethersulfone hollow-fiber membranes, J. Mem. Sci. 133(2) (1997) 161–175. https://doi.org/10.1016/S0376-7388(97)00101-4

[34] A. F. Ismail, I. R. Dunkin, S. L. Gallivan, and S. J. Shilton, Production of super selective polysulfone hollow fiber membranes for gas separation, Polymer 40(23) (1999) 6499–6506. https://doi.org/10.1016/S0032-3861(98)00862-3

[35] D. Wang, K. Li, and W. K. Teo, Polyethersulfone hollow fiber gas separation membranes prepared from NMP/alcohol solvent systems, J. Memb. Sci. 115(1) (1996) 85–108. https://doi.org/10.1016/0376-7388(95)00312-6

[36] D. Wang, K. Li, and W. K. Teo, Highly permeable polyethersulfone hollow fiber gas separation membranes prepared using water as non-solvent additive, J. Memb. Sci. 176(2) (2000) 147–158. https://doi.org/10.1016/S0376-7388(00)00419-1

[37] F. Mohamed, H. Hasbullah, W. N. R. Jamian, A. R. A. Rani, M. F. K. Saman, W. N. H. Salleh, & R. R. Ali, Gas Permeation Performance of Poly(lactic acid) Asymmetric Membrane for O2/N2 Separation, Springer (2015) 149–156. https://doi.org/10.1007/978-981-287-505-1_18

[38] A. F. Ismail, and L. P. Yean, Effects of Shear Rate on Morphology and Gas Separation Performance as Asymetric Polysulfone Membranes, J. Chem. Eng. 2(1) (2008) 67. https://doi.org/10.22146/ajche.50805

[39] R. V. Kulkarni, S. Z. Inamdar, K. K. Das, and M. S. Biradar, Polysaccharide-based stimuli-sensitive graft copolymers for drug delivery, Polysac. Car. Drug Del., Elsevier (2019) 155–177. https://doi.org/10.1016/B978-0-08-102553-6.00007-6

[40] I. Khan, M. Mansha, and M. A. Jafar Mazumder, Pol. Blends, Springer Cham, 2019 513–549. https://doi.org/10.1007/978-3-319-95987-0_16

[41] A. C. Shi, and B. Li, Block Copolymers under Confinement, Pol. Sci., Elsevier 10(7) (2012) 71–81. https://doi.org/10.1016/B978-0-444-53349-4.00186-2

[42] C. Camacho-Zuñiga, F. A. Ruiz-Treviño, S. Hernández-López, M. G. Zolotukhin, F. H. J. Maurer, and A. González-Montiel, Aromatic polysulfone copolymers for gas separation membrane applications, J. Memb. Sci. 340(1–2) (2009) 221–226. https://doi.org/10.1016/j.memsci.2009.05.033

[43] T. Komatsuka, A. Kusakabe, and K. Nagai, Characterization and gas transport properties of poly(lactic acid) blend membranes, Desalination 234(1–3) (2008) 212–220. https://doi.org/10.1016/j.desal.2007.09.088

[44] E. Rudnik, Compostable Polymer Properties and Packaging Applications, in Plastic Films in Food Packaging: Materials, Technology and Applications, Elsevier Inc (2012) 217–248. https://doi.org/10.1016/B978-1-4557-3112-1.00013-2.

[45] S. C. George, K. N. Ninan, and S. Thomas, Permeation of nitrogen and oxygen gases through styrene-butadiene rubber, natural rubber and styrene-butadiene rubber/natural rubber blend membranes, Eur. Polym. J. 37(1) (2001) 183–191. https://doi.org/10.1016/S0014-3057(00)00083-5

[46] P. Gibson, H. Schreuder-Gibson, and D. Rivin, Transport properties of porous membranes based on electrospun nanofibers, Colloids and Surfaces A: Physicochemical and Engineering Aspects 187(188) (2001) 469–481. https://doi.org/10.1016/S0927-7757(01)00616-1

[47] C. M. Zimmerman, and W. J. Koros, Polypyrrolones for membrane gas separations. I. Structural comparison of gas transport and sorption properties, Journal of Polymer Science, Part B: Polymer Physics (1999)

[48] F. Marken, M. Carta, and N. B. McKeown, Polymers of intrinsic microporosity in the design of electrochemical multicomponent and multiphase interfaces, Anal. Chem. 93(3) (2021) 1213–1220. https://doi.org/10.1021/acs.analchem.0c04554

[49] A. Arabi Shamsabadi, M. Rezakazemi, F. Seidi, H. Riazi, T. Aminabhavi, and M.

Soroush, Next generation polymers of intrinsic microporosity with tunable moieties for ultrahigh permeation and precise molecular CO2 separation, Prog. Ener. Comb. Sci., Elsevier 48 (2021) 100903. https://doi.org/10.1016/j.pecs.2021.100903

[50] D. Meis et al., Thermal rearrangement of: Ortho -allyloxypolyimide membranes and the effect of the degree of functionalization, Polym. Chem. 9(29) (2018) 3987–3999. https://doi.org/10.1039/c8py00530c

[51] S. Kim, and Y. M. Lee, Rigid and microporous polymers for gas separation membranes, Progress, Pol. Sci., Elsevier 43 (2015) 1–32, https://doi.org/10.1016/j.progpolymsci.2014.10.005

[52] K. Balani, V. Verma, A. Agarwal, and R. Narayan, Physical, Thermal, and Mechanical Properties of Polymers, Biosurfaces, John Wiley & Sons (2015) 329–344. https://doi.org/10.1002/9781118950623.app1

[53] P. M. Budd, N. B. McKeown, B. S. Ghanem, K. J. Msayib, D. Fritsch, L. Starannikova, N. Belov, O. Sanfirova, Y. Yampolskii, & V. Shantarovich, Gas permeation parameters and other physicochemical properties of a polymer of intrinsic microporosity: Polybenzodioxane PIM-1. J. Memb. Sci. 325(2) (2008) 851–860. https://doi.org/10.1016/j.memsci.2008.09.010

[54] C. L. Staiger, S. J. Pas, A. J. Hill, and C. J. Cornelius, Gas separation, free volume distribution, and physical aging of a highly microporous spirobisindane polymer, Chem. Mat. 20(8) (2008) 2606–2608. https://doi.org/10.1021/cm071722t

[55] N. Du, G. P. Robertson, J. Song, I. Pinnau, S. Thomas, and M. D. Guiver, Polymers of intrinsic microporosity containing trifluoromethyl and phenylsulfone groups as materials for membrane gas separation, Macromolecules 41(24) (2008) 9656–9662, https://doi.org/10.1021/ma801858d

[56] B. S. Ghanem, N. B. McKeown, P. M. Budd, J. D. Selbie, and D. Fritsch, High-performance membranes from polyimides with intrinsic microporosity, Adv. Mater. 20(14) (2008) 2766–2771, https://doi.org/10.1002/adma.200702400

[57] B. Comesaña-Gándara, A. Hernández, J. G. de la Campa, J. de Abajo, A. E. Lozano, and Y. M. Lee, Thermally rearranged polybenzoxazoles and poly(benzoxazole-co-imide)s from ortho-hydroxyamine monomers for high performance gas separation membranes, J. Memb. Sci. 493 (2015) 329–339. https://doi.org/10.1016/j.memsci.2015.05.061

[58] T. V. Ngo, Microfinance Complementarity and Trade-Off between Financial

Polymeric Membranes for Water Purification and Gas Separation   Materials Research Forum LLC
Materials Research Foundations **113** (2021) 171-202   https://doi.org/10.21741/9781644901632-6

Performance and Social Impact, Int. J. Econ. Financ. 7(11) (2015) 128.
https://doi.org/10.5539/ijef.v7n11p128

[59] N. Sazali, W. N. Wan Salleh, A. F. Ismail, N. H. Ismail, and K. Kadirgama, A brief review on carbon selective membranes from polymer blends for gas separation performance, Rev. Chem. Eng., De Gruyter, 37(3) (2012) 339–362.
https://doi.org/10.1515/revce-2018-0086

[60] K. C. Chong, S. O. Lai, H. S. Thiam, & W. J. Lau, The progress of polymeric membrane separation technique in O2/N2 separation. Key Eng. Mat., 701, (2016) 255–259, http://doi.org/10.4028/www.scientific.net/KEM.701.255

[61] K. Hunger, N. Schmeling, H. B. T. Jeazet, C. Janiak, C. Staudt, and K. Kleinermanns, Investigation of cross-linked and additive containing polymer materials for membranes with improved performance in pervaporation and gas separation, Membranes 2(4) (2012) 727–763. https://doi.org/10.3390/membranes2040727

[62] Q. Xu, A. Sigen, P.McMichael, J. Creagh-Flynn, D. Zhou, X. Li.Y. Gao, X. Wang, and W. Wang, Double-Cross-Linked Hydrogel Strengthened by UV Irradiation from a Hyperbranched PEG-Based Trifunctional Polymer, ACS Macro Lett. 7(5) (2018) 509–513. https://doi.org/10.1021/acsmacrolett.8b00138

[63] A. Morelli, and M. J. Hawker, Utilizing Radio Frequency Plasma Treatment to Modify Polymersic Materials for Biomedical Applications, ACS Biomat. Sci. Eng. (2021). https://doi.org/10.1021/acsbiomaterials.0c01673

[64] Z. A. Fekete, E. Wilusz, F. E. Karasz, and C. Visy, Ion beam irradiation of conjugated polymers for preparing new membrane materials-A theoretical study, Sep. Purif. Technol. 57(3) (2007) 440–443. https://doi.org/10.1016/j.seppur.2006.04.014

[65] L. Upadhyaya, X. Qian, and S. Ranil Wickramasinghe, Chemical modification of membrane surface — overview, Current Opinion in Chem. Eng. 20 (2018) 13–18. https://doi.org/10.1016/j.coche.2018.01.002

[66] S. Dilpazir, M. Usman, S. Rasul, and S. N. Arshad, A simple UV-ozone surface treatment to enhance photocatalytic performance of TiO2 loaded polymer nanofiber membranes, RSC Adv. 6(18) (2016) 14751–14755.
https://doi.org/10.1039/c5ra22903k

[67] Q. Liu et al., Effect of UV irradiation and physical aging on O 2 and N 2 transport properties of thin glassy poly(arylene ether ketone) copolymer films based on tetramethyl bisphenol A and 4,4'-difluorobenzophenone, 2019

[68] B. Comesañ A-Gá Ndara et al., Redefining the Robeson upper bounds for CO 2 / CH 4 and CO 2 /N 2 separations using a series of ultrapermeable benzotriptycene-based polymers of intrinsic microporosity, Energy Environ. Sci. 2 (2019) 2733. https://doi.org/10.1039/c9ee01384a

[69] M. Ulbricht, Advanced functional polymer membranes, Polymer, Elsevier BV 47 (7), (2006) 2217–2262. https://doi.org/10.1016/j.polymer.2006.01.084

[70] L. M. Robeson, Polymer Membranes, in Polymer Science: A Comprehensive Referencn 10(8) ((2012) 325–347. https://doi.org/10.1016/B978-0-444-53349-4.00211-9

[71] C. A. Dennett Gołda, M. Brzychczy-Włoch, M. Faryna, K. Engvall, and A. Kotarba, Listening to Radiation Damage In Situ: Passive and Active Acoustic Techniques, JOM 72(1) (2020) 197–209. https://doi.org/10.1007/s11837-019-03898-7

[72] M. Gołda, M. Brzychczy-Włoch, M. Faryna, K. Engvall, and A. Kotarba, Oxygen plasma functionalization of parylene C coating for implants surface: Nanotopography and active sites for drug anchoring, Mater. Sci. Eng. C 33(7) (2013) 4221–4227. https://doi.org/10.1016/j.msec.2013.06.014

[73] M. Kahoush, N. Behary, A. Cayla, B. Mutel, J. Guan, and V. Nierstrasz, Surface modification of carbon felt by cold remote plasma for glucose oxidase 3 enzyme immobilization, App. Surf. Sci. 476 (2019) 1016-1024. https://doi.org/10.1016/j.apsusc.2019.01.155

[74] G. Nageswaran, L. Jothi, and S. Jagannathan, Plasma Assisted Polymer Modifications, Non-Thermal Plasma Technology for Polymeric Materials, Elsevier (2019) 95–127. https://doi.org/10.1016/b978-0-12-813152-7.00004-4

[75] A. P. Kharitonov, Direct fluorination of polymers-From fundamental research to industrial applications, Non-Thermal Plasma Technology for Polymeric Materials 61(2–4) (2008) 192–204. https://doi.org/10.1016/j.porgcoat.2007.09.027

[76] J. Y. Park, H. Y. Chae, J. S. Sim, J. Park, H. H. Lee, and J. P. Yoo, Controlled wavelength reduction in surface wrinkling of poly(dimethylsiloxane), Soft Matt. 6(3) (2010) 677–684. https://doi.org/10.1039/b916603c

[77] S. Zare and A. Kargari, Membrane properties in membrane distillation, Emerging Technologies for Sustainable Desalination Handbook, Elsevier (2018) 107–156. https://doi.org/10.1016/B978-0-12-815818-0.00004-7

[78] Y. M. Lee, S. Y. Ha, Y. K. Lee, D. H. Suh, and S. Y. Hong, Gas separation through conductive polymer membranes Polyaniline membranes with high oxygen selectivity, Ind. Eng. Chem. Res. 38(5) (1999) 1917–1924. https://doi.org/10.1021/ie980259e

[79] G. Illing, K. Hellgardt, M. Schonert, R. J. Wakeman, and A. Jungbauer, Towards ultrathin polyaniline films for gas separation, J. Memb. Sci. 253(1–2) (2005) 199–208. https://doi.org/10.1016/j.memsci.2004.12.031

[80] I. S. S. Han Gyu Moon, Electrical and structural analysis of conductive polyaniline/polyimide blends, J. App. Pol. Sci., Wiley Online Library, Appl. Polym. Sci. (1999) 2169–2178

[81] M. Kawakami, Y. Yamashita, M. Iwamoto, and S. Kagawa, Modification of gas permeabilities of polymer membranes by plasma coating, J. Memb. Sci. 19(3) (1984) 249–258. https://doi.org/10.1016/S0376-7388(00)80228-8

[82] N. Inagaki and J. Ohkubo, Plasma polymerization of hexafluoropropene/methane mixtures and composite membranes for gas separations, J. Memb. Sci. 27(1) (1986) 63–75. https://doi.org/10.1016/S0376-7388(00)81382-4

[83] F. Huber, J. Springer, and M. Muhler, Plasma polymer membranes from hexafluoroethane/hydrogen mixtures for separation of oxygen and nitrogen, J. App. Pol. Sci. 63(12) (1997) 1517-1526. https://doi.org/10.1002/(SICI)1097-4628(19970321)63:12%3C1517::AID-APP2%3E3.0.CO;2-R

[84] S. Kim, L. Chen, J. K. Johnson, and E. Marand, Polysulfone and functionalized carbon nanotube mixed matrix membranes for gas separation: Theory and experiment, J. Memb. Sci. 294(1–2) (2007) 147–158. https://doi.org/10.1016/j.memsci.2007.02.028

[85] C. C. Hu, T. C. Liu, K. R. Lee, R. C. Ruaan, and J. Y. Lai, Zeolite-filled PMMA composite membranes: influence of coupling agent addition on gas separation properties, Desalination 193(1–3) (2006) 14–24. https://doi.org/10.1016/j.desal.2005.04.137

[86] J. T. Chen et al., Zeolite-filled porous mixed matrix membranes for air separation, Ind. Eng. Chem. Res. 53(7) (2014) 2781–2789. https://doi.org/10.1021/ie403833u

[87] D. Q. Vu, W. J. Koros, and S. J. Miller, Mixed matrix membranes using carbon molecular sieves: I. Preparation and experimental results, J. Memb. Sci. 211(2) (2003) 311–334. https://doi.org/10.1016/S0376-7388(02)00429-5

[88] P. S. Goh, A. F. Ismail, S. M. Sanip, B. C. Ng, and M. Aziz, Recent advances of

inorganic fillers in mixed matrix membrane for gas separation, Sep. Purif. Tech., Elsevier 81(3) (2011) 243–264. https://doi.org/10.1016/j.seppur.2011.07.042

[89] B. Zornoza, B. Seoane, J. M. Zamaro, C. Téllez, and J. Coronas, Combination of MOFs and zeolites for mixed-matrix membranes, ChemPhysChem 12 (15) (2011)

[90] M. F. A. Wahab, A. F. Ismail, and S. J. Shilton, Studies on gas permeation performance of asymmetric polysulfone hollow fiber mixed matrix membranes using nanosized fumed silica as fillers, Sep. Purif. Technol. 86 (2012) 41–48. https://doi.org/10.1016/j.seppur.2011.10.018

[91] A. Fernández-Barquín, C. Casado-Coterillo, S. Valencia, and A. Irabien, Mixed matrix membranes for O2/N2 separation: The influence of temperature, Membranes 6(2) (2016) 28. https://doi.org/10.3390/membranes6020028

[92] R. Adams, C. Carson, J. Ward, R. Tannenbaum, and W. Koros, Metal organic framework mixed matrix membranes for gas separations, Micro. Meso. Mat. 131(1–3) (2010) 13–20. https://doi.org/10.1016/j.micromeso.2009.11.035

[93] S. S. Madaeni, E. Enayati, and V. Vatanpour, Separation of nitrogen and oxygen gases by polymeric membrane embedded with magnetic nano-particle, Pol. Adv. Tech. 22(12) (2011) 2556–2563. https://doi.org/10.1002/pat.1800

[94] M. Ulbricht, Smart Polymeric Membranes with Magnetic Nanoparticles for Switchable Separation, RSC Smart Mat. 35 (2019) 297–328. https://doi.org/10.1039/9781788016377-00297

[95] O. Philippova, A. Barabanova, V. Molchanov, and A. Khokhlov, Magnetic polymer beads: Recent trends and developments in synthetic design and applications, Eur. Pol. J. 47(4) 2011 542–559. https://doi.org/10.1016/j.eurpolymj.2010.11.006

[96] X. Feng et al., Scalable fabrication of polymer membranes with vertically aligned 1 nm pores by magnetic field directed self-Assembly, ACS Nano 8(12) (2014) 11977–11986. https://doi.org/10.1021/nn505037b

[97] I. Csetneki, G. Filipcsei, and M. Zrínyi, Smart nanocomposite polymer membranes with on/off switching control, Macromolecules 39(5) (2006) 939–1942. https://doi.org/10.1021/ma052189a

[98] P. W. Majewski, M. Gopinadhan, W. S. Jang, J. L. Lutkenhaus, and C. O. Osuji, Anisotropic ionic conductivity in block copolymer membranes by magnetic field alignment, J. Am. Chem. Soc. 132(49) (2010) 17516–17522.

https://doi.org/10.1021/ja107309p

[99] Y. Li, X. F. Yin, F. R. Chen, H. H. Yang, Z. X. Zhuang, and X. R. Wang, Synthesis of magnetic molecularly imprinted polymer nanowires using a nanoporous alumina template, Macromolecules 39(3) (2006) 4497–4499. https://doi.org/10.1021/ma0526185

[100] A. Rybak, G. Dudek, M. Krasowska, A. Strzelewicz, Z. J. Grzywna, and P. Sysel, Magnetic mixed matrix membranes in air separation, Chem. Pap. 68(10) (2013) 1332–1340. https://doi.org/10.2478/s11696-014-0587-x

[101] P. Bernardo and G. Clarizia, 30 years of membrane technology for gas separation, Chem. Eng. Trans. 32 (2013) 1999–2004. https://doi.org/10.3303/CET1332334

[102] S. S. Hosseini, S. Najari, P. K. Kundu, N. R. Tan, and S. M. Roodashti, Simulation and sensitivity analysis of transport in asymmetric hollow fiber membrane permeators for air separation, RSC Adv. 5 (2015) 86359–86370. https://doi.org/10.1039/c5ra13943k

[103] H. M. Ettouney, H. T. El-Dessouky, and W. Abou Waar, Separation characteristics of air by polysulfone hollow fiber membranes in series, J. Memb. Sci. 148(1) (1998) 105–117. https://doi.org/10.1016/S0376-7388(98)00144-6

[104] R. C. Ruaan, S. H. Chen, and J. Y. Lai, Oxygen/nitrogen separation by polycarbonate/Co(SalPr) complex membranes, J. Memb. Sci. 35(1) (1997) 9–18. https://doi.org/10.1016/S0376-7388(97)00129-4

[105] O. Choi, Y. Kim, J. D. Jeon, and T. H. Kim, Preparation of thin film nanocomposite hollow fiber membranes with polydopamine-encapsulated Engelhard titanosilicate-4 for gas separation applications, J. Memb. Sci. 620 (2021) 118946. https://doi.org/10.1016/j.memsci.2020.118946

[106] H. Nishide, M. Ohyanagi, O. Okada, and E. Tsuchida, Dual-Mode Transport of Molecular Oxygen in a Membrane Containing a Cobalt Porphyrin Complex as a Fixed Carrier, Macromolecules, 20(2) (1987) 417–422. https://doi.org/10.1021/ma00168a032

[107] R. Rautenbach, A. Struck, T. Melin, and M. F. M. Roks, Impact of operating pressure on the permeance of hollow fiber gas separation membranes, J. Memb. Sci. 146(2) (1998) 217–223. https://doi.org/10.1016/S03767388(98)00119-7

[108] S. Sridhar, B. Smitha, and T. M. Aminabhavi, Separation of carbon dioxide from

natural gas mixtures through polymeric membranes - A review, Sep. Purif. Rev., 36(2) (2007) 113–174.https://doi.org/10.1080/15422110601165967

[109]   G. Dong, H. Li, and V. Chen, Challenges and opportunities for mixed-matrix membranes for gas separation, J. Mat. Chem. A 1(15) (2013) 4610–4630. https://doi.org/10.1039/c3ta00927k

[110]   T. A. Zangle, A. Mani, and J. G. Santiago, Theory and experiments of concentration polarization and ion focusing at microchannel and nanochannel interfaces, Chem. Soc. Rev. (2005). https://doi.org/ 10.1039/b902074h

Polymeric Membranes for Water Purification and Gas Separation   Materials Research Forum LLC
Materials Research Foundations **113** (2021) 203-242   https://doi.org/10.21741/9781644901632-7

Chapter 7

Polymeric Membrane for CO_2/CH_4 Separation

N. Sazali[1], W.N.W. Salleh[1,2]*, A.F. Ismail[1,2]*

[1]Advanced Membrane Technology Research Centre (AMTEC), Universiti Teknologi Malaysia, 81310 Skudai, Johor Darul Takzim, Malaysia

[2]School of Chemical and Energy Engineering, Faculty of Engineering, Universiti Teknologi Malaysia, 81310 Skudai, Johor Darul Takzim, Malaysia

hayati@petroleum.utm.my and afauzi@utm.my

Abstract

This chapter presents a critical overview of polymeric membrane applications for CO_2/CH_4 separation. Comparative summary of availability and practice of different gas separation methods are outlined to give a state-of-the-art view of this technology. Detailed discussions on polymer-based membranes are also discussed in this work, highlighting the mechanism of selective gas permeation through the membranes. Future direction is discussed for possible new experimental design to maximize the membrane performances in separation of CO_2/CH_4.

Keywords

Polymeric Membranes, Gas Separation, Membrane Technology, CO_2 Capture, Methane

Contents

1.   Introduction

Currently, the volume of greenhouse gases (GHG) in the air has increased dramatically [1]. Consequently, the accumulation of CO_2 in the atmosphere has generated much interest in the development of various techniques for the separation of CO_2 from fuel gases. Firstly, the gas separation process through a polymer has been discovered in 1892 by Thomas Graham. The polymeric membrane has been utilized in gas separation process for energy efficiency and affordability. Despite that, its incompetence to withstand high temperatures restricts their utilization in the industries [2]. In addition, polymeric membranes with permeances above 1,000 GPU and selectivity index over 30 are excellent properties for CO_2 capture from flue gases [3-5]. However, the separation of CO_2/CH_4 is likewise a notable industrial procedure. It exhibits an approximately 20% of the annual natural gas in United States (U.S.). It has unmistakably demonstrated the cost of funding and operational separation process, where numerous drawbacks have been observed at a higher concentration of CO_2 (over 2%), especially in the pipeline; for example, pipeline corrosion and declining heat rate of natural gas [6].

Some polymer membranes can function at certain levels of unconducive surrounding, yet the energy with the economic performance are counterbalanced by the need to reduce the temperature of hot streams severely. On several occasions, similar performance is inconceivable or entirely unfeasible. Thus, various researches have stressed upon the alteration of polymers to produce synthetic polymeric membranes that can endure extreme temperatures [7-9]. Fig. 1 shows the different approaches to improve polymeric membrane materials [10].

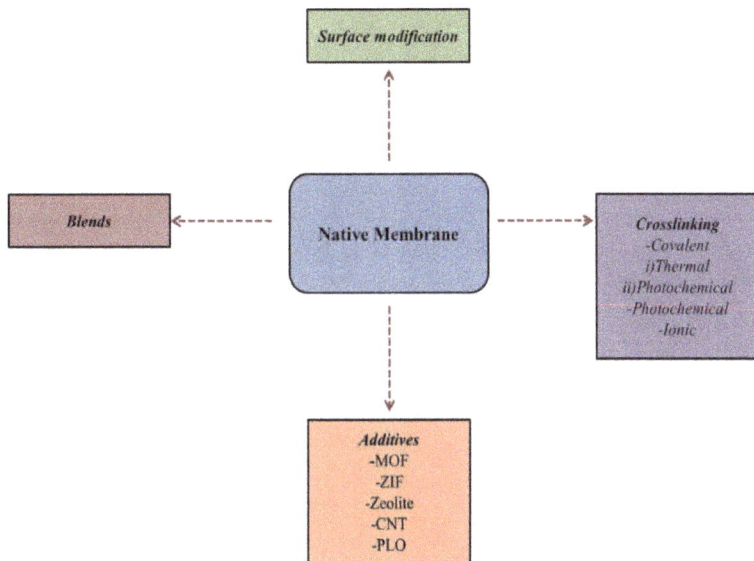

Figure 1. Few methods of membrane modification to enhance properties of polymeric membrane.

Referring to Fig. 1, there are several techniques that can be implemented onto polymeric membranes, such as surface modification, blending process, introduction of new additives, and cross-linking methods. For example, in blending process, polymeric membrane can be simply prepared by the dispersion of inorganic materials into organic solvents, such as N-methyl-2-pyrrolidone (NMP) and dimethylformamide (DMF) [11]. The blending of these two materials will yield larger polymeric/particles interfacial area that will be beneficial towards the percentage loading of the inorganic nanomaterials into the polymer matrix. Researcher presented different thermal stable polymers, especially on carboxylic and heterocyclic aromatic polymers, which showed an increase in temperature stability [2]. Fundamentally low probability of the polymers has shown significant issue because of their insolubility or long-term conversion temperature. Consequently, by far, small accomplishment has been observed in the handling of high thermal resistance polymers to produce membranes with adequate separation and performance attributes.

Issues such as expensive inorganic membranes, porous hybrid materials, and modularization lead to numerous researches that focus on modifying the polymers to develop synthetic polymeric membranes with high thermal resistance. Polymeric membranes are cheap and it can be further fabricated within vast-scale modules. Subsequently, employing polymeric membranes for gas treatment has partaken an everyday practice in numerous industries [12]. Nonetheless, shortage of resilience toward high temperature is one of the real constraints of the modern polymeric material. Due to some of these streams which cannot be exposed to the course of membrane separation, many hot gas streams in each chemical industry have to be held at an elevated temperature during the gas separation process. In particular, the process flow temperature is reduced merely to adjust the cycle gas separation membrane, then reheated and coupled with heating and cooling, all of which leads to a great deal of energy abuse. In considering offshore structures and reservoir comprising large amounts of CO_2, the use of this innovation was observed to be entirely beneficial, as it is a proficient framework with a basic moderate procedure in spite of the motion conditions. All energy sources until the middle of 21st century, which is in conjunction with the projection of world energy consumption, has presented natural gas, yet it remains outdistanced from the others. It includes the synchronous confrontation of CO_2 separation from natural/biogas (acid gas removal), along with succeeding trap of CO_2 transmitted from power plants (to mitigate climatic change due to CO_2 emission) [3].

Using membrane technology, gas separation has significant benefits amid the progress of other CO_2 separation techniques, such as refining, adsorption, and cryogenic absorption. Membrane production with permeability and selectivity is one of the major motivations behind membrane-based gas separation studies. Hasebe and co-workers [13] reported the permeability of CO_2 and selectivity of CO_2/N_2 separations via mixed matrix membranes (MMMs) comprised of polymer and nanosized silica particle, having nano-space gas permeance (Fig. 2). Although polymeric membranes offer various benefits, such as the ability to produce significant medium-sized membrane areas with minimal cost, some issues still exist, which are related to the trade-off relation between gas selectivity and gas permeance [14, 15]. As a result, the membranes usually have an upper limit. Although there are many possible uses for polymer membranes in gas separation, most membrane materials are out of phase due to the expensive materials used. Therefore, to this day, continuous efforts have been made to find alternative materials for effective CO_2 separation. Various techniques have been examined to improve the performance of specific membranes; for instance, blending, cross-linking, grafting, mixing with suitable additives, and etcetera [16].

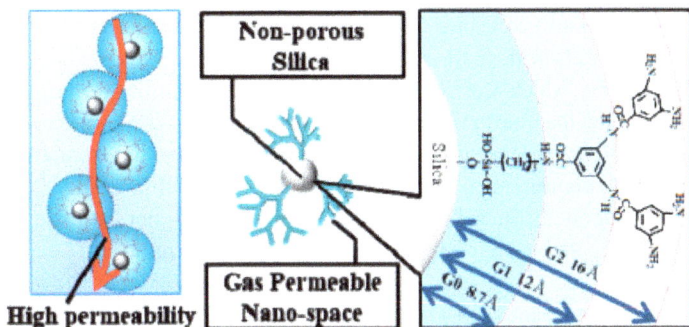

Figure 2. Selectivity of gas pairs CO_2/N_2 using MMMs made up of polymer and silica nanoparticles. Reproduced with permission from [13].

A membrane is portrayed explicitly of perm-selectivity of molecules passing through [17]. It is important to note the permeation flux is influenced by membrane structure and gas transport process driven by its potential chemical gradient. Thus, the selectivity is taken into account as an inconsistency of flux between the penetrants [18]. Membrane technology for gas separation has become drastically expended in recent decades; hence, the development owed to the rapid growth of progressively effective membrane with higher perm-selectivity. Economic advantages of the hydrogen recovery are prompted quickly in the establishment of a few frameworks around the globe [19]. Separation of gas using membrane has been established in different fields, for example, in the removal of CO_2 from natural gas, N_2 from air, as well as H_2 from various plant refineries and petroleum processing streams [20]. To add in the list, the separation of CH_4 from biogas, O_2-enrichment for further applications in metallurgy or medical area, and the removal of volatile organic liquid (VOLs) from exhaust streams are also utilized in various membrane technologies [21]. The advantage of choosing a membrane in the field of gas separation is that it is cost effective, has high efficiency, easy to use, has modularity, and is easy to scale up. Most importantly, membrane separation is a continuous process for ensuring efficient gas separation. Furthermore, classical materials could be used in the industry even though there are many numbers of new materials utilized in the membrane separation innovation. Processes of gas separation by membrane are usually considered to be in a steady-state.

It is essential to test the gas flowrate and gas inlet composition amid the task. In such event, the composition of the end-product needs to be maintained consistently during the dynamic procedure control framework throughout the separation procedure. In most cases, steady membrane permeance frequently opted for effortlessness for the simulation. Bounaceur et al. [22], in their work studied the permeance of gas mixture at different processing pressure, which was dependent on the solution-diffusion model. From the results, it was found that the consistency of permeance can cause significant problem for the membrane during the separation of CO_2/CH_4. The Bounaceur groups stated that if the difference in CO_2 concentration between the gases in and out is greater than 1%, the assumption made on the membrane permeability model is not accurate enough to estimate the gas separation performance for certain types of membranes.

2.   Polymeric membrane for CO_2/CH_4 separation

2.1   Polyimides

In general, polyimide (PI) can potentially act as a membrane material with promising separation performance. They have high chemical stability and are stable at high temperature condition, compared with other polymers, making them preferable for utilization in commercial applications [23]. However, as mentioned by Favvas et al., PI, which is classified as glassy polymers tends to suffer physical aging similar to other polymers in the same group [24]. In addition, the existence of CO_2 in gas mixtures will create strong plasticization of this polymer, thus directing to low selectivity in gas separation process. To address this issue, cross-linking or fabrication of hyperbranched polymer structure is needed. Hyperbranched of 4,4'(hexafluoroisopropylidene)diphthalic anhydride (6FDA) and 4,4'-triaminotriphenylmethane (TTM) on PI membrane had improved the CO_2/CH_4 separation selectivity value of 137 [25]. In another work by Zhang et al., the cross-linked PI has been successfully fabricated with 6FDA-carboxylic acid-containing diamines (CADA1), having superior CO_2/CH_4 separation performance with a selectivity value of 48 [26]. On top of that, the cross-linked 6FDA-PEG on PI membrane yielded high selectivity of CO_2/CH_4 with a value of 46 [27].

(a)   Matrimid®

Matrimid® 5218 is one of the commercial PI that is currently used in CO_2/CH_4 separation due to its good thermal and mechanical stability, high solubility in solvents, and excellent processability for membrane casting process [28]. Additionally, it has good selectivity but poor CO_2 permeability. However, the permeability can be improved by the incorporation of polyethylene glycol (PEG) as an additive and via the addition of nanoparticles, such as zeolitic imidazolate framework (ZIF). Compared with pristine Matrimid®, the selectivity CO_2/CH_4 of Matrimid®/PEG/ZIF was higher, with the value of 25.08 (pristine Matrimid® = 17.40) [29]. The selectivity of CO_2/CH_4 was enhanced at about 66% (selectivity = 36.3) and 57% (selectivity = 57.1) for Matrimid®/zeolite-Y and

Matrimid®/NaY-zeolite, respectively [30, 31]. The blended Matrimid® membrane and 1-ethyl-3-methyl imidazolium bis(trifluoromethylsulfonyl)imide ([Emim][Tf$_2$N]) as an ionic liquid (IL), as presented in Fig. 3, has been fabricated for separation of gas mixture CO_2/CH_4 [32]. Permselectivity of CO_2/CH_4 had improved to 96% (from 32 to 63). The IL is believed to have improved the diffusivity coefficient, which resulted in higher permeability and permselectivity, compared with bare Matrimid®. In another work by Rajati et al., the MIL-101(Cr) was added as a filler in the Matrimid®/PVDF blended membrane [33]. From the final results, selectivity of CO_2/CH_4 of blended MIL 101(Cr)/Matrimid®/PVDF was found to be higher, with the value of 61.95, compared with the unblended Matrimid® (selectivity = 34.90). The uniform dispersion of MIL 101(Cr) in Matrimid®/PVDF could have enhanced the permeability of CO_2/CH_4 separation.

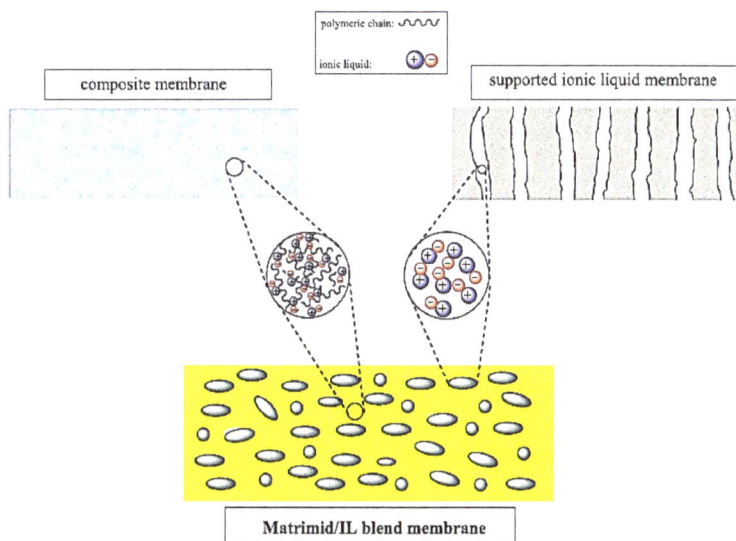

Figure 3. The preparation of blended Matrimid®/IL membrane. Reproduced with permission [32].

2.2   Polysulfones

There are three main types of polysulfone (PSF)-based membranes that are commercially used in CO_2/CH_4 separation, which is dense, asymmetric, and composite polysulfones. Both dense and asymmetric-based PSF membrane consist only PSF. Meanwhile, composite PSF-based membrane is made up of PSF incorporated with other types of polymers. The fabrication of PSF-based membrane can be done via dry, wet or dry/wet process [34]. In brief, the preparation of the PSF-based membrane involve the immersion, coagulation, and phase separation process of binary mixture between a polymer and selected solvents [35]. In terms of CO_2/CH_4 separation performance, PSF-based composite membrane can significantly improve the selectivity of the gas separation. As studied by Hatami et al., by using Pebax/PSF membrane with ZIF-90 as additives, the selectivity of CO_2/CH_4 was able to reach a value of 75.05 [36]. Mohamad et al. prepared PSF membrane incorporated with zeolite-T via priming method [37]. They found that the addition of 4 wt% loading of zeolite-T into polysulfone content a highest CO_2/CH_4 ideal selectivity of 3.37 from 2.63 due to the agglomeration of zeolite-T in polymer matrix. Polydimethylsiloxane (PDMS)/PSF membrane has been fabricated via dip coating method for CO_2/CH_4, resulting in a selectivity value of 85 [38]. Additionally, Suleman et al. mentioned that the water swelling in PDMS/PSF membrane had initiated contraction in the membrane pores, thus dramatically increasing the CO_2/CH_4 selectivity. Report from another researchers, Kiadehi et al. stated that carbon nanofibers (CNF)/PSF mixed matrix membrane exhibited a selectivity value of 12.17 for CO_2/CH_4 [39].

2.3   Polyetherimides

Other than PI, as visualized in Fig. 4, polyetherimides (PEI) is the one of the common classes of polymers that is widely studied for separation of gas by membrane due to their superior properties [40]. However, both of these polymers are facing serious problems attributed to the plasticization effects. These effects can be minimized by introducing several techniques, including cross-linking and surface modification via the addition of certain functional groups at the polymer backbones to alter the chain packing, which subsequently results in free volume increment. The effect of PEI-backbone modification has been investigated by Madzarevic et al., proving that the modified 3,3′,4,4′-biphenyltetracarboxylic dianhydride (BPDA)/PEI membrane had improved the CO_2/CH_4 selectivity value of 70 compared with that of pristine PEI [41]. In another modification, zeolite SAPO-34 incorporated with PEI membrane presented an ideally high CO_2/CH_4 selectivity of 60 [42]. Recently, the modification of PEI membrane was done via the blending technique with carboxylated carbon nanosheets [43]. Interestingly, the carbon nanosheets are produced from biomass-derived of monkeypod tree leaves. Based on the final analysis, the gas separation had a selectivity value of 42.73 for CO_2/CH_4, compared with bare PEI membrane at 23.30 [44].

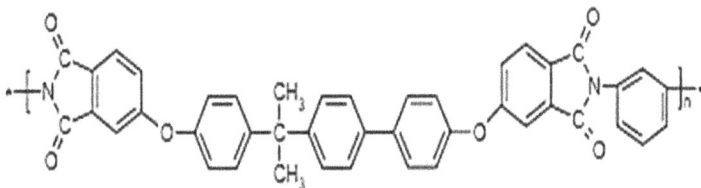

Figure 4. The chemical structure of PEI. Reproduced with permission from [40].

2.4 Polyethylene glycol

Polyethylene glycol (PEG) acts as a stabilizer of the liquid polymer in the polymer phase. The previous studies proved that an ethylene oxide unit in PEG as plasticizer is beneficial, as it can increase CO_2 permeability and therefore improve the selectivity of CO_2/nonpolar gases. The addition of nanoparticles into PEG-based membrane was able to increase the adsorption capacity of CO_2. As reported by Azizi et al., the addition of TiO_2 nanoparticles into PEBAX-1074/PEG-400 had a superior selectivity of CO_2/CH_4 value at 23.61 [45]. Noroozi et al. investigated amino- TiO_2 incorporated within blended PEBAX-PEG membrane, which then resulted in CO_2/CH_4 selectivity enhancement by 16.6 [46]. Gharibi et al. prepared a hybrid organic-inorganic poly(urethane-siloxane) membrane with built-in PEG for CO_2/CH_4 separation performance [47]. Two different contents of PEG (55 wt%, 71 wt%) were used, and the work reported that PEG of 71 wt% showed a better selectivity of CO_2/CH_4 at 20.8, compared with that shown by PEG of 55 wt%.

2.5 Classical membranes and their disadvantages

In comparison with the typical process, membrane gas separation uncovers particular benefits, as well as high proficiency, inexpensiveness, and ecological effect [48]. Despite that, the principle issues of current polymeric membranes are the confinement of permselectivity (possibilities to process low gaseous volume at under 30 million of standard cubic feet daily), with poor thermal-chemical balance, and surrender to plasticization effects. Thus, every small advance in the manufacture of gas separation membranes incorporates a significant amount of money savings, which makes the membrane a prominent presence in the industrial gas separation [4]. Presently, there is a rise in the mass uptake (adsorption) of CO_2, H_2O, H_2S, and higher hydrocarbons referred to as condensable penetrants. Swelling within the conventional polymer membrane will thus expand the polymer free volume (dilation) with its segmental versatility. A phenomenon subjected as "penetrant-induced plasticization" or "plasticization" shall either enhance or more commonly diminishes the gas diffusivity selectivity [49]. Plasticization frequently leads in higher CO_2 gas flux but lower CO_2/CH_4 and CO_2/N_2 gas selectivity of conventional membrane materials polysulfides, especially at high pressure. For plasticized and glassy polymeric membranes, typically, the gas permeability is reduced at a higher feed pressure up until plasticization takes place. Subsequently, it will begin to expand with increasing pressure. Induced plasticization process in glassy polymers also known as "plasticization pressure", is carried out between 10 to 35 bars for polymers related to gas separation with regard to CO_2. These measurements are different from the equilibrium value for low upstream pressure, which include increments in time. By now, dependent relaxation process of the polymer stated to contain the basics of non-equilibrium glassy condition of a polymer. It is also expressed that the penetrated molecules swell and consequently, slackens the matrices of polymer, which brought about the suppression of glass transition temperature, T_g [50, 51].

As shown in Fig. 5, it is seen that it portrays the external plasticization and internal plasticization, where (a) shows plasticizers that are not bound to any polymers, and (b) shows plasticizers that are bound to polymers. Thermal treating has been aforementioned [52], hence, the remaining two will be considered thereafter. Apart from reducing the plasticization effects, frequent blending enhances the properties of polymeric membranes in terms of their mechanical and thermal strength, as well as their separation performance [53]. Moreover, it encourages the transportation of a particular gas, which can be accomplished by altering the blend composition and observing the resulting morphologies of the membrane [54, 55]. In several cases, a blend that is able to separate the different phase material is needed for the sake of species transport [56]. Miscible/homogeneous type of blends is better suited for expanding the gas selectivity while drastically improving the permeability performance. Free volume polymer extension takes place amid the mixing procedure of immiscible blends, causing higher gas permeation rates. However, a significant negative volume change in the mixing of homogeneous miscible polymer blend had led to a notable reduction in gas permeation rates and therefore, an expansion in gas selectivity [57, 58].

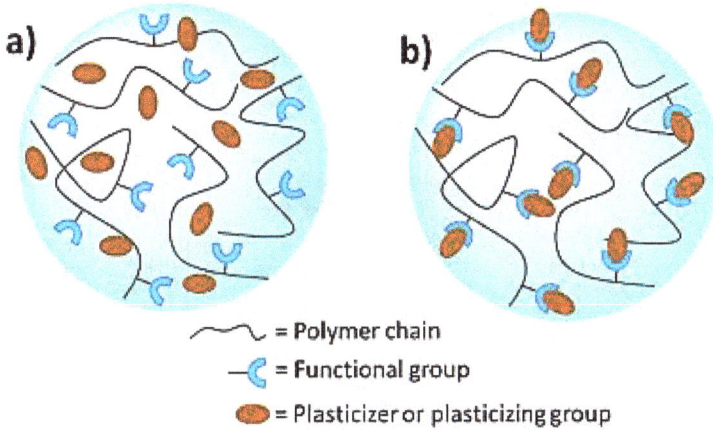

Figure 5. External plasticization and internal plasticization where (a) shows plasticizers that are not bound to any polymers and (b) is plasticizers that are bound to polymer. Reproduced with permission from [52].

2.6   Polymeric membrane for gas separation: mechanism and efficiency

The separation of CO_2/CH_4 across polymeric membrane can occur through several transport mechanisms. Fig. 6 exhibits five main types of mechanisms of gas transport through the membrane, namely: capillary condensation, Knudsen diffusion, molecular sieving, solution-diffusion mechanism and surface diffusion [59]. These mechanisms are applicable for various types of gases that are highly dependent on the characteristics of the membrane, especially on the size of membrane pores. Occasionally, mixing of cross-linking and blending is opted to accomplish a superior quality. For instance, the CO_2-selective membrane of PVA blended with PVP incorporated with appropriate amine carriers has been synthesized by Mondal and Mandal [60]. Formaldehyde was applied as the cross-linking agent, where it resulted in high CO_2/N_2 selectivity of 370, with CO_2 permeability of 1396 Barrer at a temperature of 100°C and 2.8 amp. The latest methodology was established as hybrid ternary blends, whereby it implies the suitability of the blend constituents by the add-on of specific additives.

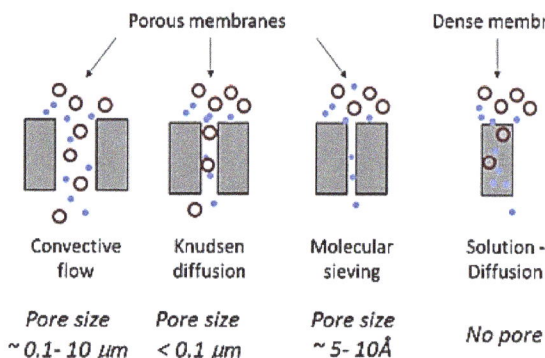

Figure 6. Gas separation mechanism through polymeric membrane. Reproduced with permission from [59].

Apart from having the ability to improve the mechanical properties, another aspect of the specific fillers is that is able to induce a great decline in plasticization of the blends [61]. Cross-linking of membrane material at high temperatures (~15°C above T_g) for polyimide-based membranes was demonstrated by Qiu et al. (2011) [51]. The composite is capable of balancing out the membranes in regard to the swelling and plasticization effects in aggressive feed streams. It should be noted that at a raised temperature, it could lead to the breakdown inside the matrix membrane, which includes the substructure and transition layers in membranes. The sub-T_g cross-linking outcome demonstrated that the cross-linked membranes selectivity could be kept even at tough CO_2 working condition, which is impractical without cross-linking. Additionally, the plasticization pressure denotes the maximized feed pressure beyond which glass polymer-typed membrane can no longer operate profitably [62]. Mixing of cross-linking (GC) and grafting was further revealed as an enhanced procedure to inhibit plasticization. The altered membranes conserve a precisely similar selectivity to that of CO_2/CH_4, which is in contrast with the unaltered form. Nonetheless, the pure CO_2 permeability of 139 Barrer was acquired in an excess level of magnitude, as compared with the neat CA polymer.

The CO_2 separation research industry has its focus on evolving the separation techniques of polymeric membranes. Since membranes are competing with other separation industries, efficiency will be dependent on safety, environment, economic, and technologies prospects [63, 64]. Hence, enhancing polymeric membranes by increasing the permselectivity can shorten the high expense of membrane innovation. Thus, one needs to concentrate on the present challenges by presenting new polymeric membranes that can carry higher CO_2 permeability, combined with adequate selectivity and has great physical and chemical strength. By far, in manufacturing gas separation membranes, various polymers have been utilized, especially poly(pyrrolone), polyacetylenes, poly(ethylene oxide), polysulfones, polyaniline, polyacrylates, poly(arylene ether), polyimides, polycarbonates, polyetherimides, and poly(phenylene oxide) [65]. The utilization of these materials has drawn much research attention. Several primary issues in the area are thermal and pressure of the system, plasticization, and traces of contaminants, permeation hysteresis, casting solvent, and physical and chemical aging. Current studies are committed to clarifying concerning matters influencing the CO_2 separation properties of established polymeric membranes. Apart from that, further classes of CO_2 separation performance polymeric membrane are outlined. Polymers having characteristics such as microporosity (PIM) [66], thermally-rearranged polymers [67], PI [68], and polyurethanes are a few examples of membranes that display higher efficacy for CO_2 separation [69]. Consequently, the industry is convinced that the above-mentioned elite membranes will be able to isolate the different CO_2-containing gas mixtures. In the future, CO_2 permeable membranes shall concentrate on upgrading these economically effective polymers by increasing the awareness on industrial applications.

Nevertheless, the advancement of CO_2 permeable materials is beginning to become a concern towards conventional polymeric membranes. Even though cellulose acetate membranes are widely utilized, more recent membranes, for example, PI membranes, have been manufactured and marketed by the Air Liquide Medal Company. One of the causes is the presence of flux within cellulose acetate membranes, which diminishes considerably with time due to plasticization (response among CO_2 and polymeric chain) and compaction of material (poor mechanical quality) [70, 71]. In reality, the existence of differential pressure inside the membrane will prompt the breakdown of the pores, reducing the bare regions to flow, which later influence the performance of membrane. PI membranes are less responsive but have higher resistance; the mentioned membranes are gradually substituting cellulose acetate. It is only essential to show proof of product contribution to the industry involved when introducing new technologies to the market [53]. To verify an enduring and productive future of an alternative system, one must ensure it can be viewed as a sustainable technology. On the subject related to sustainable development, knowledge on the general definition of concept is vital. In spite of obvious words such as "green" or "eco-friendly", which may have been utilized in the early stage of developing a sustainable technology, the present definition has been used for the ecological sphere enclosing economic and social concepts. Nowadays, innovation is not determined only based on its effectiveness, cost, and other performance-oriented problems, rather, on its environmental effect and future social impact, which are deemed to be likely very significant factors [71]. Those three spheres (economic, ecological, and social impacts) are related, can be observed with a substantially intricate investigation on the consequences of a new project. In the modern world, it is without a doubt that there may be occasions that prevail one sphere over the other, for instance, over the economic sphere; nonetheless, the remainder elements still bear a considerable weight in a decision-making process.

It is conceivable to distinguish research patterns oriented to every three parameters concerning relatively latest membranes innovation. One would be regarded as oblivious to expect that the oil industry would put its attention to the system while considering its benefit involvement with environmental and social ideas [72]. However, there are some slight advantages that can be observed; unmistakably, the industry target is aimed towards profit increment and the making of significant worth. The membranes innovation employed to CO_2 productive oil production wells will allow an effective gas separation into two essential streams: methane-rich stream and carbon dioxide-rich stream. The later stream is utilized to improve the oil recovery method, which will extend the life cycle of an oil reservoir through improved oil flow, flow properties, and/or its water-rock interaction. A major advancement in gas separation via polymeric materials for membrane was observed over the last decade [73, 74]. About several polymeric materials have been designed because of their exceptional transport properties, which makes them capable of being an energy-effective course for large-scale gas separation. These deliberate research endeavors are catalyzed by the critical need to lower CO_2 emission from the burning of fossil fuels and flue gas combustion by various plants (i.e., coal-fired power plant flue gas). It is considered a viable technological solution to alleviate the impact of climate change due to carbon dioxide emission to both economy and environment [8, 75]. The membrane, in contrast with another CO_2 separation method, is favorable due to its simplicity to operate, compactness, low energy consumption, and ability to prevent limitation of thermodynamic solubility [76]. These highlighted advantages are derived from the use the membrane as a thin, intermediate layer that acts as a selective barrier to separate the two different phases [77]. This slimness commonly scales to around 100 nanometers to some micrometers, giving a practically universal stage to actualize sophisticatedly engineered macromolecular structures. To maintain the pressure difference across membrane, feed stream compressing as in Fig. 7 can be done or alternatively, installing the permeate side with vacuum pump. Spiral-wound composite membranes have reported the recovery of 90% to 99.99%.

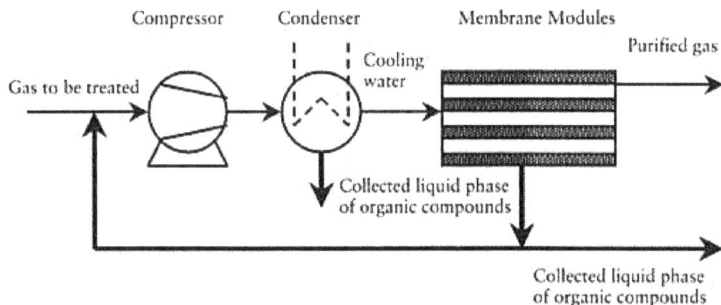

Figure 7. The pressure difference across the membrane can be maintained either by compressing the feed stream [78].

2.6.1 Surface functionalization

Improvement of polymeric materials with increasing permeability and CO_2 selectivity has been widely sought in the past decade. However, an important obstacle is the permeability or selectivity exchange, where high-permeability polymers generally have low selectivity and vice versa [79].

Various polymeric materials were employed for the synthesis of membranes in gas separation application, such as poly(pyrrolone), polyacetylenes, poly(ethylene oxide), polysulfones, polyaniline, polyacrylates, poly(arylene ether), polyimides, polycarbonates, polyetherimides, and poly(phenylene oxide). Several difficulties concerning the utilization of these materials have gained the interest of researchers. Some of the issues faced in the field are the thermal and pressure of the system, plasticization, traces of contaminants, permeation hysteresis, casting solvent, and physical and chemical aging [12, 16]. An abundant free volume in glassy polymers matrix is subject to change due to the current state of production, thermal history, and exposure to sorbing materials. The so-called "physical aging" that drives the nature of membrane transport depends on time mainly articulated in thin films [80]. Huang and Paul suggested a technique for monitoring physical aging of thin glassy polymer films associated with gas separation [81]. Due to physical aging the packing density will elevate and the permeability will drop drastically over the time [82].

Lately, for thickness of selective membrane that is lower than 1 micron, the job of the membrane thickness on physical aging was likewise perceived. Membranes used in gas separation frequently has a thickness below 0.1 mm, in which the effect of thickness on aging has a significant interest in the field of study mentioned [83]. The physical and chemical aging rely upon time and temperature, as chemical aging causes changes within the chemistry of the polymeric membrane, which typically prompts membrane degradation with concurrently weakening of the properties (for example, coloration and embrittlement). Physical aging adjusts the local packing of the chains, yet results in no adjustments in the polymer structure. Hence, only the measurements and physical properties of the polymer, such as T_g, tensile strength, and brittleness, were modified. In this manner, the fundamental contrast among both impacts is that physical aging is a reversible procedure, but chemical aging is not [50, 84, 85]. It compromises the two "anti-aging" methods, including:

❖ Solidifying the free volume state setup employing inorganic or organic non-porous or porous fillers in enhancing free-transport pathways; and

❖ Utilizing cross-linking, copolymerization, and surface plasma treatment to strengthen the setup of first polymer structure.

The added material cannot adequately compensate for FFV to minimize aging, although cross-linking can effectively stop aging; however, drastic decrease in gas permeance was observed [86]. The incompetency to prevent aging in available glassy polymers causes interest to new rigid and porous materials, known as super glassy polymers with intrinsic microporosity (PIMs). The PIMs were developed by rigid segments, which are particularly internally flexible [86]. Another concept presented from mixing both of the techniques mentioned in the above by solving the issue of polymer aging through the particular supplement of microporous microparticle, porous aromatic framework towards super glassy polymers.

Solvent casting technique may influence the crystallinity or amorphous region of the polymeric membrane along with subsequently modifies the permeability. Furthermore, the distinctions in the molecular sizes of the solvents are able to present several free volumes in the membrane manufacture process, namely, distinct solubility coefficient esteems [87, 88]. Casting solvent has an important part in the morphology of polymeric blend membranes. Demonstration of fully dense ethylene vinyl acetic acid derivation (EVA)/liquid poly(ethylene glycol) (PEG) blend membrane changes via evaporation process of solvent was done by Zamiri et al. [89]. However, a few voids had showed up in the membrane frameworks. In their study, the solvent and non-solvent for EVA were chloroform and PEG, respectively. The mixing of chloroform and PEG had dissolved the EVA, and high EVA and non-solvent PEG composition were obtained in the arrangement by continuous evaporation of chloroform upon the casted solution. Hence, the polymer precipitated together while displaying cavity. They additionally showed examples with

20 wt% of PEG; the cavity where it amplified and interconnected, hence, producing a noteworthy drop in the (CO_2/N_2) gas selectivity.

The impact of impurities at high CO_2 permeability membranes is suggestive of a sensitive problem. It usually occurs in the field condition that impacts the separation of various gas separation membranes, which will incorporate the gas flow with compressor oils from pump and presence of the water vapor in the inlet stream [90]. The event is viewed as "penetrant-induced plasticization" or in short, "plasticization" [50,51]. Often, plasticization results in increment of carbon dioxide flux but lowers the selectivity of the mixed gas (e.g., CO_2/CH_4 and CO_2/N_2) of conventional membrane materials (e.g., polysulfones, CA, polyimides), especially at high pressure. Commonly, for plasticized, glassy polymeric membranes, gas permeability diminishes with expanding feed pressure until plasticization occurs. Next, it will begin to rise with respect to pressure. For example, in CO_2-induced plasticization in glassy polymers, regular plasticization pressures are between 10 to 35 bar for suitable polymers in gas separation applications [91]. These qualities are not equivalent to the equilibrium value for low upstream pressure with increments in time. Time-based relaxation procedure of the polymer asserted that it sources from among the non-equilibrium glassy conditions of the polymer. This demonstrates the swelling of the sorbed penetrant, thus easing the polymer matrix composite and thereby causing the suppression of glass transition temperature, T_g [92].

Three elective techniques can stifle plasticization of glassy polymers, which are temperature treatment, cross-linking of polymers, and combining with less plasticizable polymer [22, 53]. Since recent discussion on thermal treating, the second of the two will be considered thereinafter. Blending usually enhances the overall properties of the polymeric membranes while decreasing the plasticization effects [93]. Transportation of a particular gas can be facilitated by tuning the blend composition which result in morphologies that are shaped [56]. On numerous occasions, a phase-separated blend is necessary for greater transport of species [94]. To increase selectivity, homogenous blends are favorable while for enhancement of permeability, appropriate type of blends is the immiscible blends. Amid the mixing procedure of immiscible blends, polymer free volume expansion caused expanded gas permeation rates. However, miscible polymer blend demonstrates a noteworthy negative volume changes in mixing. The outcome reveals a huge reduction in the rate of permeation for common gas and therefore, an expansion in gas selectivity [95, 96].

Sometimes, a union of cross-linking and blending is seen as a good achievement. For instance, the CO_2-selective cross-linked thin-film composite poly(vinyl liquor) (PVA)/polyvinylpyrrolidone (PVP) blend membranes doped with appropriate amine carriers were combined by Mondal and Mandal [60]. Another methodology was created as hybrid ternary blends, which implies a compatibilization of the blend constituents through the extra targeted fillers. The aspect of appropriate fillers, besides improving the mechanical properties, is inducing an impressive decline in blends' plasticization [97]. Qiu et al. [51] demonstrated that cross-linking at raised temperatures (~15°C above T_g) for polyimides are able to balance out membrane in swelling and plasticization. Additionally, the plasticization pressure showed an important shift to higher pressures [98]. Mixing of GC and grafting was further revealed as one of the approaches to inhibit plasticization. Among these techniques, Achoundong et al. [99] was able to alter the CA membranes through the grafting of VTMS into hydroxyl groups to produce a cross-linked structure of the polymer.

Anti-plasticization is another phenomenon that disrupts the separation performance of polymeric membranes. The contrary conduct of plasticization impacts the polymeric membranes at low concentrations of a particular penetrant. Rather than expanding the segmental movements of polymer chains as accomplished for the plasticization, the penetrant molecule shows an induced drag in chain fluctuation, thus diminishing the diffusion of all different permeation in the membrane [100]. For example, either a decrease of segmental portability or an expanded stiffness can diminish the gas permeability, while the related selectivity was changed alternately [49]. Anti-plasticization has impacts on the TCP, PNA, and DDS of PSF and PPO membranes [101]. These extra diluents or plasticizers to the polymers have a low molecular weight that was joined by critical decreases in absorption level and the permeabilities for gases such as He, CO_2, and CH_4. Upon low anti-plasticizer concentration, the selectivity expanded through surpassing from an established concentration, causing a decline in the selectivity because of the substitution of anti-plasticization against the anti-plasticization impact. Another major complication in commercial glass-type polymers is the hysteresis that appears in sorption, permeation, and dilation isotherms of CO_2. Amid feed pressurising, these properties are primarily estimated to be lower compared with those deliberated while feed depressurising. Among the features in the hysteresis, it can be considered because of a late polymer chain rearrangement in CO_2 pressure growth [102 – 104].

2.6.2 Hybrid and composite membranes

Recently, the technology of membrane gas separation technology is rapidly developing due to its diverse benefits compared with the conventional methods (primarily absorption by amines). Particularly, membrane gas separation demonstrates flexibility in operation, low operating costs, easy scale-up, and high quality of product and compact design. The most significant implementation of membrane gas separation is CO_2 capture and recovery from the gas streams. Atmospheric concentration of CO_2 is raising significantly over the last decades. The average CO_2 concentration was identified at 399 ppm in 2015, which is around 40%, higher compared to the years of mid-1800s. Intensified carbon dioxide release was detected in the previous era with 2 ppm/year of average growth [105]. This strong rising trend is caused by the increasing demand for energy production from fossil fuels. One-third of the global energy utilization, as well as 40% of carbon dioxide emission come from the industrial sector [106]. Significant greenhouse gases impact with carbon dioxide emission as the primary perpetrator on global warming, as well as climate change, forces the capture and recovery of carbon dioxide to be a worldwide interest [107-109].

There are several types of membrane materials, namely, inorganic substance (metals, ceramics, carbons, zeolites), organic substances (glassy, polymer blends and rubbery), porous hybrid materials (zeolitic imidazolate frameworks (ZIFs), and metal-organic framework (MOFs). The MMMs were discovered by Chung et al. [110], in which they reported on an additional path for solution-diffusion transport. According to Powell and Qiao [111], nanoparticles assimilation comes with three primary impacts on the permeability of membrane, for example, the ability to modify permeability by acting as molecular sieves, the ability to increase permeability by disturbing the structure of the membrane matrix, and the ability to lower the permeability by acting as a barrier. Other relations are also suggested for the purpose of describing the properties of gas transport in MMMs, for example, the Higuchi model [112] and the three-phase Maxwell model [113]. The mentioned models are naturally complex without providing significant improvement in their results compared with the two-phase Maxwell equation. On a small scale, restrictions of selectivity and permeability can be overcome via the use of inorganic membranes; however, it requires high costs for the production of inorganic membranes (some magnitude higher than the cost for polymeric membranes).

In addition, inorganic membranes production in a large scale requires enhancements in its reproducibility. The principle of MMMs merges the selectivity and size/shape of nanoporous materials with mechanical stability and processability of polymers [114]. The advantages of the combination of polymeric and inorganic components are that MMMs can be taken into account as an evolutionary method to elevate both the selectivity and the permeability. Gas permeance via MMM is determined by both polymer and inorganic filler properties of intrinsic molecular sieve [115]. Benefit of MMMs can be found in its greater properties of gas separation contributed by the molecular inorganic filler that provides the ability to sieve molecules. Inorganic particles utilized in MMMs are divided into two which are porous and non-porous; for example, CMS, metal oxides, non-porous silica and porous zeolites. Zeolites are well known to be recognized as the leading edge molecular sieve family utilized in MMMs [116]. Zeolites are crystal-like particles with three-dimensional (3D) and uniform micropore structures, and interconnected cages and tunnels in order for the gas molecules to be sieved according to the desired shapes and sizes [117]. Zeolite addition in the discerning skin layer improves the selectivity of CO_2/CH_4 to about half, compared with the neat hollow fibers made up of PSf/Matrimid® [118].

Incorporation of CMSs were also done [114], whereby its construction were carried out with highly porous ingredients to allow for high permeability of gas; however, its molecular structure of sieving pore offers gas penetrants discrimination to give high selectivity [73, 119]. Studies by Barsema et al. [120] introduced the incorporation of CMS fiber into two different polymer, namely Ultem®1000 and Matrimid®5218. For membrane made up of Matrimid® 5218-CMS, CO_2/CH_4 selectivity increases by 45% in accordance with the pristine Matrimid® 5218 membrane. Meanwhile, Ultem®1000-CMS observed 40% increment for selectivity of CO_2/CH_4, which was in accordance to Ultem® 1000 membrane. Moreover, MMM with carbon nanotubes (CNTs) has also become an attractive research field.

Another significant inorganic particle type is if the metal oxide nanoparticles have a great affinity to CO_2 [121]. Hosseini et al. [122] incorporated nanosized MgO with Matrimid® polymer prior to ten days of treatment with silver ion in nanocomposite membrane preparation. The best separation performance of CO_2/CH_4 increased to 50%, which is in accordance to the neat membrane of Matrimid®. Separation of CO_2/CH_4 in nanocomposite membrane are through solution-diffusion mechanism combination in the polymer matrices with size exemption by silver covered MgO, as well as assisted the transportation through MgO/Ag and CO_2 interaction [123]. The MOFs are anticipated to favor the organic phase over the inorganic fillers, as organic linkers in MOFs have better affinity with polymer chains. The ZIF-90 possesses sodalite cage-like structures with size of pore of 0.35 nm. In addition, carboxyl groups in particle ZIF-90 have a good CO_2 affinity via a non-covalent chemical interaction. Permeability of CO_2 increased from 390 to 720 Barrer when polyimide 6FDA-DAM membrane was incorporated with ZIF-90. Selectivity of CO_2/CH_4 increased from 27 to 37, which indicates that the MMM is defect-free [117]. Meanwhile, the ZIF-8 has a facile synthesis, as well as good thermal and chemical stability, making it an attractive member of MOFs. Both had improved the CO_2/CH_4 selectivity as well as the CO_2 permeability, as perceived for the membranes of ZIF-8/Matrimid® [124].

Inorganic particles that are non-porous, for instance, silica, are used in membranes of nanocomposite structures [53, 86]. Previous works stated that the additional void located at the interface between silica agglomerates and the polymers has the probability to induce permeability enhancement [114, 125, 126]. Sadeghi et al. [127] studied CH_4 and CO_2 properties of transport through composite membrane made of silica and polybenzimidazole (PBI). They identified the solubility enhancement of the condensable gases as the hydrated silica particles was incorporated due to an increment in the amount of hydroxyl groups in the polymer matrix. Ruling mechanism of gas permeance transformed from the mechanism of diffusion in the neat PBI membrane into mechanism of solution in the hybrid membrane. Permeability for non-condensable gas was also found to be decreased. This is due to gas diffusivity reduction in the polymer matrix affected by limited gas molecules motion in the company of silica particles. One primary challenge in MMMs development is the ability for the inorganic nanoparticles to disperse across the polymer matrices.

Generally, nanoparticles scatter below par in the polymer matrix, making them to be more likely to aggregate in the MMM formed which leads to the establishment of various stress under the act of forces from outside, hence weakening the stability performance of the materials, particularly at high concentrations of inorganic filler [128]. Zeolites addition to glassy polymer usually causes defects formation at the polymer-zeolite interface due to poor polymer matrix and zeolites compatibility. This issue has the ability to seriously deteriorate the performance of membrane separation. As an example, MMM constructed from Matrimid®-zeolite exhibited higher permeability and no selectivity enhancement in accordance to the neat membrane of Matrimid® caused by voids formation which is delaminating away. In the application for sweetening of natural gas, inlet gas is normally gained from the gas well at 20-70 bar of typical pressure and less than 8% of CO_2 content. The content of CO_2 needs to be decreased to around 2% in order to condone to pipeline requirements and specifications for transportation.

Although gas permeation membranes usage in improved recovery of oil process at elevated partial pressure of CO_2 is considered to be successful, the usage of dense membranes and certain membranes of porous inorganic structure, for example, CMS at low partial pressure of CO_2 is deemed as unsuccessful. This is mainly caused by significant hydrocarbon loss within the CO_2-rich permeate, as permeability are in contrast with selectivity. Hydrocarbon loss causes the process to be economically unattractive. However, combination of a hollow fiber gas liquid membrane contactor and acid gas absorption as a substitute to dense membranes obtains higher selectivity of separation and is deemed as more attractive in the applications of natural gas sweetening; particularly to minimize the loss of hydrocarbon in situations with low concentration of CO_2. Various studies were conducted through the utilization of this process of membrane and chemical absorption in order to separate the CO_2. However, most works were done under low membrane pressure (≤ 2 bar), primarily to highlight the separation of CO_2 from gaseous blends in the applications of flue gas without the involvement of pressurized gas stream [129]. This is because the membranes of microporous polymeric structures are naturally delicate, which limits their utilization in harsh pressure condition. Thus, in the case of natural gas sweetening involving operation under high feed pressure, numerous research on membrane contactor module construction, operation condition and materials are needed to review the membrane gas-liquid contactors potential for high pressure usage.

Nowadays, available works on high pressure membrane contactors are not enough with only two reported works can be found, which is by Dindore et al. [130] over a decade ago and by Marzouk et al. [131] in 2010. Both researches described the gas side pressure as having extended impact on carbon dioxide flux for physical absorption compared with chemical absorption, making physical absorption more suitable for high pressure usage than its use in low pressure usage. At high pressure and temperature, inorganic membranes are favored due to its exceptional stability in terms of chemical, thermal, and mechanical aspects, high pressure stability, and good erosion resistance over the conventional polymeric membranes. In addition, inorganic membranes also possess regeneration potential after fouling primarily caused by its relatively robust chemical and mechanical structures. In general, pore plugging is a normal concern for all membrane types. However, plugging of pore is not seen as a primary issue in small scale experiment, where cleaned gas is utilized. However, gas feed usually contains particles in industrial operations, which in numerous cases require the pretreatment of feed gas.

3.   Research gaps

Commercialization of polymeric membranes for gas separation began years ago. These systems have gained increased research interest due to its simplicity and low-energy intake. Gas separation is required in numerous processes in the refinery, for example, in the capture of carbon dioxide, treatment of natural gas, separation of hydrocarbons and purification of hydrogen. Membranes in these processes have potential to be substituted with the existing techniques of pressure swing adsorption, amine scrubbing, as well as cryogenic distillation. This chapter discusses the implementations of polymer membranes in the refinery through commercialized units and current materials reviews. Economical assessment of the membranes involved in contrast to the traditional processes is also included. Average permeance calculation offers reasonable results as only one of the gases' permeance is altered. If alteration of permeance for each gas was done during the separation process, proper extraction of the required membrane area cannot be attained. Thus, the effects of permeance parameters need to be considered during the minimization of overall and operation costs, and energy usage, if the behavior of variable permeance was perceived in the module of the membrane.

It is assumed that the separation process of membrane will be regularly utilized as the future practical biogas enhancement technology. High purity and high CH_4 gas recovery can be attained using a two-step membrane process. Permeance and CO_2/CH_4 selectivity enhancement could further decrease the capital cost of gas separation by membrane. Certain conundrums are required to be investigated or resolved in the future, which are:

> Systematic investigation on the effect of membrane materials and operation conditions on the operating cost with predetermined purity and recovery of products that are still lacking. Thus, enhancement of systematic investigation for biogas involving optimal operating cost and fixed separation target are immediately required;

> More upgrades in membrane permeance or selectivity might not result to extra advantages for biogas enhancement. Systematical investigations on the effect of membrane permeance, membrane selectivity, operating pressure towards energy consumption, operational cost. and capital cost are essential in understanding the required permeance and selectivity of a membrane with optimal cost of operation;

> Detailed investigation should be conducted on two-step membrane process, as it is the optimum process that can be done to achieve CH_4 concentration above 97% and above 98% for CH_4 recovery, both of which are attractive parameters for the production of biomethane;

> If any significant changes of the gases' permeance during membrane separation process are identified, calculation of membrane permeance in membrane process needs to be attuned as per the precise operating condition.

4.   Future perspectives

The technology of membrane separation has the most potential in capturing CO_2 from flue gas according to the advantageous of membrane processes, such as operation simplicity, small footprint, modular construction, and zero emission of hazardous by-product. Nevertheless, utilization of membrane process to CO_2 capture from flue gas is still under further discussion, as the separation aims suggested by the United States Department of Energy for product with recovery of more than 90% and purity of 95% and higher are very hard to obtain through low cost of capture and energy consumption. Low concentration of CO_2 and low flue gas pressure are accountable for the struggle in separation process primarily due to high consumption of energy and the cost of capture. Numerous researchers have optimized the membrane-based processes to separate the carbon dioxide from flue gas mixture with the intention to search for optimum operating parameters and appropriate materials with the lowest cost and energy consumption [132-135]. CO_2 concentration in the flue gas of power plants is low (approximately 13% to

15% for the usual coal fired power plant) brought the most vital challenges for the capture of CO_2 process. Castel et al. [136] reported 10% of feed gas CO_2 concentration requires the amount of energy intake for membrane system that is too high to be acknowledged, although having selectivity of higher than 120. As concentration of CO_2 in inlet feed surpassed 20%, the energy intake for the membrane system is reduced making the membrane process to be more economical than the old absorption process. As for the usual case of 60 selectivity with 30 mbar downstream pressure, membrane system with single stage is able to fullfil 0.7 MJ/kg energy requirement for 20% of CO_2 feed gas. Thus, decreasing concentration of CO_2 in flue gas is the primary blockage in realizing the low capture cost of 20 \$/ton CO_2. Systematic analysis of techno-economic was conducted in order to establish an ideal membrane material offering optimal operating and capital costs for the membrane system.

Membrane permeance and selectivity were identified according to the Robeson upper limit for separation of CO_2/CH_4 via membrane. Usually, attractive area is found in the vicinity of the upper bound or surpassing the Robeson upper limit. As for the two-stage membrane process, minimum cost of operation for fixed CH_4 purity of 96%, as well as fixed CH_4 recovery of 85% was detected. Increment in the selectivity of CO_2/CH_4 initially reduces the operating cost followed by increasing operating cost during the latter half. Cost of 0.037 \$/Nm^3 for separation process is calculated in the event that the optimal membrane with selectivity range of 10 to 25 and operating pressure ratio of 10 under high permeance of CO_2 of above 2000 GPU. As the optimal membrane is confirmed, the cost of biogas upgrades is controlled by the compressor, comprising the utility expenses and capital depreciation, which totaled to roughly 85% of the total cost. Furthermore, more reduction in the membrane skid cost reduces the operating cost and increases the CO_2/CH_4 selectivity with minimized cost of operation. Lastly, single step gas separation membrane has higher operating cost than two-stage membrane with optimal selectivity.

It is ideal for membranes to possess high CO_2 permeability with sufficient CO_2/CH_4 selectivity (10–25) [137]. Polyethersulfone (PESU) membrane with 35.5 selectivity of CO_2/CH_4 is sufficient for biogas enhancement of the membrane process with two-step. The effect of CO_2/CH_4 selectivity as well as the area of membrane towards separation performance was investigated by Makaruk et al. [138]. Investigation on biogas enhancement with the flow rate of feed gas of 1000 Nm^3/h, methane concentration of feed gas of 60%, as well as product gas of 98% methane was conducted. Energy usage in biogas upgrades to natural gas replacement was about 0.3 kWh/Nm^3 which is equal to 1.08 MJ/Nm^3. Results produced indicates increment in membrane area which contributes to relatively higher recovery of product. Nevertheless, if a very high recovery of nearly

100% were to be attained, it is anticipated that more areas of the membrane will be utilized. As lower selectivity of CO_2/CH_4 membrane is employed, higher consumption of compression power can be observed to obtain similar recovery of methane. Utilization of CO_2/CH_4 with selectivity greater than 50 could only slightly reduce the energy consumed. In this reference, the process of membrane gas separation has the ability to offer sufficient flexibility to maintain high recovery of product while the biogas process is upgraded.

Membrane system application is rising in the industry due to unique features provided by the membrane. Membrane separation process has simple installation procedure with minimum supervision required compared to other separation processes [139]. Moreover, it utilizes less space with no moving parts making maintenance as not one of its crucial needs [140]. Additionally, it consumes less energy as well as is deemed as a technology that is environmentally friendly due to the involvement of no gases emission and solvents [18]. Scaling up the membrane for commercialization purposes is also an easy task. According to its material, membranes are divided into polymeric, inorganic, and metallic. Platinum or palladium made metallic membranes have exceptional performance; however, the precious metals cost highly affect the selection of the membranes. Inorganic membranes are recognized as good substitutes, having better chemical stability and lower cost of fabrication. However, elevated temperature between 200°C and 900°C is required in the operation of inorganic membranes [8]. Lately, polymeric membranes are a hit in the industries due to its competitive performance and outstanding economy [107]. Operation of the membranes is done at environment temperature and they possess excellent properties. In 1960s, polymeric membranes revolution began as Loeb and Sourirajan [141], which fabricated a membrane using CA for water desalination through reverse osmosis. The membrane was 0.2 µm thick, supported on a porous substrate with capability of transforming seawater into potable water.

Conclusion

Most of the polymer membranes utilized in the gas separation processes that is membrane-based exhibit low thermal resistance. In various implementations, working at elevated temperatures is essential. Alternatively, high stability membranes could offer new application fields through prolonged membrane lifetime at high temperatures. In fact, in terms of energy and cost saving, it is important for the operation of the membranes to be conducted at the highest permissible temperatures. Presently, most systems of gas separation need an effective heat exchanger to reduce the influence of the membrane temperature to less than 50°C. Such requirement indicates operating cost and extra capital demands during operation with a hot stream. Furthermore, expensive cost related to the novel raw materials and low processing ability of thermally stable polymers restricted their commercialization to the comparatively dedicated gas separation field. Development of polymeric membranes separation performance for CO_2 separation industry is broad research area due to membranes competition with other separation technology in terms of technological, economic, environmental, and safety aspects. Certainly, one the most significant aspect in membrane gas separation processes is proper material selection. Improvement of polymer membranes with higher selectivity and permeance reassures the concerns on high membrane technology cost. Hence, existing efforts should focus on eliminating them through the introduction of new polymer membranes with high CO_2 permeability, combined with good chemical and physical stability, as well as acceptable level of selectivity. Certain stimulating concerns in utilization of these materials for membrane fabrication has gained interests from academician. Primary concerns in this matter are the pressure and thermal conditioning, chemical and physical aging, permeation hysteresis, plasticization, impurities or existence of trace contaminants, and casting solvent. In a nutshell, this chapter compiles all of the reviewed papers that particularly focused on enlightening the challenging issues that influence the CO_2 separation properties of conventional polymeric membranes.

References

[1] N.C. Mat, G.G. Lipscomb, Membrane process optimization for carbon capture, Int. J. Greenhouse Gas Control 62 (2017) 1-12. https://doi.org/10.1016/j.ijggc.2017.04.002

[2] M. Rezakazemi, M. Sadrzadeh, T. Matsuura, Thermally stable polymers for advanced high-performance gas separation membranes, Prog. Energy Combust. Sci. 66 (2018) 1-41. https://doi.org/10.1016/j.pecs.2017.11.002

[3] Y. Han, W.S.W. Ho, Recent advances in polymeric membranes for CO_2 capture, Chin. J. Chem. Eng. 26 (2018) 2238-2254. https://doi.org/10.1016/j.cjche.2018.07.010

[4] J. Liu, X. Hou, H.B. Park, H. Lin, High-Performance Polymers for Membrane CO_2/N_2 Separation, Chem. - Eur. J. 22 (2016) 15980-15990. https://doi.org/10.1002/chem.201603002

[5] L. Yang, Z. Tian, X. Zhang, X. Wu, Y. Wu, Y. Wang, D. Peng, S. Wang, H. Wu, Z. Jiang, Enhanced CO_2 selectivities by incorporating CO_2-philic PEG-POSS into polymers of intrinsic microporosity membrane, J. Membr. Sci. 543 (2017) 69-78. https://doi.org/10.1016/j.memsci.2017.08.050

[6] H. Shabgard, M.J. Allen, N. Sharifi, S.P. Benn, A. Faghri, T.L. Bergman, Heat pipe heat exchangers and heat sinks: Opportunities, challenges, applications, analysis, and

state of the art, Int. J. Heat Mass Transfer 89 (2015) 138-158. https://doi.org/10.1016/j.ijheatmasstransfer.2015.05.020

[7] S. Haider, A. Lindbråthen, J.A. Lie, I.C.T. Andersen, M.-B. Hägg, CO2 separation with carbon membranes in high pressure and elevated temperature applications, Sep. Purif. Technol. 190 (2018) 177-189. https://doi.org/10.1016/j.seppur.2017.08.038

[8] N. Sazali, W.N.W. Salleh, A.F. Ismail, N.H. Ismail, F. Aziz, N. Yusof, H. Hasbullah, Effect of stabilization temperature during pyrolysis process of P84 co-polyimide-based tubular carbon membrane for H2/N2 and He/N2 separations, IOP Conf. Ser.: Mater. Sci. Eng. 342 (2018) 012027. https://doi.org/10.1088/1757-899X/342/1/012027

[9] N.H. Ismail, W.N.W. Salleh, N. Sazali, A.F. Ismail, N. Yusof, F. Aziz, Disk supported carbon membrane via spray coating method: Effect of carbonization temperature and atmosphere, Sep. Purif. Technol. 195 (2018) 295-304. https://doi.org/10.1016/j.seppur.2017.12.032

[10] K. Hunger, N. Schmeling, H.B.T. Jeazet, C. Janiak, C. Staudt, K. Kleinermanns, Investigation of cross-linked and additive containing polymer materials for membranes with improved performance in pervaporation and gas separation, Membranes 2 (2012) 727-763. https://doi.org/10.3390/membranes2040727

[11] W.J. Lau, C.S. Ong, N.A.H.M. Nordin, N.A.A. Sani, N.M. Mokhtar, R.J. Gohari, D. Emadzadeh, A.F. Ismail, Surface Modification of Polymeric Membranes for Various Separation Processes, in: M. Gürsoy, M. Karaman (Eds.), Surf. Treat. Biol., Chem., Phys. Appl., Wiley-VCH, Weinheim, 2017, pp. 115-180. https://doi.org/10.1002/9783527698813.ch4

[12] X.Q. Cheng, Z.X. Wang, X. Jiang, T. Li, C.H. Lau, Z. Guo, J. Ma, L. Shao, Towards sustainable ultrafast molecular-separation membranes: From conventional polymers to emerging materials, Prog. Mater. Sci. 92 (2018) 258-283. https://doi.org/10.1016/j.pmatsci.2017.10.006

[13] S. Hasebe, S. Aoyama, M. Tanaka, H. Kawakami, CO2 separation of polymer membranes containing silica nanoparticles with gas permeable nano-space, J. Membr. Sci. 536 (2017) 148-155. https://doi.org/10.1016/j.memsci.2017.05.005

[14] S. Yuan, F. Shen, C.K. Chua, K. Zhou, Polymeric composites for powder-based additive manufacturing: Materials and applications, Prog. Polym. Sci. 91 (2019) 141-168. https://doi.org/10.1016/j.progpolymsci.2018.11.001

[15] X. Wang, E.N. Kalali, J.-T. Wan, D.-Y. Wang, Carbon-family materials for flame retardant polymeric materials, Prog. Polym. Sci. 69 (2017) 22-46. https://doi.org/10.1016/j.progpolymsci.2017.02.001

[16] B. Notario, J. Pinto, M.A. Rodriguez-Perez, Nanoporous polymeric materials: A new class of materials with enhanced properties, Prog. Mater. Sci. 78-79 (2016) 93-139. https://doi.org/10.1016/j.pmatsci.2016.02.002

[17] O. Heinz, M. Aghajani, A.R. Greenberg, Y. Ding, Surface-patterning of polymeric membranes: fabrication and performance, Curr. Opin. Chem. Eng. 20 (2018) 1-12. https://doi.org/10.1016/j.coche.2018.01.008

[18] C. Castel, L. Wang, J.P. Corriou, E. Favre, Steady vs unsteady membrane gas separation processes, Chem. Eng. Sci. 183 (2018) 136-147. https://doi.org/10.1016/j.ces.2018.03.013

[19] M. Takht Ravanchi, T. Kaghazchi, A. Kargari, Application of membrane separation processes in petrochemical industry: a review, Desalination 235 (2009) 199-244. https://doi.org/10.1016/j.desal.2007.10.042

[20] H. Nakajima, P. Dijkstra, K. Loos, The Recent Developments in Biobased Polymers toward General and Engineering Applications: Polymers that are Upgraded from Biodegradable Polymers, Analogous to Petroleum-Derived Polymers, and Newly Developed, Polymers 9 (2017) 523. https://doi.org/10.3390/polym9100523

[21] A.M. Abdalla, S. Hossain, O.B. Nisfindy, A.T. Azad, M. Dawood, A.K. Azad, Hydrogen production, storage, transportation and key challenges with applications: A review, Energy Convers. Manage. 165 (2018) 602-627. https://doi.org/10.1016/j.enconman.2018.03.088

[22] R. Bounaceur, E. Berger, M. Pfister, A.A. Ramirez Santos, E. Favre, Rigorous variable permeability modelling and process simulation for the design of polymeric membrane gas separation units: MEMSIC simulation tool, J. Membr. Sci. 523 (2017) 77-91. https://doi.org/10.1016/j.memsci.2016.09.011

[23] M.W. Anjum, F. de Clippel, J. Didden, A.L. Khan, S. Couck, G.V. Baron, J.F.M. Denayer, B.F. Sels, I.F.J. Vankelecom, Polyimide mixed matrix membranes for CO2 separations using carbon–silica nanocomposite fillers, J. Membr. Sci. 495 (2015) 121-129. https://doi.org/10.1016/j.memsci.2015.08.006

[24] E.P. Favvas, F.K. Katsaros, S.K. Papageorgiou, A.A. Sapalidis, A.C. Mitropoulos, A review of the latest development of polyimide based membranes for CO2 separations, React. Funct. Polym. 120 (2017) 104-130. https://doi.org/10.1016/j.reactfunctpolym.2017.09.002

[25] M. Lanč, P. Sysel, M. Šoltys, F. Štěpánek, K. Fónod, M. Klepić, O. Vopička, M. Lhotka, P. Ulbrich, K. Friess, Synthesis, preparation and characterization of novel hyperbranched 6FDA-TTM based polyimide membranes for effective CO2

separation: Effect of embedded mesoporous silica particles and siloxane linkages, Polymer 144 (2018) 33-42. https://doi.org/10.1016/j.polymer.2018.04.033

[26] C. Zhang, P. Li, B. Cao, Decarboxylation crosslinking of polyimides with high CO2/CH4 separation performance and plasticization resistance, J. Membr. Sci. 528 (2017) 206-216. https://doi.org/10.1016/j.memsci.2017.01.008

[27] I. Hossain, A.Z. Al Munsur, O. Choi, T.H. Kim, Bisimidazolium PEG-mediated crosslinked 6FDA-durene polyimide membranes for CO2 separation, Sep. Purif. Technol. 224 (2019) 180-188. https://doi.org/10.1016/j.seppur.2019.05.014

[28] C. Atalay-Oral, M. Tatlier, Effects of structural properties of fillers on performances of Matrimid® 5218 mixed matrix membranes, Sep. Purif. Technol. 236 (2020) 116277. https://doi.org/10.1016/j.seppur.2019.116277

[29] R. Castro-Muñoz, V. Fíla, V. Martin-Gil, C. Muller, Enhanced CO2 permeability in Matrimid® 5218 mixed matrix membranes for separating binary CO2/CH4 mixtures, Sep. Purif. Technol. 210 (2019) 553-562. https://doi.org/10.1016/j.seppur.2018.08.046

[30] A.E. Amooghin, M. Omidkhah, A. Kargari, The effects of aminosilane grafting on NaY zeolite–Matrimid®5218 mixed matrix membranes for CO2/CH4 separation, J. Membr. Sci. 490 (2015) 364-379. https://doi.org/10.1016/j.memsci.2015.04.070

[31] A.E. Amooghin, M. Omidkhah, H. Sanaeepur, A. Kargari, Preparation and characterization of Ag+ ion-exchanged zeolite-Matrimid®5218 mixed matrix membrane for CO2/CH4 separation, J. Energy Chem. (2016). https://doi.org/10.1016/j.jechem.2016.02.004

[32] S. Abdollahi, H.R. Mortaheb, A. Ghadimi, M. Esmaeili, Improvement in separation performance of Matrimid®5218 with encapsulated [Emim][Tf2N] in a heterogeneous structure: CO2/CH4 separation, J. Membr. Sci. 557 (2018) 38-48. https://doi.org/10.1016/j.memsci.2018.04.026

[33] H. Rajati, A.H. Navarchian, S. Tangestaninejad, Preparation and characterization of mixed matrix membranes based on Matrimid/PVDF blend and MIL-101(Cr) as filler for CO2/CH4 separation, Chem. Eng. Sci. 185 (2018) 92-104. https://doi.org/10.1016/j.ces.2018.04.006

[34] H. Julian, I.G. Wenten, Polysulfone membranes for CO2/CH4 separation: State of the art, IOSR J. Eng. 2 (2012) 484-495. https://doi.org/10.9790/3021-0203484495

[35] L.G. Tiron, S. Pintilie, M. Vlad, I. Bîrsan, Ş. Baltă, Characterization of Polysulfone Membranes Prepared with Thermally Induced Phase Separation Technique, IOP

Conf. Ser.: Mater. Sci. Eng. 209 (2017) 012013. https://doi.org/10.1088/1757-899X/209/1/012013

[36] A. Hatami, I. Salahshoori, N. Rashidi, D. Nasirian, The effect of ZIF-90 particle in Pebax/PSF composite membrane on the transport properties of CO2, CH4 and N2 gases by molecular dynamics simulation method, Chin. J. Chem. Eng. 28 (2020) 2267-2284. https://doi.org/10.1016/j.cjche.2019.12.011

[37] M.B. Mohamad, Y.Y. Fong, A. Shariff, Gas Separation of Carbon Dioxide from Methane Using Polysulfone Membrane Incorporated with Zeolite-T, Procedia Eng. 148 (2016) 621-629. https://doi.org/10.1016/j.proeng.2016.06.526

[38] M.S. Suleman, K.K. Lau, Y.F. Yeong, Characterization and Performance Evaluation of PDMS/PSF Membrane for CO2/CH4 Separation under the Effect of Swelling, Procedia Eng. 148 (2016) 176-183. https://doi.org/10.1016/j.proeng.2016.06.525

[39] A.D. Kiadehi, A. Rahimpour, M. Jahanshahi, A.A. Ghoreyshi, Novel carbon nano-fibers (CNF)/polysulfone (PSf) mixed matrix membranes for gas separation, J. Ind. Eng. Chem. 22 (2015) 199-207. https://doi.org/10.1016/j.jiec.2014.07.011

[40] A. Rusli, N.S.M. Raffi, H. Ismail, Solubility, Miscibility and Processability of Thermosetting Monomers as Reactive Plasticizers of Polyetherimide, Procedia Chem. 19 (2016) 776-781. https://doi.org/10.1016/j.proche.2016.03.084

[41] Z.P. Madzarevic, S. Shahid, K. Nijmeijer, T.J. Dingemans, The role of ortho-, meta- and para-substitutions in the main-chain structure of poly(etherimide)s and the effects on CO2/CH4 gas separation performance, Sep. Purif. Technol. 210 (2019) 242-250. https://doi.org/10.1016/j.seppur.2018.08.006

[42] S. Belhaj Messaoud, A. Takagaki, T. Sugawara, R. Kikuchi, S.T. Oyama, Mixed matrix membranes using SAPO-34/polyetherimide for carbon dioxide/methane separation, Sep. Purif. Technol. 148 (2015) 38-48. https://doi.org/10.1016/j.seppur.2015.04.017

[43] M.Y. Khan, A. Khan, J.K. Adewole, M. Naim, S.I. Basha, M.A. Aziz, Biomass derived carboxylated carbon nanosheets blended polyetherimide membranes for enhanced CO2/CH4 separation, J. Nat. Gas Sci. Eng. 75 (2020) 103156. https://doi.org/10.1016/j.jngse.2020.103156

[44] S. Saimani, M.M. Dal-Cin, A. Kumar, D.M. Kingston, Separation performance of asymmetric membranes based on PEGDa/PEI semi-interpenetrating polymer network in pure and binary gas mixtures of CO2, N2 and CH4, J. Membr. Sci. 362 (2010) 353-359. https://doi.org/10.1016/j.memsci.2010.06.045

[45] N. Azizi, T. Mohammadi, R.M. Behbahani, Synthesis of a new nanocomposite membrane (PEBAX-1074/PEG-400/TiO2) in order to separate CO2 from CH4, J. Nat. Gas Sci. Eng. 37 (2017) 39-51. https://doi.org/10.1016/j.jngse.2016.11.038

[46] Z. Noroozi, O. Bakhtiari, Preparation of amino functionalized titanium oxide nanotubes and their incorporation within Pebax/PEG blended matrix for CO2/CH4 separation, Chem. Eng. Res. Des. 152 (2019) 149-164. https://doi.org/10.1016/j.cherd.2019.09.030

[47] R. Gharibi, A. Ghadimi, H. Yeganeh, B. Sadatnia, M. Gharedaghi, Preparation and evaluation of hybrid organic-inorganic poly(urethane-siloxane) membranes with build-in poly(ethylene glycol) segments for efficient separation of CO2/CH4 and CO2/H2, J. Membr. Sci. 548 (2018) 572-582. https://doi.org/10.1016/j.memsci.2017.11.058

[48] S. Roy, S. Ragunath, Emerging Membrane Technologies for Water and Energy Sustainability: Future Prospects, Constraints and Challenges, Energies 11 (2018) 2997. https://doi.org/10.3390/en11112997

[49] C. Zhang, L. Fu, Z. Tian, B. Cao, P. Li, Post-crosslinking of triptycene-based Tröger's base polymers with enhanced natural gas separation performance, J. Membr. Sci. 556 (2018) 277-284. https://doi.org/10.1016/j.memsci.2018.04.013

[50] R. Swaidan, B. Ghanem, E. Litwiller, I. Pinnau, Physical Aging, Plasticization and Their Effects on Gas Permeation in "Rigid" Polymers of Intrinsic Microporosity, Macromolecules 48 (2015) 6553-6561. https://doi.org/10.1021/acs.macromol.5b01581

[51] W. Qiu, C.-C. Chen, L. Xu, L. Cui, D.R. Paul, W.J. Koros, Sub-Tg Cross-Linking of a Polyimide Membrane for Enhanced CO2 Plasticization Resistance for Natural Gas Separation, Macromolecules 44 (2011) 6046-6056. https://doi.org/10.1021/ma201033j

[52] M. Klähn, R. Krishnan, J.M. Phang, F.C.H. Lim, A.M. van Herk, S. Jana, Effect of external and internal plasticization on the glass transition temperature of (Meth)acrylate polymers studied with molecular dynamics simulations and calorimetry, Polymer 179 (2019) 121635. https://doi.org/10.1016/j.polymer.2019.121635

[53] S.A. Stern, Y. Mi, H. Yamamoto, A.K.S. Clair, Structure/permeability relationships of polyimide membranes. Applications to the separation of gas mixtures, J. Polym. Sci., Part B: Polym. Phys. 27 (1989) 1887-1909. https://doi.org/10.1002/polb.1989.090270908

[54] Z. Ahmad, N.A. Al-Awadi, F. Al-Sagheer, Thermal degradation studies in poly(vinyl chloride)/poly(methyl methacrylate) blends, Polym. Degrad. Stab. 93 (2008) 456-465. https://doi.org/10.1016/j.polymdegradstab.2007.11.019

[55] A. Choudhury, A. Balmurulikrishnan, G. Sarkhel, Polyamide 66/EPR-g-MA blends: mechanical modeling and kinetic analysis of thermal degradation, Polym. Adv. Technol., 19 (2008) 1226-1235. https://doi.org/10.1002/pat.1116

[56] P.R. Couchman, Compositional Variation of Glass-Transition Temperatures. 2. Application of the Thermodynamic Theory to Compatible Polymer Blends, Macromolecules 11 (1978) 1156-1161. https://doi.org/10.1021/ma60066a018

[57] Y. Li, X. Yu, H. Li, Q. Guo, Z. Dai, G. Yu, F. Wang, Detailed kinetic modeling of homogeneous H2S-CH4 oxidation under ultra-rich condition for H2 production, Appl. Energy 208 (2017) 905-919. https://doi.org/10.1016/j.apenergy.2017.09.059

[58] L. Yaning, K. Xinting, T. Huiping, W. Jian, Synthesis of Pd-Ag Membranes by Electroless Plating for H2 Separation, Rare Met. Mater. Eng. 46 (2017) 3688-3692. https://doi.org/10.1016/S1875-5372(18)30058-4

[59] E. Lasseuguette, M.C. Ferrari, Polymer Membranes for Sustainable Gas Separation, in: G. Szekely, A. Livingston (Eds.), Sustainable Nanoscale Eng., Elsevier Inc., 2020, pp. 265-296. https://doi.org/10.1016/B978-0-12-814681-1.00010-2

[60] A. Mondal, B. Mandal, Synthesis and characterization of crosslinked poly(vinylalcohol)/poly(allylamine)/2-amino-2-hydroxymethyl-1,3-propanediol/polysulfone composite membrane for CO2/N2 separation, J. Membr. Sci. 446 (2013) 383-394. https://doi.org/10.1016/j.memsci.2013.06.052

[61] J.B. Faisant, A. Aït-Kadi, M. Bousmina, L. Desche^nes, Morphology, thermomechanical and barrier properties of polypropylene-ethylene vinyl alcohol blends, Polymer, 39 (1998) 533-545. https://doi.org/10.1016/S0032-3861(97)00313-3

[62] M.G. Kamath, S. Fu, A.K. Itta, W. Qiu, G. Liu, R. Swaidan, W.J. Koros, 6FDA-DETDA: DABE polyimide-derived carbon molecular sieve hollow fiber membranes: Circumventing unusual aging phenomena, J. Membr. Sci. 546 (2018) 197-205. https://doi.org/10.1016/j.memsci.2017.10.020

[63] N.H. Ismail, W.N.W. Salleh, N. Sazali, A.F. Ismail, Development and characterization of disk supported carbon membrane prepared by one-step coating-carbonization cycle, J. Ind. Eng. Chem. 57 (2018) 313-321. https://doi.org/10.1016/j.jiec.2017.08.038

[64] N. Sazali, W.N.W. Salleh, N.I. Mahyoun, Z. Harun, K. Kadirgama, Precursor Selection for Carbon Membrane Fabrication: A Review, J. Appl. Membr. Sci. Technol. 22 (2018) 131-144. https://doi.org/10.11113/amst.v22n2.122

[65] N. Sazali, W.N.W. Salleh, A.F. Ismail, N.H. Ismail, K. Kadirgama, A brief review on carbon selective membranes from polymer blends for gas separation performance, Rev. Chem. Eng. 37 (2019) 339-362. https://doi.org/10.1515/revce-2018-0086

[66] W.F. Yong, F.Y. Li, T.S. Chung, Y.W. Tong, Highly permeable chemically modified PIM-1/Matrimid membranes for green hydrogen purification, J. Mater. Chem. A 1 (2013) 13914-13925. https://doi.org/10.1039/c3ta13308g

[67] D. Popov, K. Fikiin, B. Stankov, G. Alvarez, M. Youbi-Idrissi, A. Damas, J. Evans, T. Brown, Cryogenic heat exchangers for process cooling and renewable energy storage: A review, Appl. Therm. Eng. 153 (2019) 275-290. https://doi.org/10.1016/j.applthermaleng.2019.02.106

[68] M. Inagaki, N. Ohta, Y. Hishiyama, Aromatic polyimides as carbon precursors, Carbon 61 (2013) 1-21. https://doi.org/10.1016/j.carbon.2013.05.035

[69] O. Salinas, X. Ma, E. Litwiller, I. Pinnau, Ethylene/ethane permeation, diffusion and gas sorption properties of carbon molecular sieve membranes derived from the prototype ladder polymer of intrinsic microporosity (PIM-1), J. Membr. Sci. 504 (2016) 133-140. https://doi.org/10.1016/j.memsci.2015.12.052

[70] R.J. Lee, Z.A. Jawad, A.L. Ahmad, J.Q. Ngo, H.B. Chua, Improvement of CO2/N2 separation performance by polymer matrix cellulose acetate butyrate, IOP Conf. Ser.: Mater. Sci. Eng. 206 (2017) 012072. https://doi.org/10.1088/1757-899X/206/1/012072

[71] A. Kaboorani, B. Riedl, P. Blanchet, M. Fellin, O. Hosseinaei, S. Wang, Nanocrystalline cellulose (NCC): A renewable nano-material for polyvinyl acetate (PVA) adhesive, Eur. Polym. J. 48 (2012) 1829-1837. https://doi.org/10.1016/j.eurpolymj.2012.08.008

[72] W.N.W. Salleh, A.F. Ismail, T. Matsuura, M.S. Abdullah, Precursor Selection and Process Conditions in the Preparation of Carbon Membrane for Gas Separation: A Review, Sep. Purif. Rev. 40 (2011) 261-311. https://doi.org/10.1080/15422119.2011.555648

[73] N. Sazali, W.N.W. Salleh, N.A.H.M. Nordin, A.F. Ismail, Matrimid-based carbon tubular membrane: Effect of carbonization environment, J. Ind. Eng. Chem. 32 (2015) 167-171. https://doi.org/10.1016/j.jiec.2015.08.014

[74] W.N.W. Salleh, A.F. Ismail, Fabrication and characterization of PEI/PVP-based carbon hollow fiber membranes for CO2/CH4 and CO2/N2 separation, AIChE J. 58 (2012) 3167-3175. https://doi.org/10.1002/aic.13711

[75] N. Sazali, W.N.W. Salleh, A.F. Ismail, K. Kadirgama, F.E.C. Othman, N.H. Ismail, Impact of stabilization environment and heating rates on P84 co-polyimide/nanocrystaline cellulose carbon membrane for hydrogen enrichment, Int. J. Hydrogen Energy 44 (2018) 20924-20932. https://doi.org/10.1016/j.ijhydene.2018.06.039

[76] M. Hong, E.Y.X. Chen, Chemically recyclable polymers: a circular economy approach to sustainability, Green Chem. 19 (2017) 3692-3706. https://doi.org/10.1039/C7GC01496A

[77] S. Saeidi, N.A.S. Amin, M.R. Rahimpour, Hydrogenation of CO2 to value-added products—A review and potential future developments, J. CO2 Util. 5 (2014) 66-81. https://doi.org/10.1016/j.jcou.2013.12.005

[78] P. Sjöholm, D. Ingham, M. Lehtimäki, L. Perttu-Roiha, H. Goodfellow, H. Torvela, Gas-Cleaning Technology, in: H. Goodfellow, E. Tähti (Eds.), Ind. Vent. Des. Guideb., Academic Press, 2001, pp. 1197-1316. https://doi.org/10.1016/B978-012289676-7/50016-3

[79] H.H. Tseng, C.-T. Wang, G.-L. Zhuang, P. Uchytil, J. Reznickova, K. Setnickova, Enhanced H2/CH4 and H2/CO2 separation by carbon molecular sieve membrane coated on titania modified alumina support: Effects of TiO2 intermediate layer preparation variables on interfacial adhesion, J. Membr. Sci. 510 (2016) 391-404. https://doi.org/10.1016/j.memsci.2016.02.036

[80] A. Jha, A.K. Bhowmick, Thermal degradation and ageing behaviour of novel thermoplastic elastomeric nylon-6/acrylate rubber reactive blends, Polym. Degrad. Stab. 62 (1998) 575-586. https://doi.org/10.1016/S0141-3910(98)00044-5

[81] C. Wang, L. Ling, Y. Huang, Y. Yao, Q. Song, Decoration of porous ceramic substrate with pencil for enhanced gas separation performance of carbon membrane, Carbon 84 (2015) 151-159. https://doi.org/10.1016/j.carbon.2014.12.003

[82] M. Aguilar-Vega, D.R. Paul, Gas transport properties of polycarbonates and polysulfones with aromatic substitutions on the bisphenol connector group, J. Polym. Sci., Part B: Polym. Phys. 31 (1993) 1599-1610. https://doi.org/10.1002/polb.1993.090311116

[83] M. Kiyono, P.J. Williams, W.J. Koros, Effect of pyrolysis atmosphere on separation performance of carbon molecular sieve membranes, J. Membr. Sci. 359 (2010) 2-10. https://doi.org/10.1016/j.memsci.2009.10.019

[84] S.S. Stivala, L. Reich, Structure vs stability in polymer degradation, Polym. Eng. Sci. 20 (1980) 654-661. https://doi.org/10.1002/pen.760201003

[85] D.S. Achilias, C. Roupakias, P. Megalokonomos, A.A. Lappas, E.V. Antonakou, Chemical recycling of plastic wastes made from polyethylene (LDPE and HDPE) and polypropylene (PP), J. Hazard. Mater. 149 (2007) 536-542. https://doi.org/10.1016/j.jhazmat.2007.06.076

[86] S. Matteucci, Y. Yampolskii, B.D. Freeman, I. Pinnau, Transport of Gases and Vapors in Glassy and Rubbery Polymers, in: Y. Yampolskii, I. Pinnau, B.D. Freeman (Eds.), Mater. Sci. Membr. Gas Vap. Sep., John Wiley & Sons, Ltd., 2006, pp. 1-47. https://doi.org/10.1002/047002903X.ch1

[87] J.R. Khurma, D.R. Rohindra, R. Devi, Miscibility study of solution cast blends of poly(lactic acid) and poly(vinyl butyral), South Pac. J. Nat. Appl. Sci. 23 (2005) 22-25. https://doi.org/10.1071/SP05004

[88] M.M. Reddy, S. Vivekanandhan, M. Misra, S.K. Bhatia, A.K. Mohanty, Biobased plastics and bionanocomposites: Current status and future opportunities, Prog. Polym. Sci. 38 (2013) 1653-1689. https://doi.org/10.1016/j.progpolymsci.2013.05.006

[89] M.A. Zamiri, A. Kargari, H. Sanaeepur, Ethylene vinyl acetate/poly(ethylene glycol) blend membranes for CO2/N2 separation, Greenhouse Gases Sci. Technol. 5 (2015) 668-681. https://doi.org/10.1002/ghg.1513

[90] C.J. Anderson, W. Tao, C.A. Scholes, G.W. Stevens, S.E. Kentish, The performance of carbon membranes in the presence of condensable and non-condensable impurities, J. Membr. Sci. 378 (2011) 117-127. https://doi.org/10.1016/j.memsci.2011.04.058

[91] S. Kanehashi, G.Q. Chen, D. Danaci, P.A. Webley, S.E. Kentish, Can the addition of carbon nanoparticles to a polyimide membrane reduce plasticization?, Sep. Purif. Technol. 183 (2017) 333-340. https://doi.org/10.1016/j.seppur.2017.04.013

[92] J.H. Petropoulos, A comparative study of approaches applied to the permeability of binary composite polymeric materials, J. Polym. Sci., Part B: Polym. Phys. 23 (1985) 1309-1324. https://doi.org/10.1002/pol.1985.180230703

[93] R.J. Swaidan, X. Ma, I. Pinnau, Spirobisindane-based polyimide as efficient precursor of thermally-rearranged and carbon molecular sieve membranes for enhanced propylene/propane separation, J. Membr. Sci. 520 (2016) 983-989. https://doi.org/10.1016/j.memsci.2016.08.057

[94] M. Naffakh, G. Ellis, M.A. Gómez, C. Marco, Thermal decomposition of technological polymer blends 1. Poly(aryl ether ether ketone) with a thermotropic liquid crystalline polymer, Polym. Degrad. Stab. 66 (1999) 405-413. https://doi.org/10.1016/S0141-3910(99)00093-2

[95] D.R. Paul, J.W. Barlow, A binary interaction model for miscibility of copolymers in blends, Polymer 25 (1984) 487-494. https://doi.org/10.1016/0032-3861(84)90207-6

[96] P. Pötschke, D.R. Paul, Formation of Co-continuous Structures in Melt-Mixed Immiscible Polymer Blends, J. Macromol. Sci., Polym. Rev. 43 (2003) 87-141. https://doi.org/10.1081/MC-120018022

[97] L. Botta, M.C. Mistretta, S. Palermo, M. Fragalà, F. Pappalardo, Characterization and Processability of Blends of Polylactide Acid with a New Biodegradable Medium-Chain-Length Polyhydroxyalkanoate, J. Polym. Environ. 23 (2015) 478-486. https://doi.org/10.1007/s10924-015-0729-4

[98] B. Singh, N. Sharma, Mechanistic implications of plastic degradation, Polym. Degrad. Stab. 93 (2008) 561-584. https://doi.org/10.1016/j.polymdegradstab.2007.11.008

[99] C.S.K. Achoundong, N. Bhuwania, S.K. Burgess, O. Karvan, J.R. Johnson, W.J. Koros, Silane Modification of Cellulose Acetate Dense Films as Materials for Acid Gas Removal, Macromolecules 46 (2013) 5584-5594. https://doi.org/10.1021/ma4010583

[100] V. Siracusa, P. Rocculi, S. Romani, M.D. Rosa, Biodegradable polymers for food packaging: a review, Trends Food Sci. Technol. 19 (2008) 634-643. https://doi.org/10.1016/j.tifs.2008.07.003

[101] C. Ma, J. Yu, B. Wang, Z. Song, J. Xiang, S. Hu, S. Su, L. Sun, Chemical recycling of brominated flame retarded plastics from e-waste for clean fuels production: A review, Renewable Sustainable Energy Rev. 61 (2016) 433-450. https://doi.org/10.1016/j.rser.2016.04.020

[102] C.T. Nguyen, F. Desgranges, G. Roy, N. Galanis, T. Maré, S. Boucher, H. Angue Mintsa, Temperature and particle-size dependent viscosity data for water-based nanofluids – Hysteresis phenomenon, Int. J. Heat Fluid Flow 28 (2007) 1492-1506. https://doi.org/10.1016/j.ijheatfluidflow.2007.02.004

[103] N. Wang, F. Niu, S. Wang, Y. Huang, Catalytic activity of flame-synthesized Pd/TiO2 for the methane oxidation following hydrogen pretreatments, Particuology 41 (2018) 58-64. https://doi.org/10.1016/j.partic.2018.01.005

[104] Z. Said, R. Saidur, A. Hepbasli, N.A. Rahim, New thermophysical properties of water based TiO2 nanofluid—The hysteresis phenomenon revisited, Int. Commun. Heat Mass Transfer 58 (2014) 85-95. https://doi.org/10.1016/j.icheatmasstransfer.2014.08.034

[105] N. Sazali, W.N.W. Salleh, A.F. Ismail, N.H. Ismail, CO2/CH4 Separation by Using Carbon Membranes, in: A. Basile, E.P. Favvas (Eds.), Curr. Trends Future Dev. (Bio-) Membr., Elsevier Inc., 2018, pp. 209-234. https://doi.org/10.1016/B978-0-12-813645-4.00007-6

[106] M.R. Rahimpour, F. Samimi, A. Babapoor, T. Tohidian, S. Mohebi, Palladium membranes applications in reaction systems for hydrogen separation and purification: A review, Chem. Eng. Process. 121 (2017) 24-49. https://doi.org/10.1016/j.cep.2017.07.021

[107] C.Z. Liang, T.S. Chung, J.Y. Lai, A review of polymeric composite membranes for gas separation and energy production, Prog. Polym. Sci. 97 (2019) 101141. https://doi.org/10.1016/j.progpolymsci.2019.06.001

[108] L. Li, R. Xu, C. Song, B. Zhang, Q. Liu, T. Wang, A Review on the Progress in Nanoparticle/C Hybrid CMS Membranes for Gas Separation, Membranes 8 (2018) 134. https://doi.org/10.3390/membranes8040134

[109] Z. Dai, L. Ansaloni, L. Deng, Recent advances in multi-layer composite polymeric membranes for CO2 separation: A review, Green Energy Environ. 1 (2016) 102-128. https://doi.org/10.1016/j.gee.2016.08.001

[110] T.S. Chung, L.Y. Jiang, Y. Li, S. Kulprathipanja, Mixed matrix membranes (MMMs) comprising organic polymers with dispersed inorganic fillers for gas separation, Prog. Polym. Sci. 32 (2007) 483-507. https://doi.org/10.1016/j.progpolymsci.2007.01.008

[111] C.E. Powell, G.G. Qiao, Polymeric CO2/N2 gas separation membranes for the capture of carbon dioxide from power plant flue gases, J. Membr. Sci. 279 (2006) 1-49. https://doi.org/10.1016/j.memsci.2005.12.062

[112] T. Higuchi, Some physical chemical aspects of suspension formulation, J. Am. Pharm. Assoc. 47 (1958) 657-660. https://doi.org/10.1002/jps.3030470913

[113] W.J. Koros, R. Mahajan, Pushing the limits on possibilities for large scale gas separation: which strategies?, J. Membr. Sci. 175 (2000) 181-196. https://doi.org/10.1016/S0376-7388(00)00418-X

[114] Y. Zhao, D. Zhao, C. Kong, F. Zhou, T. Jiang, L. Chen, Design of thin and tubular MOFs-polymer mixed matrix membranes for highly selective separation of H2 and

CO2, Sep. Purif. Technol. 220 (2019) 197-205.
https://doi.org/10.1016/j.seppur.2019.03.037

[115] A. Mundstock, S. Friebe, J. Caro, On comparing permeation through Matrimid®-based mixed matrix and multilayer sandwich FAU membranes: H2/CO2 separation, support functionalization and ion exchange, Int. J. Hydrogen Energy 42 (2017) 279-288. https://doi.org/10.1016/j.ijhydene.2016.10.161

[116] C. Feng, K.C. Khulbe, T. Matsuura, R. Farnood, A.F. Ismail, Recent Progress in Zeolite/Zeotype Membranes, J. Membr. Sci. Res. 1 (2015) 49-72.

[117] M.J.C. Ordoñez, K.J. Balkus, J.P. Ferraris, I.H. Musselman, Molecular sieving realized with ZIF-8/Matrimid® mixed-matrix membranes, J. Membr. Sci. 361 (2010) 28-37. https://doi.org/10.1016/j.memsci.2010.06.017

[118] L.Y. Jiang, T.-S. Chung, S. Kulprathipanja, Fabrication of mixed matrix hollow fibers with intimate polymer-zeolite interface for gas separation, AIChE J. 52 (2006) 2898-2908. https://doi.org/10.1002/aic.10909

[119] N. Sazali, W.N.W. Salleh, N.A.H.M. Nordin, Z. Harun, A.F. Ismail, Matrimid-based carbon tubular membranes: The effect of the polymer composition, J. Appl. Polym. Sci. 132 (2015). https://doi.org/10.1002/app.42394

[120] J.N. Barsema, S.D. Klijnstra, J.H. Balster, N.F.A. van der Vegt, G.H. Koops, M. Wessling, Intermediate polymer to carbon gas separation membranes based on Matrimid PI, J. Membr. Sci. 238 (2004) 93-102. https://doi.org/10.1016/j.memsci.2004.03.024

[121] C.C. Hu, Y.-J. Fu, S.-W. Hsiao, K.-R. Lee, J.-Y. Lai, Effect of physical aging on the gas transport properties of poly(methyl methacrylate) membranes, J. Membr. Sci. 303 (2007) 29-36. https://doi.org/10.1016/j.memsci.2007.06.004

[122] S.S. Hosseini, M.M. Teoh, T.S. Chung, Hydrogen separation and purification in membranes of miscible polymer blends with interpenetration networks, Polymer 49 (2008) 1594-1603. https://doi.org/10.1016/j.polymer.2008.01.052

[123] S. Friebe, B. Geppert, F. Steinbach, J. Caro, Metal–Organic Framework UiO-66 Layer: A Highly Oriented Membrane with Good Selectivity and Hydrogen Permeance, ACS Appl. Mater. Interfaces 9 (2017) 12878-12885. https://doi.org/10.1021/acsami.7b02105

[124] X. Gong, Y. Wang, T. Kuang, ZIF-8-Based Membranes for Carbon Dioxide Capture and Separation, ACS Sustainable Chem. Eng. 5 (2017) 11204-11214. https://doi.org/10.1021/acssuschemeng.7b03613

[125] S.N.A. Shafie, W. X. Liew, N.A.H. Md. Nordin, M. Roil Bilad, N. Sazali, Z. Adi Putra, M.D.H. Wirzal, CO2-Philic [EMIM][Tf2N] Modified Silica in Mixed Matrix Membrane for High Performance CO/CH4 Separation, Adv. Polym. Technol. 2019 (2019). https://doi.org/10.1155/2019/2924961

[126] G.L. Zhuang, M.-Y. Wey, H.-H. Tseng, The density and crystallinity properties of PPO-silica mixed-matrix membranes produced via the in situ sol-gel method for H2/CO2 separation. II: Effect of thermal annealing treatment, Chem. Eng. Res. Des. 104 (2015) 319-332. https://doi.org/10.1016/j.cherd.2015.08.020

[127] M. Sadeghi, M.A. Semsarzadeh, H. Moadel, Enhancement of the gas separation properties of polybenzimidazole (PBI) membrane by incorporation of silica nano particles, J. Membr. Sci. 331 (2009) 21-30. https://doi.org/10.1016/j.memsci.2008.12.073

[128] Z. Yang, X.-H. Ma, C.Y. Tang, Recent development of novel membranes for desalination, Desalination 434 (2018) 37-59. https://doi.org/10.1016/j.desal.2017.11.046

[129] W. Li, H. Wang, X. Jiang, J. Zhu, Z. Liu, X. Guo, C. Song, A short review of recent advances in CO2 hydrogenation to hydrocarbons over heterogeneous catalysts, RSC Adv. 8 (2018) 7651-7669. https://doi.org/10.1039/C7RA13546G

[130] V. Dindore, W. Brilman, F. Geuzebroek, G. Versteeg, Membrane–solvent selection for CO2 removal using membrane gas–liquid contactors, Sep. Purif. Technol. 40 (2004) 133-145. https://doi.org/10.1016/j.seppur.2004.01.014

[131] S. Marzouk, M. Al-Marzouqi, M. El-Naas, N. Abdullatif, Z. Ismail, Removal of carbon dioxide from pressurized CO2–CH4 gas mixture using hollow fiber membrane contactors, J. Membr. Sci. 351 (2010) 21-27. https://doi.org/10.1016/j.memsci.2010.01.023

[132] J. Lee, J. Kim, H. Kim, K.S. Lee, W. Won, A new modeling approach for a CO2 capture process based on a blended amine solvent, J. Nat. Gas Sci. Eng. 61 (2019) 206-214. https://doi.org/10.1016/j.jngse.2018.11.020

[133] Q. He, G. Yu, S. Yan, L.F. Dumée, Y. Zhang, V. Strezov, S. Zhao, Renewable CO2 absorbent for carbon capture and biogas upgrading by membrane contactor, Sep. Purif. Technol. 194 (2018) 207-215. https://doi.org/10.1016/j.seppur.2017.11.043

[134] S.A. Wassie, S. Cloete, V. Spallina, F. Gallucci, S. Amini, M. van Sint Annaland, Techno-economic assessment of membrane-assisted gas switching reforming for pure H2 production with CO2 capture, Int. J. Greenhouse Gas Control 72 (2018) 163-174. https://doi.org/10.1016/j.ijggc.2018.03.021

[135] I.M. Bernhardsen, H.K. Knuutila, A review of potential amine solvents for CO2 absorption process: Absorption capacity, cyclic capacity and pKa, Int. J. Greenhouse Gas Control 61 (2017) 27-48. https://doi.org/10.1016/j.ijggc.2017.03.021

[136] C. Castel, E. Favre, Membrane separations and energy efficiency, J. Membr. Sci. 548 (2018) 345-357. https://doi.org/10.1016/j.memsci.2017.11.035

[137] W.F. Yong, T.S. Chung, M. Weber, C. Maletzko, New Polyethersulfone (PESU) Hollow Fiber Membranes for CO2 Capture, J. Membr. Sci. 552 (2018) 305-314. https://doi.org/10.1016/j.memsci.2018.02.008

[138] A. Makaruk, M. Miltner, M. Harasek, Biogas desulfurization and biogas upgrading using a hybrid membrane system – modeling study, Water Sci. Technol. 67 (2013) 326-332. https://doi.org/10.2166/wst.2012.566

[139] N. Thomas, M.O. Mavukkandy, S. Loutatidou, H.A. Arafat, Membrane distillation research & implementation: Lessons from the past five decades, Sep. Purif. Technol. 189 (2017) 108-127. https://doi.org/10.1016/j.seppur.2017.07.069

[140] S. Ayadi, I. Jedidi, S. Lacour, S. Cerneaux, M. Cretin, R.B. Amar, Preparation and characterization of carbon microfiltration membrane applied to the treatment of textile industry effluents, Sep. Sci. Technol. 51 (2016) 1022-1029. https://doi.org/10.1080/01496395.2016.1140201

[141] S. Loeb, L. Titelman, E. Korngold, J. Freiman, Effect of porous support fabric on osmosis through a Loeb-Sourirajan type asymmetric membrane, J. Membr. Sci. 129 (1997) 243-249. https://doi.org/10.1016/S0376-7388(96)00354-7

Polymeric Membranes for Water Purification and Gas Separation   Materials Research Forum LLC
Materials Research Foundations **113** (2021) 243-334   https://doi.org/10.21741/9781644901632-8

Chapter 8

Polymeric Membranes for H_2 and N_2 Separation

J. Wu[1‡], S. Japip[2‡], T.S. Chung[1,2*]

[1] NUS Graduate School for Integrative Sciences and Engineering, National University of Singapore, 117456, Singapore

[2] Department of Chemical and Biomolecular Engineering, National University of Singapore, 117585, Singapore

‡ These authors contributed equally

* chencts@nus.edu.sg

Abstract

H_2 and N_2 separations are of paramount importance to the global development of clean energy and environment. While the traditional thermal-driven processes are often deemed overly energy-intensive, the polymeric membrane technology presents an energy-efficient and potentially cost-effective alternative that comes with many operational and environmental advantages to offer. However, a variety of key challenges revolving around the membrane performance and stability issues require new material innovations in order to be eventually overcome. This chapter provides the background for polymeric gas separation membranes and also a detailed evaluation on state-of-the-art polymeric membrane materials for H_2 and N_2 separations. Performance enhancement strategies will also be discussed in the later parts.

Keywords

Membrane Gas Separation, Polymeric Membranes, Glassy Polymers, Hydrogen Recovery, CO_2 Capture, N_2/CH_4 Separation, Polymers of Intrinsic Microporosity, Mixed-Matrix Membranes

Contents

1.   Introduction

1.1   Overview of membranes for H_2 and N_2 separations

As the world increasingly recognizes the threats of rising carbon dioxide (CO_2) emissions, the transition away from fossil-fuel-based power sources is now fast becoming a global endeavor, which sparks the prospect of an economy based on clean and more efficient fuel alternatives, especially hydrogen (H_2) [1, 2]. Not only is hydrogen considered as a zero-emission fuel because its sole combustion product is water which is perfectly nonpolluting, hydrogen also offers a higher energy density than most hydrocarbon fuels [3]. However, the separation of hydrogen from a gas mixture is theoretically an entropy-decreasing, non-spontaneous process that requires an external energy input to occur. As such, the development of efficient hydrogen separation technologies is of critical importance to the production of high-purity, cost-effective hydrogen fuels. On the other hand, the significant role of nitrogen (N_2) separation for environmental development lies not particularly in the utilization or production of N_2 gas, but rather, it lies in the efficient capture of environmentally problematic gas components, such as CO_2 and methane (CH_4). The N_2 separation process is also being increasingly demanded in the recovery and purification of valuable gas components, for instance, methane (CH_4), from refinery and petrochemical gas streams that usually contain considerable amounts of nitrogen [4-6].

Traditional thermal-driven technologies for H_2 and N_2 separations, such as pressure swing adsorption (PSA), fractional/cryogenic distillation or amine absorption, involve extremely energy-intensive processes and are generally deemed cost-ineffective, even though they are still dominating the gas separation (GS) industry [7]. In comparison, membrane GS, driven only by an applied transmembrane pressure, has no requirement for phase change, thermal regeneration, or any active moving parts during its operation, enabling a much smaller energy and maintenance input than the traditional methods [8]. This key advantage of energy efficiency, along with its other attractive features, such as small footprint, modular design, and simple and continuous operation, renders membrane GS the most promising technology for realizing energy-saving and potentially more cost-effective H_2 and N_2 separations. It also fits nicely with the process design principles delineated by the process intensification strategy which is central to all current chemical engineering developments [8]. Among various membrane materials including metals, inorganics, porous carbons and purely organic polymers, organic polymers are by far the most developed and the most commercially feasible choice of material for GS owing to their easy solution-processability, wide functioning temperature range as well as a balanced combination of cost, performance, mechanical and chemical properties [2].

1.2   History and key applications of polymer membranes for H_2 and N_2 separations

Modern polymeric membrane technology for gas separation was actually first developed upon the asymmetric cellulose acetate (CA) membrane developed by Loeb and Sourirajan for reverse osmosis (RO) seawater desalination in the early 1960s, which they later applied to gas separation [9]. However, the existence of only sub-Angstrom-level size difference between common industrial gas molecules renders the design of GS membranes technically much more challenging, because any defects formed during the membrane fabrication, though as small as at a nanometer-scale, could significantly compromise the membrane performance [10]. For example, as shown in Table 1, the differences between the kinetic diameters of H_2, N_2, CH_4 and CO_2, which are 2.89, 3.64, 3.80 and 3.30 Å respectively, are all smaller than 1.0 Å [11].

Table 1. Kinetic diameters of gas species of interest to gas separation

Gas Species	H_2	CO_2	O_2	N_2	CH_4
Kinetic Diameter (Å)	2.89	3.30	3.46	3.64	3.80
Critical Temperature, T_c (K)	33.2	304.2	154.8	126.2	190.6

In the mid- to late-1970s, DuPont pioneered the first generation of small-diameter hollow fiber membranes (HFM) for GS which was groundbreaking, but the low manufacturing efficiency and gas permeance limited their economic viability for industrial applications [12, 13]. The key breakthrough occurred in 1980 when Monsanto first developed and commercialized the asymmetric polysulfone (PSF) HFMs coated by a thin silicone rubber layer that plugged the defects for separating H_2 from ammonia plant purge gases [14]. Inspired by Monsanto's success, Separex and Cynara soon applied similar design concepts to develop the spiral-wound CA membranes for H_2 and natural gas separations [15, 16], and Ube from Japan later introduced and commercialized the polyimide (PI) GS membranes which demonstrated one of the best thermal properties and solvent resistance at that time [17]. A brief timeline of the key developments in polymeric GS membranes is shown in Fig. 1.

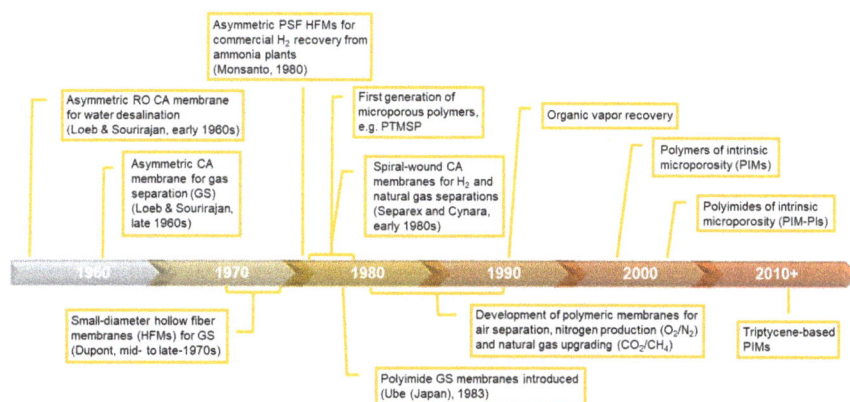

Figure 1. Timeline of polymeric gas separation membranes

Thereafter, membrane GS technology underwent rapid growth and had spanned a great variety of commercial applications, such as H_2/N_2 and H_2/CH_4 separations for hydrogen recovery from ammonia purge gases, H_2/CO ratio adjustment in synthesis gas (syngas) production, and H_2/light HCs separation in refineries or petrochemical processes (e.g. off-gases from hydrotreaters and hydrocrackers) [8]. Relatively new applications involving H_2 and N_2 separations also emerged, which included H_2/CO_2 separation for both pre-combustion CO_2 capture and syngas production, CO_2/N_2 separation for CO_2 capture from flue gas of power plants [18], N_2/CH_4 separation for N_2 rejection to enhance the heat value of natural gas [5], and also the recovery of organic vapors, like light olefins (e.g. C_2H_4, C_3H_6), from N_2-containing off-gases [6]. After a significant expansion over the past three decades, the current global membrane GS market reaches an estimated total sales value of around 1.0 to 1.5 billion US dollars each year [7]. Some subsidiary H_2 and N_2 separation markets, such as hydrogen recovery and organic vapor recovery as mentioned above, are now worth about US$200 and US$100 million per year, respectively. Table 2 provided a summary of the target applications and the common choices of polymer materials for some industrially important membrane-based H_2 and N_2 separations.

1.3   Performance and Mechanism of Polymeric Gas Separation Membranes

Evaluating the gas separation performance of a polymeric membrane is based on its two critical gas transport characteristics – permeability and selectivity. Permeability (P) is a thickness- and pressure-normalized measure of the flux of a particular gas permeant, which determines the throughput or productivity of the membrane gas separation process [2]. It is an intrinsic property of the polymer material and is generally expressed in units of *Barrer* ($1\ Barrer = 1 \times 10^{-10}\ cm^3 (STP)\ cm/cm^2\ s\ cmHg$). When the membrane thickness is factored in the measurement of gas flux, permeance is used instead, which is calculated as the permeability divided by the thickness of the membrane (P/l) and is typically expressed in gas permeance units (GPU, $1\ GPU = 1 \times 10^{-6}\ cm^3 (STP)/ cm^2\ s\ cmHg$). Selectivity, on the other hand, measures the effectiveness of the membrane in preferentially permeating one particular gas over the other, which determines the purity of the target gas after separation. The ideal selectivity ($\alpha_{i/j}$) is typically used for assessing the pure-gas separation efficiency of a membrane, which is expressed as a ratio between the pure-gas permeabilities of the membrane for two different gas species ($\alpha_{i/j} = P_i/P_j$) [2, 19].

Table 2. Common applications of H_2 and N_2 separation and their corresponding polymer materials [30]

Application	Separation	Traditional Technology	Membrane Materials
Hydrogen recovery from ammonia plants	H_2/N_2, H_2/CH_4	PSA	polysulfone, polyimide
Adjustment of H_2/CO ratio in syngas plants	H_2/CO	PSA	polysulfone, polyimide
Hydrogen purification and CO_2 capture from syngas production	H_2/CO_2	Amine absorption	polybenzimidazole (PBI)
Hydrogen recovery in refineries (e.g. hydrocrackers)	H_2/hydrocarbons	Cryogenic distillation	silicon rubber, polyimide
CO_2 capture from flue gas	CO_2/N_2	Amine absorption	polyimide
Nitrogen removal from natural gas	N_2/CH_4	Cryogenic distillation	silicon rubber
Vapor recovery	C_2H_4/N_2, C_3H_6/N_2	Condensation	silicon rubber

Polymeric membranes are generally dense and nonporous such that the gas transport through them follows a well-established solution-diffusion mechanism that comprises three major processes: 1) gas molecules adsorb onto the membrane surface on the high-pressure upstream side, 2) then they diffuse into and through the polymer matrix, 3) and eventually desorb away from the low-pressure downstream side [3]. More specifically, from a microscopic point of view, process 2) actually does not occur as a continuous passage. Instead, it takes place by permeants making 'jumps', in the direction of the applied pressure gradient, from one transient gap to another, which is formed by the thermally agitated segment motions of the polymer backbone. An illustrative schematic of the solution-diffusion transport mechanism is shown in Fig. 2. As such, the permeability of polymeric membranes for a particular gas permeant can be expressed as the product of its solubility (S) and diffusivity (D) coefficients as shown in Eq. 1 below:

$$P = S \times D \tag{1}$$

Therefore, the ideal selectivity can be expressed as shown in Eq. 2:

$$\alpha_{i/j} = \frac{P_i}{P_j} = \frac{S_i}{S_j} \times \frac{D_i}{D_j} \tag{2}$$

There are some key distinctions between solubility and diffusivity. The former is associated with the thermodynamic phenomenon of gas adsorption onto the membrane and thus demonstrates a strong correlation to the condensability of gas permeants as well as the chemical interaction and affinity between the permeants and polymer matrix [20]. The latter, in contrast, is essentially a kinetic factor strongly influenced by the spatial extents of polymer chain random motions and is found closely related to the size of gas permeants, the polymer chain mobility and also the amount of free volume in the membrane (defined as the space unoccupied by the entangled polymer chains due to conformational constraints). As processes requiring an activation energy, both the solution and diffusion of gas permeants are usually temperature-dependent. Therefore, a successful polymeric GS membrane design must carefully consider both these gas transport parameters.

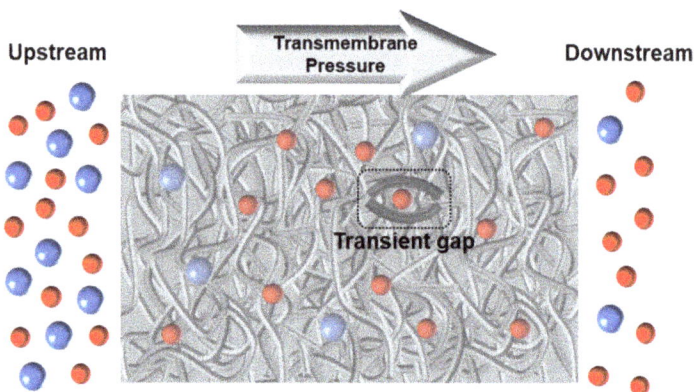

Figure 2. The membrane gas separation process and the solution-diffusion mechanism of gas transport in dense polymeric membranes

1.4   Difference between Glassy and Rubbery Polymers

Polymeric membranes can be categorized as glassy or rubbery depending on the operation temperature. When the operation temperature is below the glass transition temperature, T_g, of the polymer, the polymer exhibits behaviors of rigid glass and typically possesses a stiffened polymer backbone with a small amount of free volume that restricts large-scale cooperative chain movements [3]. When the operation temperature is kept above its T_g, the polymer displays rubbery characteristics and obtains a relatively larger amount of free volume owing to the motion of flexible polymer chains that gives rise to greater transient voids. As a result, glassy polymers generally tend to exhibit a lower gas permeability, higher selectivity and also better mechanical properties than rubbery polymers [8]. Moreover, glassy polymers often produce GS membranes that are diffusivity-selective (i.e., the overall selectivity (permselectivity) is dominantly influenced by the diffusivity selectivity, D_i/D_j) while membranes fabricated from rubbery polymers are usually solubility-selective (i.e., the permselectivity is dominantly influenced by the solubility selectivity, S_i/S_j) [21, 22]. Therefore, glassy polymers demonstrate a stronger size-selective effect towards the separation of gas molecules by preferentially permeating the smaller ones, like H_2 and CO_2, over the larger ones, like N_2 and CH_4. In contrast, rubbery polymers tend to preferentially permeate the more condensable gases that possess higher solubility in the membranes [23]. This often results in a reverse-selective system whereby larger but more soluble gas molecules transport faster through the membrane than smaller, non-condensable ones. Examples of reverse-selective membranes include those that can enable high CO_2/H_2 or light $HCs/(H_2$ or $N_2)$ selectivities. Industrially speaking, glassy polymers, including CA, PSF and PI, have achieved the widest commercial implementation over the past few decades in the GS industry, especially for H_2 separation, because of their good size-selectivity and mechanical robustness. On the other hand, rubbery polymers, such as poly(dimethyl siloxane) (PDMS) and poly(ethylene oxide) (PEO), have gained increasing attention recently for the reverse-selective H_2 and N_2 separations, including CO_2/H_2 and CO_2/N_2 separations for CO_2 capture as well as HCs/H_2 and HCs/N_2 separations for organic vapor recovery, because of their preferential affinity, solubility, and hence permeability for condensable gases.

1.5   Challenges in Polymeric Gas Separation Membrane Design

Despite the early commercial success, the ongoing growth and advancement of polymeric GS membranes are not without challenges. Firstly, by analyzing the huge pool of performance data from the available polymeric GS membrane materials, Robeson discovered in 1991 that there existed a characteristic trade-off relationship between the permeability and selectivity of polymeric GS membranes such that a high permeability was typically accompanied by a low selectivity and vice versa [24]. Based on the transition state theory, a simultaneous increase in both polymer chain stiffness (i.e., lower intrasegmental mobility) and interchain spacing (i.e., larger free volume) could systematically improve either permeability or selectivity without sacrificing the other, or in some occasions, improve both [25]. However, such improvement would quickly hit a limit when the interchain spacing becomes so large that the motion of thermally agitated polymer chains is no longer governing the diffusion process of gas permeants. Under such circumstances, without a substantial enhancement in solubility selectivity, this limit would become the asymptotic end-point for the separation performance of polymeric membranes whose gas transport properties are ruled by the solution-diffusion mechanism. For example, certain glassy polymers can achieve an unprecedentedly high gas permeability (in the order of ~10^4 *Barrer*) due to their possession of an extraordinary amount of free volume arising from their microporous polymer structures, such as poly(trimethylsilyl propyne) (PTMSP) [26]. However, their extremely low selectivity renders them unfit for any large-scale commercial GS applications. This well-defined trade-off limit on the polymeric GS membranes is famously known as the upper bound (first proposed by Robeson) and has been widely used as the 'gold standard' for the comparison of membrane performance for several important H_2 and N_2 separations, including H_2/N_2, H_2/CH_4, H_2/CO_2, CO_2/N_2 separations and so on [24]. A huge part of polymeric GS membrane research has been focused on overcoming this inherent performance limit, and detailed discussions on the material breakthroughs that have accelerated this cause will be given in later sections. A simplified illustration of the upper bound was shown in Fig. 3.

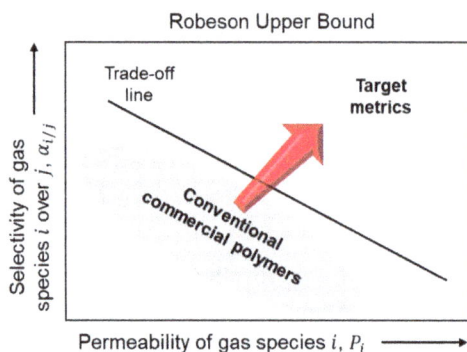

Figure 3. Illustrative diagram of the Robeson upper bound

Besides the challenge of intrinsic performance, the critical issues of plasticization and physical aging also put the operation durability of glassy polymers to the test. Plasticization occurs when the sorption of highly condensable gas components, like CO_2 and HCs, significantly swells the membrane, increases the free volume, and subsequently results in substantial losses of size-selectivity, especially under high operating pressures when the amount of irreversibly dissolved condensable gases in the membrane becomes significant [3]. In some occasions, the mechanical stability could also be compromised due to swelling. Physical aging is the decrease of gas permeability with time caused by the gradual relaxation and subsequent collapse of non-equilibrium excess free volume elements in glassy polymers [7]. Its effect is more significant for highly permeable polymers like PTMSP and PIMs because the amount of thermodynamically unstable free volume in these polymers is particularly large. These critical durability problems, plus other stability issues (e.g. chemical, thermal or mechanical) and the challenge of large-scale fabrication of defect-free thin-film membranes, are the key impeding forces against the progress of the commercialization of new membrane materials. Consequently, current commercial membrane-based H_2 and N_2 separations are still relying heavily on the few polymers developed about 30 years ago. New polymeric membrane materials with not only better performance but also promising stability against plasticization, aging, harsh operating and fabrication conditions are what the industry needs to eventually break this stagnation.

The following sections will cover the key properties and performances of a variety of conventional and advanced polymeric membrane materials for H_2 and N_2 separation. The focus is primarily on glassy polymers because of the size-selective nature in these separations, but some rubbery polymers will also be discussed for solubility-selectivity-dominated separations, such as CO_2/H_2, light HCs/H_2 and CO_2/N_2. Some discussions on the various material strategies for gas transport enhancement and the importance of membrane configuration will also be included in later parts of this chapter.

2.   Polymeric Membranes for H_2 Separation

Hydrogen is highly permeable owning to its small molecular size so that it transports through size-selective membranes much faster than other larger gases that do not have high condensability, such as nitrogen (N_2), methane (CH_4) and carbon monoxide (CO). Therefore, the separation of hydrogen from these gases is relatively straightforward and was enabled as the first set of commercial applications [3]. For example, the earliest successful industrial-scale GS application based on polymeric membranes was the recovery of hydrogen from ammonia purge gases that contained nitrogen and methane (i.e., H_2/N_2 and H_2/CH_4 separations). These purge streams have been produced at high pressures and are 'clean' from condensable gas molecules, which make them ideal target mixtures to be separated by polymeric membranes. Similarly, the separation of hydrogen from carbon monoxide for the adjustment of syngas ratio (i.e., H_2/CO separation) was also among the first few commercially successful GS processes because CO has a similar size and non-condensability to N_2. Glassy polymers are naturally preferred for these size-discrimination-based separations because of the dominance of diffusivity selectivity in their gas transport properties (Table 3). For example, the earliest commercial polymeric membrane systems, installed for the hydrogen recovery in ammonia plants or for the syngas ratio adjustment, were developed based on asymmetric glassy polysulfone (PSF) hollow fiber membranes (HFMs) by Monsanto under the trade name of Prism® membrane. These PSF HFMs were coated by silicone rubbers to plug the 'leaky' defects formed during fabrication.

It is also of rising industrial interest and environmental importance to recover hydrogen from refineries to meet the increasing demand of hydrogen as a feedstock for hydrotreating, hydrocracking or other processes [27]. The target separations are H_2/CH_4 and H_2/light HCs separations in the purge or off gases from the refinery processes, and size-selective glassy polymers, such as PSFs (Prism®) and PIs, have been used industrially as the membrane materials. Now, hydrogen separation by polymeric membranes in ammonia plants, syngas production, and refineries and petrochemical processes is considered as a tried and true technology, and many plants have been installed and in operation since more than 30 years ago [7].

A major source of hydrogen for refineries, power production or chemical synthesis comes from the syngas produced primarily from the steam-methane reforming (SMR) reaction or coal gasification as a mixture of mainly H_2, CO and CO_2. An additional water-gas shift (WGS) reaction can convert CO into more hydrogen and CO_2, and the target separation becomes essentially the H_2/CO_2 separation for fulfilling both the recovery of high purity hydrogen and the capture of greenhouse gas, CO_2 [2]. The reaction equations were shown below as reaction 1 to 3.

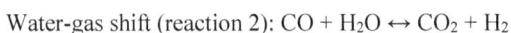

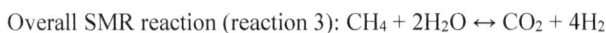

$$\text{Initial steam reforming (reaction 1): } CH_4 + H_2O \leftrightarrow CO + 3H_2$$

$$\text{Water-gas shift (reaction 2): } CO + H_2O \leftrightarrow CO_2 + H_2$$

$$\text{Overall SMR reaction (reaction 3): } CH_4 + 2H_2O \leftrightarrow CO_2 + 4H_2$$

As an easily condensable gas with a high solubility in polymers, CO_2 is also very permeable in polymeric membranes due to its large solubility coefficient. Therefore, the success of using glassy polymers for the H_2/CO_2 separation is relatively limited due to the competition between the diffusivity selectivity-favored H_2 permeation and the solubility selectivity-favored CO_2 permeation as well as the CO_2-induced plasticization effect, especially under high operating pressures. On the other hand, gas transport in flexible rubbery polymers, like PDMS and PEO, or glassy polymers with an extremely high free volume, like PTMSP, is dominated by the solubility selectivity, which enables the reverse-selective separation whereby larger but more condensable gases could permeate faster than smaller, non-condensable gases [23]. One of the key process advantages offered by reverse-selective separations is that the desirable gases can be retained at the high-pressure feed side, avoiding or reducing the need for their recompression and hence additional energy consumption. Also, the recovery of hydrogen from light hydrocarbons in refineries and petrochemical processes generally involves high-throughput separations so that the use of reverse-selective membranes with an extremely high permeability for the condensable hydrocarbons can minimize the required membrane areas, lowering the capital cost. Meanwhile, the adverse effect of plasticization on the diffusivity selectivity of polymeric membranes is arguably less detrimental to reverse-selective separations because these separations actually become more efficient when the solubility selectivity gains more dominance. Nevertheless, plasticization is still undesirable because it could compromise mechanical stability.

Table 3. Difference of gas transport characteristic between size-selective and reverse-selective membranes

Type of Selectivity	H_2-selective (size-selective)	H_2-rejective (reverse-selective)
$\dfrac{D_{H_2}}{D_{other\ gas}}$	>>> 1	>> 1
$\dfrac{S_{H_2}}{S_{other\ gas}}$	< 1	<<< 1
$\dfrac{P_{H_2}}{P_{other\ gas}}$	>> 1	< 1

In this section, a variety of glassy polymeric membrane materials used commercially or under research for H_2/N_2, H_2/CH_4 and H_2/CO_2 separations will be discussed and evaluated, followed by other polymers that are applicable to the reverse-selective separations, including CO_2/H_2 and light HCs/H_2 separations.

2.1 Polymers for H_2/N_2 and H_2/CH_4 Separations

2.1.1 Conventional Polymers

Early studies on the structure-property relationship based on polysulfone (PSF), polycarbonate (PC) and polyimide (PI) polymers revealed a key design principle for the concomitant enhancements in both permeability and selectivity for light gases over bulkier gases, which is to hinder the intersegmental/interchain packing while simultaneously restricting the intra-segmental mobility (equivalent to maximizing the polymer backbone rigidity) [28]. The chain packing efficiency is often reflected by the fractional free volume (FFV) in the membrane, while the large-scale segmental motions and the local sub-segmental mobility can be measured by the glass transition temperature

(T_g) and the sub-T_gs. Both the T_g and sub-T_gs can also serve as indications of the extent of steric effects among the polymer chains. It is important to note that the extent of condensability-dependent solubility selectivity is relatively much smaller than the size-dependent diffusivity selectivity in the overall gas transport of H_2/N_2 and H_2/CH_4 separations because the three involved gases; namely, H_2, N_2 and CH_4, all have low to less than moderate condensabilities in dense polymeric membranes. Therefore, the membranes designed for H_2/N_2 and H_2/CH_4 separations primarily target the size-selective effect.

2.1.1.1 Polysulfone

Polysulfones (PSF) are chemically and thermally stable engineering thermoplastics with attractive physical and chemical properties, including good mechanical strength, wide operating temperature range, high chemical resistance and facile processability [29]. Monsanto's PSF-based Prism® membranes were the first polymeric material being successfully commercialized for industrial gas separations. The structure of PSF is characterized by the diphenylene sulfone repeating units which effectively impart the polymer backbones with a certain degree of rigidity owing to their hindrance to the intra-segmental rotations. This bestows PSF with a high T_g of around 186 °C and a satisfactory combination of a H_2 permeability of 14 *Barrer* and H_2/N_2 and H_2/CH_4 selectivities of both 56 at 35 °C, which was fairly acceptable during the early stage of development for GS membranes [30]. The iconic structure of PSF, which is also a commercial product under the trade name of Ultrason® S, is shown in Fig. 4. There are generally two key structural modifications on PSFs that could manipulate their gas transport properties: changing the bridging moieties between the phenylene rings or changing the substitutions on the phenylene rings [29]. For example, PSF polymers can be made much more permeable by substituting the isopropylidene bridging group ($-C(CH_3)_2-$) with the bulkier hexafluoro bridging group ($-C(CF_3)_2-$) because the later significantly hinders interchain packing and enhances the FFV of polymer [31]. Also, symmetrical bulky methyl substitutions on the phenylene rings could result in a significant improvement of permeability as compared to the unsubstituted polymer because of the upset chain packing [32].

2.1.1.2 Polycarbonate

Polycarbonates (PC) are widely used engineering thermoplastics with outstanding strength and toughness. They are characterized as the polyesters of carbonic acid that are typically derived from either diphenyl carbonate or phosgene [29]. Commercial PCs that contain structural units based on the bisphenol A groups (BPA) are the most commonly used type of PCs. Shown in Fig. 2.1 is the archetypical BPA-type PC structure which possesses a reasonably high T_g of 150 °C and a comparable H_2 permeability of 13 *Barrer* and a lower H_2/CH_4 selectivity of 36 as compared to the analogous PSF [33]. Gas transport properties of PC membranes can be tuned via similar structural modifications as those for PSFs, including bridging moiety and/or aromatic ring substitutions. For example, substitution of bulkier hexafluoro bridging groups could similarly enhance the permeability, while symmetrical tetrabromo-substitution on the biphenyl rings of PC tends to reduce the diffusivity coefficient but enhances the diffusivity selectivity due to the higher interchain interaction by polar halogen groups [33, 34].

PSF (Ultrason® S)

PC (BPA-type)

CA (CTA)

PPO

Aramid
PPPT(Kevlar®)

PMPI (Nomex®)

Polyimide
Matrimid®

Upilex®

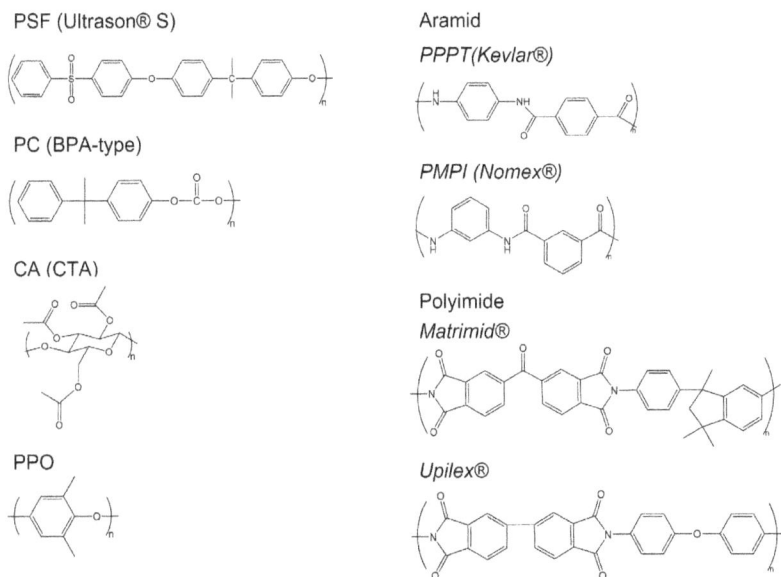

Figure 4. Structures of several well-established commercial polymers for gas separation

2.1.1.3 Cellulose Acetate

Cellulose Acetates (CA) are produced from the acetylation of cellulose using acetate esters, which make them relatively inexpensive given the abundance and the renewable nature of these raw materials [35]. The development of integrally skinned asymmetric membrane configuration, which enabled a higher process productivity by tremendously increasing the membrane area per unit volume, was initially based on CA that first targeted desalination applications and gave rise to the early commercialization of CA membranes [9]. As the technology of membrane module production that stemmed from the development of asymmetric membranes quickly developed, CA membranes were soon found promising for GS applications, too. They became one of the earliest first-generation commercial GS polymers [15, 16]. CA membranes have been reported to demonstrate a lower H_2 permeability but much higher H_2/N_2 and H_2/CH_4 selectivities than PSFs [36, 37], and the overall separation performances of CA are similar to PSF in terms of their respective distance to the upper bound lines. The gas transport properties of CA membranes are strongly influenced by the degree of acetylation of the hydroxyl groups on cellulose [29]. The greater number of hydroxyl groups being substituted by the acetyl groups, the larger the degree of acetylation and the higher the gas permeability (accompanied by lower selectivity) are to be obtained, because the polymer structure is opened up by the bulky acetyl groups. One of the most common CA structures is the cellulose triacetate (CTA, Fig. 4), which is formed when the three hydroxyl groups in each glucose unit have all been acetylated.

2.1.1.4 Poly(phenylene oxide)

Poly(phenylene oxide) (PPO) is characterized by the aromatic ether linkage (-Ar-O-Ar-) which assumes a kinked conformation. The structure of its most iconic member, the symmetrically dimethyl-substituted poly(2,6-dimethyl-1,4-phenylene oxide), is shown in Fig. 4, which possesses a T_g of 220 °C [29, 38]. Being mechanically stable while having excellent thermo-oxidative resistance owing to its high T_g, PPO became one of the earliest high-performance (i.e., high-temperature) commercial engineering thermoplastics. In fact, PPO was introduced as early as in the 1960s by General Electric as the first commercially available polyether in the market, which later gained much interest as a membrane material for industrial gas separations [38, 39]. PPO demonstrates a much higher gas permeability than most other commercial glassy polymers, such as PSF, PC, CA and various polyimides (PI), but its selectivity is much lower due to its relatively open polymer structure [28]. The kinked ether linkages, in the absence of any polar groups substituted on the phenylene rings, could prevent efficient chain packing and induce a high free volume in the membrane. Also, there is high freedom of ring rotation about the ether bonds that could give PPO high intra-segmental mobility. The iconic dimethyl PPO is able to obtain a H_2 permeability of 113 *Barrer* and H_2/N_2 and H_2/CH_4 selectivities of around 24 and 18 respectively [40]. Structural modifications on the PPO polymers can take place by substituting the aromatic hydrogens with large polar atoms or substituting the existing methyl groups with other functionalities that could tune the free volume [41].

2.1.1.5 Polyamide

Aromatic polyamides are often abbreviated as aramids and are generally the polycondensation products of aromatic diamines and aromatic diacyl chlorides via a two-phase process known as interfacial polymerization, which has been one of the most common methods for the synthesis of aramids [29]. Nomex® and Kevlar® are two very representative examples of commercial aramids and they are actually the *meta-* and *para-* isomers of the same polyamide; namely, poly(*m*-phenylene isophthalamide) (PMPI) and poly(*p*-phenylene terephthalamide) (PPPT) (Fig. 4) [42]. Aramids typically possess excellent mechanical strength and thermal stability owing to their high cohesive energy and the extremely efficient polymer packing that both arise in a large part from the interchain hydrogen bonding between amide linkages. Such a tight packing imparts aramids with a very low gas permeability but a high size-selectivity [43]. Typical structural modifications on aramids involve bulky substitutions on aromatic rings that could act as spacers to reduce the extent of interchain hydrogen bonding such that the free volume and hence gas permeability could be increased [29]. Meanwhile, the weakened interchain interactions also help enhance aramids' processability in common solvents. For example, DuPont produced ring-substituted high-performance aramids for commercial hydrogen purification in refinery feed streams and they demonstrated a H_2 permeability from 4 to 40 *Barrer* and a H_2/CH_4 selectivity of 75 to 600 for separating equimolar H_2/CH_4 mixtures at a moderately high temperature of 90 °C [44].

2.1.1.6 Polyimide

Aromatic polyimides (PI) used as GS membrane materials are generally polycondensation products of aromatic dianhydrides and aromatic diamines [28]. PIs are by far the most studied and also the most rapidly growing family of gas separation polymers owing to their broad monomer choices, attractive gas transport properties, high processability in common organic solvents as well as one of the best mechanical and thermal properties among most glassy polymers [29]. The highly rigid polymer backbones of PIs impart them with very high T_g and hence superior thermo-oxidative resistance. A well-known example of commercial PIs is Matrimid® (Fig. 4) which is synthesized from 3,3'-3,3'-benzophenone tetracarboxylic dianhydride (BTDA) and diaminophenylindane (DAPI) [45]. The bent indane structure and the bulky methyl and phenyl groups that extend outside the plane could significantly hinder the intra-segmental mobility, effectively stiffening the polymer backbone and subsequently giving rise to a high T_g of above 300 °C for Matrimid®. Meanwhile, the efficient interchain packing is also impeded by these rigid and odd-shaped backbones, leading to the higher H_2 permeability and comparable or even better H_2/N_2 and H_2/CH_4 selectivities of Matrimid®

as compared to PC, PSF and CA [29]. As a result, Matrimid®'s H_2/N_2 and H_2/CH_4 separation performances were placed much closer to the 1991 upper bounds than most other first-generation commercial glassy polymers. Another good example of commercial PIs is BPDA-based (biphenyl tetracarboxylic dianhydride) polyimides that were developed and first commercialized by Ube Industries [46]. Upilex® (Fig. 4), synthesized from BPDA and 4,4'-oxydianiline (ODA) monomers, is one representative member of the Ube polyimide family and it has demonstrated outstanding hydrogen separation performance with a H_2 permeability of 50 *Barrer* and H_2/N_2 and H_2/CH_4 selectivities of 83 and 125, respectively [30].

One key reason why PIs have attracted such broad attention is that there exists a huge number of structural variants depending on the choice of dianhydride and diamine monomers, which offer a versatile opportunity to tailor their gas transport properties via a relatively easy polycondensation synthesis scheme. For example, the dianhydride structure can vary from being single-ring to double-ring and the choice of bridging moieties for the latter is also numerous [28]. The diamine structure can also vary in similar manners. Fig. 5 illustrates some common choices of dianhydride and diamine structures constituting the corresponding PI backbones. The general trend is that the permeability tends to increase and the selectivity tends to decrease as the dianhydride and/or diamine units become bulkier. Among the various dianhydride choices, those containing the hexafluoro bridging moieties, for example, hexafluoro isopropylidene diphthalic anhydride (6FDA), are highly popular because the bulky hexafluoro group could effectively hinder the intra-segmental mobility, stiffen the polymer backbone, and inhibit polymer packing and interchain interaction as spacers, resulting in not only a significant increase in permeability, but also sometimes a simultaneous enhancement in selectivity [29]. These favorable gas transport properties observed on hexafluoro dianhydrides allowed the 6FDA-based PIs to be among the earliest researched PIs and are often referred to as the first-generation PIs.

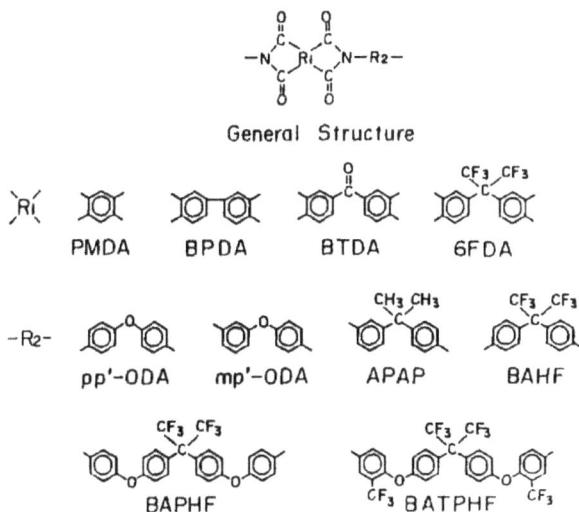

Figure 5. Some common choices of dianhydride and diamine for building polyimides. Reproduced with permission [28]. Copyright 1994, Elsevier Ltd.

2.1.1.7 Other Glassy Polymers for H_2/N_2 and H_2/CH_4 Separations

Besides the above six early commercial examples which have been relatively more extensively studied, there are many other glassy polymer choices available that have demonstrated viable H_2/N_2 and H_2/CH_4 separation performances. As discussed before, these two separations rely primarily on the high diffusivity selectivity of membranes, which is a common feature of most glassy polymers. They also involve no condensable species that tend to deteriorate the membrane performance and subsequently complicate the polymer choice and structural design. The chemical structures of additional representatives, either commercial or research glassy polymers, including polyetherimide (PEI), polyethersulfone (PES), polyetherketone (PEK), polystyrene (PS), poly(methyl methacrylate) (PMMA), polyaniline, polyamide imide (PAI), polypyrrolone (PP), polytriazole (PT) and polynorbornene (PN), are summarized in Fig. 6.

PEI (Ultem® 1000)

Polyaniline (emeraldine base)

PES (Ultrason® E)

PAI

PEK

Polypyrrolone

PEEK (Victrex®)

Polytriazole

Polystyrene

Polynorbornene

PMMA

Cardo-polymer (Cardo-PC)

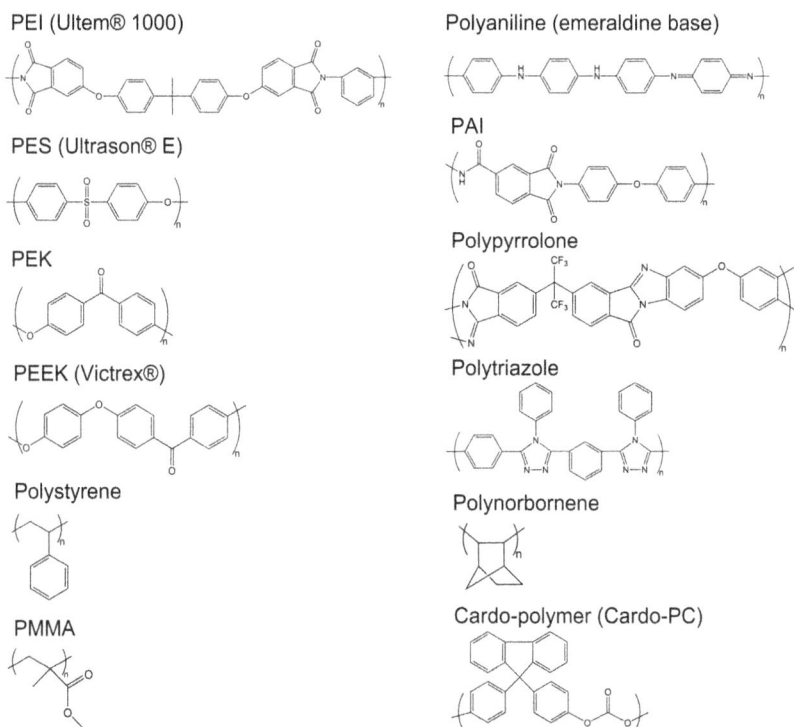

Figure 6. Structures of various other glassy polymers for gas separation

Cardo-type polymers that are based on conventional polymer backbones but modified with the fluorene structure are an interesting class of glassy polymers. They demonstrate excellent thermal stability for operations up to 200 °C and could be made from many different choices of conventional polymers, such as polyimides, polyamides, polyesters, polysulfones, polycarbonates and polyetherketones, by incorporating the fluorene structure into the repeating unit of their polymer backbones [47]. The structure of a fluorene-based cardo-polycarbonate is shown in Fig. 6. It has been reported that the cardo-type PSF and PC exhibit higher H_2 permeabilities than their respective unmodified polymers while their H_2/CH_4 selectivities remain somewhat similar [48].

Teflon AF2400 (x = 87)   Hyflon AD60X (x = 60)

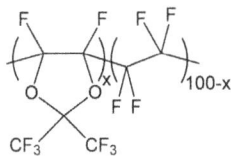

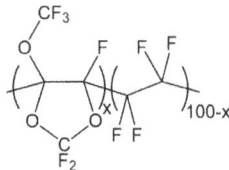

Figure 7. Structures of representative commercial perfluoropolymers for gas separation

Perfluoropolymers are a special class of glassy polymers whose hydrogen atoms have all been replaced with fluorine atoms. Besides having very good mechanical properties, perfluoropolymers generally possess extremely high chemical stability and thermal resistance owing to the high C-F bond energy [8]. One unique property of perfluoropolymers is their unusual sorption behavior which results in a very low solubility of condensable species in them. This gives rise to the high resistance of perfluoropolymers against the swelling and plasticization induced by highly condensable gases, like CO_2 and light HCs, even at high pressures. Because of that, perfluoropolymers have been actually developed initially for tackling challenging separations that involve hydrocarbons or for natural gas treatment. As the energy barrier for rotation is much higher when there are all fluorine atoms on the backbone, the rigidity of polymer chains is greatly enhanced and the tendency of efficient packing is reduced [49]. As a result, perfluoropolymers generally display very high H_2 permeabilities while maintaining good H_2/N_2 and H_2/CH_4 selectivities that allow their overall separation performances to be superior to most conventional commercial polymers discussed in previous sections. For example, Teflon AF2400 developed by DuPont has an unusually high H_2 permeability of 3300 *Barrer* and a sufficient H_2/CH_4 selectivity of 5.5 [50]. Hyflon AD60X developed by Solvay Solexis has a much lower H_2 permeability of 187 *Barrer*, which is still higher than most other conventional glassy polymers, and an excellent H_2/CH_4 selectivity of around 62 [51]. Such separation performances place these two commercial perfluoropolymers well beyond the 1991 upper bound, and in fact, the update on the upper bounds in 2008 was in part promoted by the high performance of perfluoropolymers [52]. Both the structures of Teflon AF2400 and Hyflon AD60X are shown in Fig. 7. It is interesting to note that unlike other highly permeable glassy polymers, most perfluoropolymers experience minimal physical aging. Fig. 8 summarizes the H_2/N_2 and H_2/CH_4 separation performances of most glassy polymers discussed so far on the respective 1991 and 2008 upper bounds.

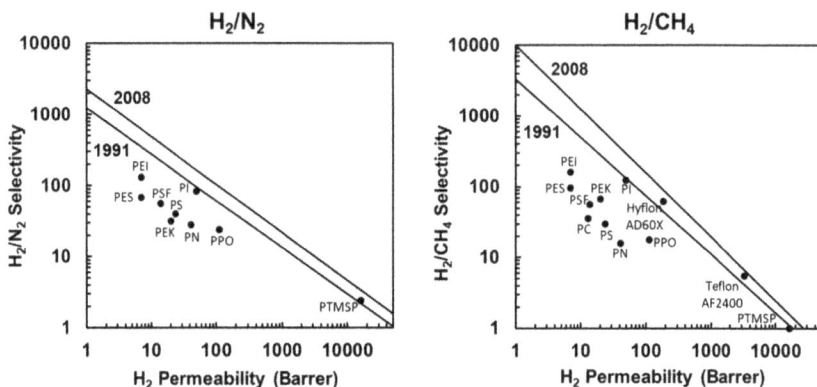

Figure 8. Hydrogen separation performance of representative glassy polymers on 1991 and 2008 Robeson upper bounds.

The separation of H_2 from N_2 and CH_4 by their size differences is a straightforward membrane separation process and its mechanism is also relatively simple to understand as the gas transport is primarily governed by the diffusivity selectivity while the solubility selectivity plays a little role in that. Virtually all glassy polymers that can be fabricated into dense membranes could provide some selectivities for H_2 over N_2 and CH_4 because of their sufficiently large size differences (still on a sub-angstrom level though) as evidenced in this section by many examples viable for these two separations. The direction for new design is then shifted to enhancing H_2 permeability via creating a microporous polymer structure with an average pore size < 2 nm in the dense membranes for faster diffusion of gases. While most literatures, including this chapter, refer to such new polymer structures that are of interest to highly permeable gas separations as 'microporous', it is important to understand that the dimension of pores within these structures are, in fact, in the nano- or sub-nano range rather than the micron range.

2.1.2 Microporous polymers

2.1.2.1 Polyacetylene

Polyacetylenes are characterized by the alternating double bonds and are synthetic glassy polymers with the highest intrinsic permeabilities for small gas molecules, which is even higher than most rubbery polymers including PDMS [49, 53]. In 1983, Masuda et al. successfully synthesized the most representative member of disubstituted polyacetylene, poly(trimethylsilyl)propyne (PTMSP), which possessed a high T_g of above 200 °C. It soon became the most widely known polyacetylene for GS and is now generally considered as the first generation of intrinsically microporous polymers [54]. The structure of PTMSP is shown in Fig. 9. The extraordinarily high gas permeability of PTMSP altered the common perception that only rubbery polymers could possess an extremely high permeability at that time.

PTMSP has very high chain stiffness due to both the alternating double bonds that rigidify the polymer backbone as well as the bulky trimethylsilyl side groups that further hinder the intra-segmental rotation and mobility [55]. The latter also serves as bulky spacers that induce looser packing among polymers and enlarge the interchain spacing. These effects collectively result in the extremely high FFV of PTMSP (up to 30%) with an average free volume size of around 0.68 nm which is more than twice as big as that of conventional glassy polymer [7, 8]. These free volume elements or nanopores also appear to be interconnected, forming a so-called interconnected microporous structure inside the PTMSP membranes. This special structure imparts PTMSP with diffusivities of several orders of magnitude higher for small gas molecules, like H_2 and CO_2, than other glassy polymers, and its H_2 permeability is in the range of ~10^4 *Barrer* [26]. However, the size-selectivity of PTMSP is among the lowest due to its extra-large FFV, which renders its commercial hydrogen separation unpreferable. The physical aging problem caused by the gradual relaxation of thermodynamically unstable excess free volumes with time further compromises its separation efficiency and attractiveness for commercial use [8]. Crosslinking the PTMSP polymers to reduce aging has been attempted in the literatures, but only limited success has been obtained so far. An extremely high H_2 permeability of 16,160 *Barrer*, coupled with very low H_2/N_2 and H_2/CH_4 selectivities of 2.4 and 1.01 respectively, has placed PTMSP at the high permeability end of the 1991 upper bounds [24].

PTMSP

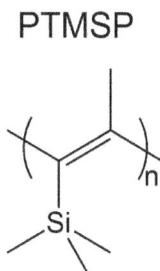

Figure 9. The chemical structure of PTMSP.

Despite being unsuitable for the diffusivity-selectivity-governed H_2/N_2 and H_2/CH_4 separations, PTMSP have actually found promising applicability for the reverse-selective H_2 and N_2 separations, such as the HCs/H_2 and HCs/N_2 separations, which primarily rely on the high solubility selectivity of PTMSP for condensable gases over light gases [56]. The large free volume of PTMSP could act as spacious sorption sites for these condensable gases, giving them very high solubility in PTMSP and allowing them to effectively compete for passage through the polymers with small, non-condensable gases like H_2 [8]. PTMSP has thus been often compared with as an industrial benchmark for vapor/gas separation, and a more detailed discussion on polymers for reverse-selective H_2 and N_2 separations will be given in a later section.

2.1.2.2 Polymers of intrinsic microporosity

Polymers of intrinsic microporosity (PIM) are a new class of microporous glassy polymers with good solution processability, film-forming ability as well as high functional tunability. They were first developed by Budd and McKeown in 2004 and have revealed a novel polymer structure comprising interconnected micropores of less than 2 nm (20 Å) owing to the presence of bulky contortion sites on their rigid backbones that prevent efficient chain packing [57, 58]. Such microporosity is considered intrinsic because it arises from the inherent polymer structure instead of the processing protocol. The possession of intrinsic microporosity imparts PIMs with a large FFV and also higher gas and vapor solubility than the traditional classy polymers, including even the most permeable PTMSP [49]. Shown in Fig. 10 is the archetypical ladder-type PIM-1 structure, which could obtain gas permeabilities about hundred times higher than traditional glassy polymers due to both high diffusivity and solubility coefficients arising from its large FFV [7]. The gas permeabilities of PIMs are generally in a decreasing order from CO_2, H_2, O_2, CH_4 to N_2 [8]. At the same time, the diffusivity selectivity of PIM-1 is not completely compromised because of the high rigidity of polymer chains such that the permselectivity of PIM-1 remains within reasonable ranges even with such high permeabilities. As a result, the H_2/N_2 and H_2/CH_4 separation performances of early PIMs were typically between the 1991 and 2008 upper bounds or, in some cases, even surpassed the 2008 upper bounds [7]. In fact, the outstanding performance demonstrated by PIM membranes has promoted the redefinition of the upper bounds for these two hydrogen separations twice, once in 2008 and the other time in 2015, and have been attracting huge research attention among the membrane community ever since the advent of PIM-1 [37, 52]. One interesting key feature of PIM membranes is that they tend to swell reversibly in non-solvents like methanol or ethanol, which makes their gas permeation properties fairly sensitive to the film-forming protocol [8]. For example, a certain duration of treatment in a non-solvent can result in a few times higher gas permeabilities of the PIM membranes as shown in Table 4.

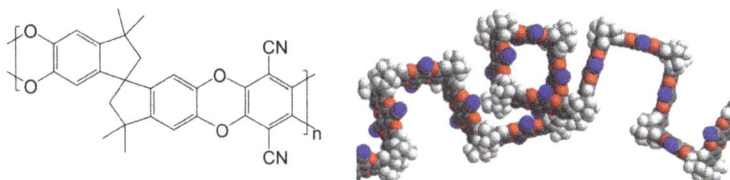

Figure 10. The polymer structure of PIM-1. Reproduced with permission [59]. Copyright 2016, Springer Publishing Group.

In terms of the nature of their polymer backbone, PIMs can be classified into two general types: ladder PIMs and polyimides of intrinsic microporosity (PIM-PIs). Ladder PIMs usually contain a characteristic sterically hindered building block in their repeating units. These building blocks provide the sites of contortion for creating the kinked ladder conformation which gives rise to the rigid and contorted PIM backbones that significantly upset the efficient chain packing. For example, ladder-type monomers consisting of two aromatic rings linked by a single tetrahedral carbon atom (also known as the spiro-carbon center), such as the spirobisindane-based monomer used in PIM-1, are classic examples for constructing early ladder PIM structures. Ladder PIMs can be also sub-categorized according to the types of their sterically hindered building blocks, including but not limited to, spirobisindane (SBI), phenazine, ethanoanthracene (EA), triptycene (Trip), spirobifluorene (SBF), Troger's base (TB), and tetraphenylethylene (TPE) [10].

PIM-PIs are polyimides with rigid and contorted PIM segments that contain the sterically hindered contortion sites. The first series of PIM-PIs, ranging from PIM-PI-1 to 8, were reported by Ghanem et al. in 2008 [60]. The structural design of PIM-PIs generally depends on the target location of contortion sites, whether in the dianhydride unit or in the diamine unit, or both. A diagram illustrating the detailed classification of PIMs according to their structures is shown in Fig. 11. It is important to note that physical aging has been a notorious problem associated with highly permeable glassy polymers including PIMs, but the focus of this section is on PIM structural designs targeting superior intrinsic gas permeation properties so that the detailed discussions on aging behavior is outside the scope here [61].

(a) Ladder PIMs

As shown in Fig. 11, the structure of ladder PIMs is generally characterized by three types of linkage groups that integrate the sterically hindered building blocks into the backbone; namely, the benzodioxane-, phenazine-, and TB-based linkers. Given the highly rigid and/or contorted nature of the phenazine- and TB-based linkers, they can also be considered as the sterically hindered building blocks themselves. In the design of PIM structures, the key to generating the intrinsic microporosity is to use at least one monomer with a rigid and contorted structure that can effectively disrupt efficient chain packing while keeping sufficient chain stiffness for good diffusivity selectivity [10]. Common choices of monomers with bulky contortion sites (i.e., the sterically hindered building blocks) include those with the spiro-carbon center (e.g. SBI and SBF), a single covalent bond with highly restricted rotation (e.g. TPE), or non-planar rigid bulky 3D structures (e.g. EA and Trip), which in combination with the three types of linkers can produce a huge variety of ladder PIM structures.

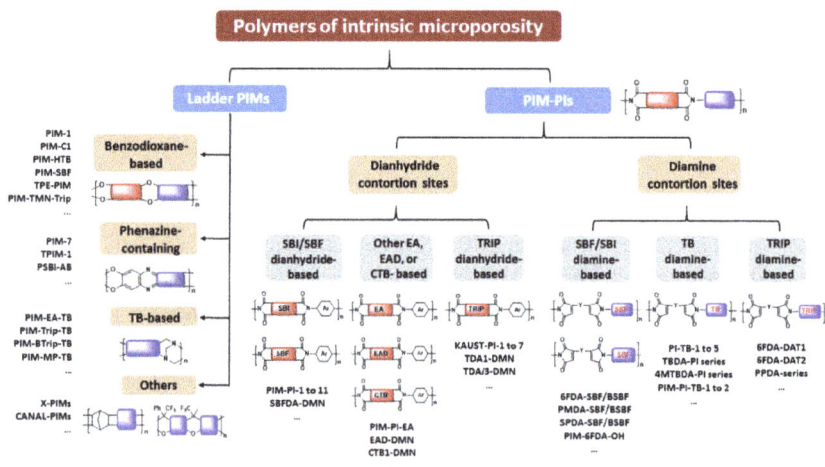

Figure 11. Classification of polymers of intrinsic microporosity. Reproduced with permission [10]. Copyright 2018, Elsevier Ltd.

(b) Ladder PIMs with Benzodioxane Linker

The archetypal PIMs, also widely recognized as the first-generation PIMs, were synthesized by the aromatic nucleophilic substitution reaction that formed the benzodioxane linkers in the resultant polymers [57, 58]. Such reactions typically occur between a spirocyclic bis-catechol monomer (e.g. spirobisindane-based) and a tetrahalo monomer (often a tetrafluoro monomer) and will give rise to a fused-ring polymer structure. For example, PIM-1 is the earliest benzoedioxane-type ladder PIM which was synthesized from equimolar 5,5',6,6'-tetrahydroxy-3,3,3,3'-tetramethyl-1,1'-spirobisindane (TTSBI) and 2,3,5,6-tetrafluoroterephthalonitrile (TFTPN), and has been by far the most extensively studied ladder PIM member for gas separation owing to its simple structure, promising performance and physical properties as well as excellent solution-processability [10]. McKeown's group recently reported a unique and interesting 2D ladder PIM structure, named PIM-TMN-Trip (TMN = tetramethyltetrahydronaphthalene), which is solution-processable and ultra-permeable [62]. In the PIM-TMN-Trip structure, two phenyl rings of the rigid Trip unit were used to construct the benzodioxane linkers in the ribbon-like, shape-persistent 2D backbones, while the third ring was fused to a TMN group to maintain the high 2D aspect ratio for obtaining the ultra-permeability by significantly impeding the efficient chain packing. The PIM-TMN-Trip membranes demonstrated an extraordinarily high H_2 permeability while keeping sufficient H_2/N_2 and H_2/CH_4 selectivities for easily surpassing the 2008 upper bounds. Various representative benzodioxane-based ladder PIM structures are shown in Fig. 12.

(c) Ladder PIMs with Phenazine Linker

Phenazine-based linkers are more rigid than the benzodioxane-based ones. Examples of phenazine-containing ladder PIMs include PIM-7, PSBI-AB and TPIMs whose structures are shown in Fig. 13 [63-65]. Among these examples, TPIM-1 developed by Pinnau's group in 2014, which comprised both phenazine linkers and Trip building blocks in its rigid, fused-ring structure, demonstrated superior hydrogen separation performances owing to its strong molecular-sieving characteristic [64]. TPIM-1 was actually one of the top-performance PIMs that was used to establish the 2015 H_2/N_2 and H_2/CH_4 upper bounds [37].

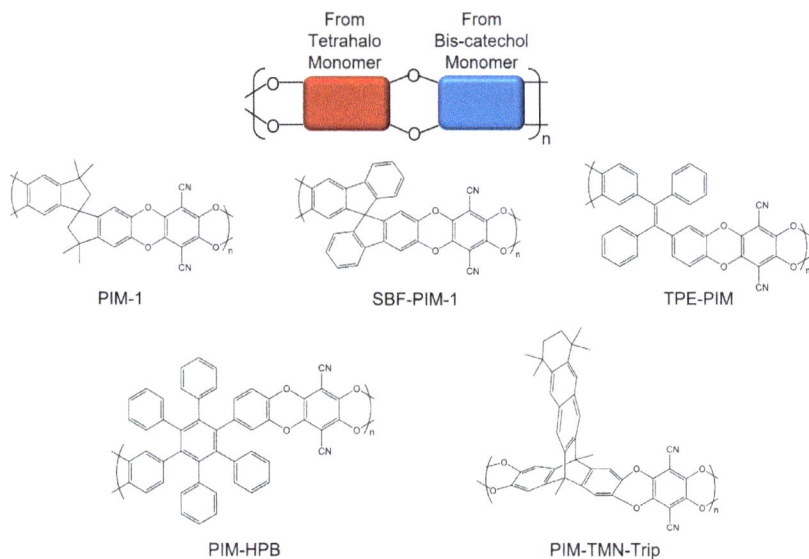

Figure 12. Structures of some representative benzodioxane-based ladder PIMs

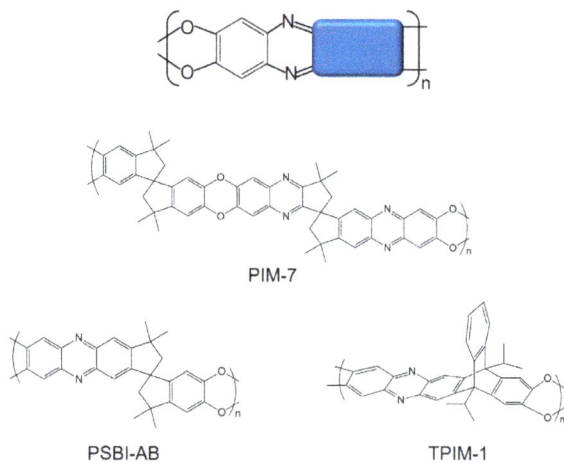

Figure 13. Structures of some representative phenazine-based ladder PIMs

(d) Ladder PIMs with Troger's Base Linker

Troger's base (TB) is a bulky and V-shape linker first introduced into PIMs and PIM-PIs in 2013 by Carta et al. [66]. They developed the novel PIM-EA-TB polymer that combined two highly rigid, bicyclic and shape-persistent groups; namely, ethanoanthracene (EA) and TB, into the same compact, fused-ring ladder structure with no spiro-carbon centers or dioxane linkers. Such an advanced polymer structure displayed a strong nature of intrinsic ultramicroporosity and also outstanding hydrogen separation performances. Various novel TB-based ladder PIM structures are shown in Fig. 14, and the hydrogen separation performances of several representative ladder PIMs are summarized in Table 4.

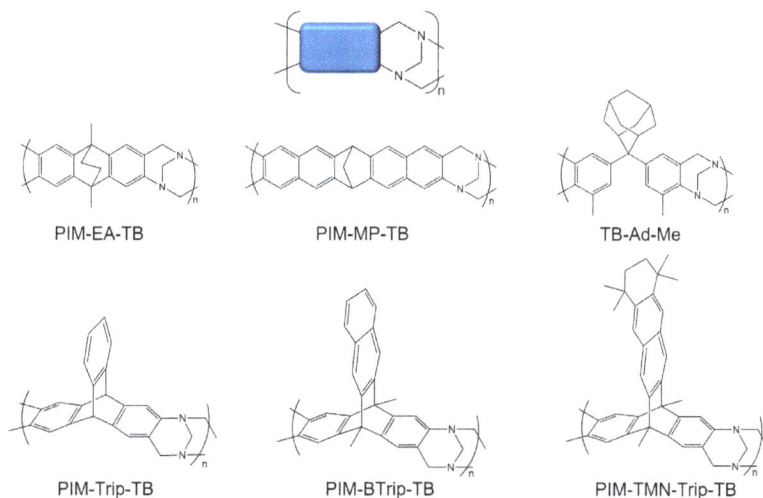

| PIM-EA-TB | PIM-MP-TB | TB-Ad-Me |

| PIM-Trip-TB | PIM-BTrip-TB | PIM-TMN-Trip-TB |

Figure 14. Structures of some representative TB-based ladder PIMs

(e) Polyimides of Intrinsic Microporosity

Highly rigid and sterically hindered bulky sites of contortion can be introduced into polyimide backbones to create polyimides of intrinsic microporosity (PIM-PI), which like conventional polyimides are typically synthesized via the polycondensation between an aromatic dianhydride and an aromatic diamine. Depending on where the introduced contortion site is (i.e., in the dianhydride monomer or the diamine monomer), the structure of PIM-PIs can be sub-categorized into those that contain dianhydride contortion sites, those that contain diamine contortion sites, or both.

Table 4. Summary of hydrogen separation performance of representative ladder PIMs[10, 37, 67]

Ladder PIMs	Permeability (Barrer)			Selectivity	
	H_2	N_2	CH_4	H_2/N_2	H_2/CH_4
Benzodioxane-based					
PIM-1[a]	1300	92	125	14.1	10.4
PIM-1[b]	5010	823	1360	6.1	3.7
SBF-PIM-1[b]	6320	786	1100	8	5.7
PIM-HPB[b]	1413	190	361	7.4	3.9
TPE-PIM[c]	604	33.4	41	18.1	14.7
PIM-TMN-Trip[b]	16900	2230	3420	7.6	4.9
Phenazine-based					
PIM-7[a]	860	42	62	20.5	13.9
PSBI-AB[c]	1630	141	257	11.5	6.3
TPIM-1[e]	2666	54	50	49.4	53.3
TB-base					
PIM-EA-TB[b]	7760	525	699	14.8	11.1
PIM-Trip-TB[b]	8039	629	905	12.8	8.9
PIM-TMN-Trip-TB[b]	6100	396	710	15.4	8.6
TB-Ad-Me[b]	1800	121	162	14.8	11.1
PIM-BTrip-TB[b]	9980	926	1440	10.8	6.9
PIM-MP-TB[b]	4050	200	264	20.3	15.3

[a] *As-cast membranes tested at 30 °C and 200 – 300 mbar*
[b] *Methanol treated and air-dried before testing at 25 °C and 1 bar*
[c] *Methanol treated, air-dried and then 120 °C dried under vacuum overnight before testing at 35 °C and 2 bar*
[d] *Methanol treated before testing at 22 °C and 4.4 bar*
[e] *Exactly same as* [b] *except that testing temperature was 35 °C*

(f) Contortion Sites in the Dianhydride

The first series of PIM-PIs were invented by McKeown's group in 2008 which utilized the classic SBI unit as the contortion site in its dianhydride monomer [60]. The choice of diamine was varied to form PIM-PI-1 to PIM-PI-8. Compared with most conventional PIs, these early PIM-PIs exhibited a much higher FFV, larger Brunauer-Emmett-Teller (BET) surface area and lower density due to inefficient packing arising from the rigid and contorted PIM-type segments [68, 69]. As a result, they could achieve much higher H_2 permeabilities but lower H_2/N_2 and H_2/CH_4 selectivities than conventional PIs. Nevertheless, as compared to ladder PIMs, especially those with superior gas transport properties, the hydrogen separation performances of early PIM-PIs were not as attractive. The structures of some representative PIM-PIs built by SBI- or SBF-containing dianhydrides are shown in Fig. 15.

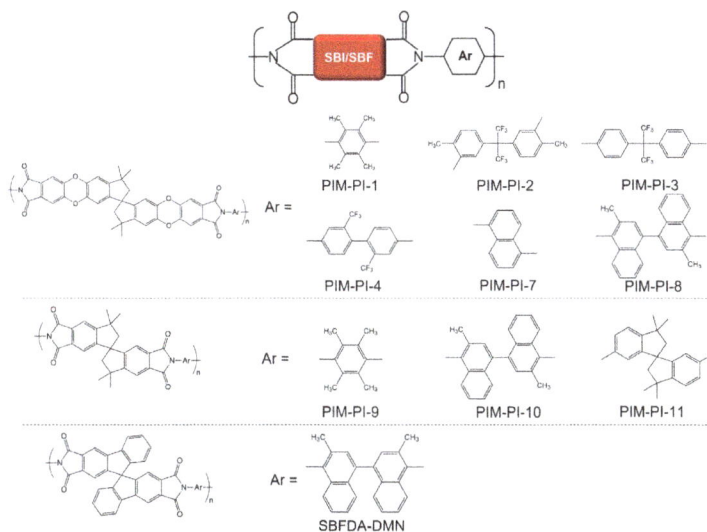

Figure 15. Structures of representative PIM-PIs built by SBI- or SBF-containing dianhydrides

The same principle in designing novel ladder PIMs, which utilized the bridged bicyclic building blocks with high rigidity and contortedness, such as TB, EA and Trip, to increase the chain rigidity and internal free volume of polymers, could be also applied to the design of advanced PIM-PI structures. For example, structures like PIM-PI-EA (also known as PIM-PI-12) and CTB-DMN (a PIM-PI based on EA and carbocyclic pseudo TB (CTB) building blocks) delivered much better H_2/N_2 and H_2/CH_4 separation performances than early PIM-PIs that utilized the comparatively less rigid SBI and SBF units. [70, 71]. Fig. 16 provides the structures of some representative PIM-PI structures with EA and TB-containing dianhydrides. One of the real breakthroughs took place when Ghanem at al. and Swaidan at al. developed a series of novel, top-performance PIM-PIs, called KAUST-PI-1 to KAUST-PI-7, using a rigid, 3D paddlewheel-structured Trip-containing dianhydride (Fig. 17) [72, 73]. With 2,3,5,6-tetramethyl-1,4-phenylene (TMPD or durene) being used as the diamine, the KAUST-PI-1 polymer reported by Ghanem et al. in 2014 achieved extraordinary H_2/N_2 and H_2/CH_4 separation performances that sit very close to or on the respective 2015 upper bounds [37, 72]. This innovative PIM-PI structure prompted the realization of an efficient strategy for designing high-performance PIM-PIs. It created a tightened bimodal interconnected microporous structure in the polymers with a unique combination of large micropores (> 10 Å) that not only facilitated a high permeability and the formation of ultrafine micropores (< 7 Å) but also maintained a good size-selectivity. As a result, KAUST-PI-1 displayed far superior performance than early ladder PIMs and PIM-PIs. The hydrogen separation performances of several representative PIM-PIs containing contorted dianhydrides are summarized in Table 5.

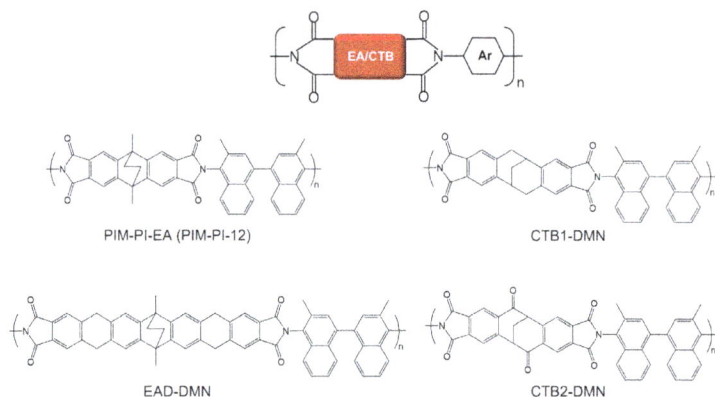

Figure 16. Structures of representative PIM-PIs built by EA- or CTB-containing dianhydrides

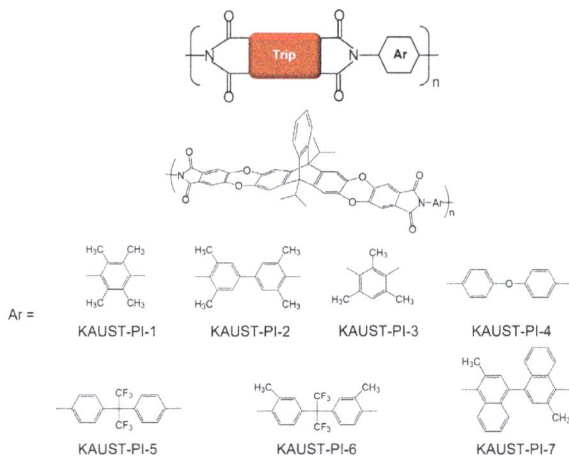

Figure 17. Structures of representative PIM-PIs built by Trip-containing dianhydrides

Table 5. Summary of hydrogen separation performance of PIM-PIs with contorted dianhydrides[10, 37, 73]

PIM-PIs with Contorted Dianhydride	Permeability (Barrer)			Selectivity	
	H_2	N_2	CH_4	H_2/N_2	H_2/CH_4
PIM-PI-1[a]	530	47	77	11.3	6.9
PIM-PI-2[a]	220	9	9	24.4	24.4
PIM-PI-3[a]	360	23	27	15.7	13.3
PIM-PI-4[a]	300	16	20	18.8	15
PIM-PI-7[a]	350	19	27	18.4	13
PIM-PI-8[a]	1600	160	260	10	6.2
PIM-PI-9[b]	840	94	170	8.9	4.9
PIM-PI-10[b]	670	84	168	8	4
PIM-PI-11[b]	624	65	129	9.6	4.8
PIM-PI-12[d] (PIM-PI-EA)	4230	369	457	11.5	9.3
SBFDA-DMN[c]	2966	226	326	13.1	9.1
EAD-DMN[e]	4703	480	707	9.8	6.7
CTB1-DMN[c]	1295	76.2	95.7	17	13.5
CTB2-DMN[c]	1150	39.9	40.4	28.8	28.5
KAUST-PI-1[c]	3983	107	105	37.2	37.9
KAUST-PI-2[c]	2368	98	101	24.2	23.4
KAUST-PI-3[c]	1625	43	43	37.8	37.8
KAUST-PI-4[c]	302	10.6	10.7	28.5	28.2
KAUST-PI-5[c]	1558	87	77	17.9	20.2
KAUST-PI-6[c]	409	14.4	11	28.4	37.2
KAUST-PI-7[c]	3198	225	354	14.2	9

[a] *Air-dried before testing at 30 °C and 200 – 500 mbar*
[b] *Methanol treated and air-dried before testing at 25 °C and 1 bar*
[c] *Methanol treated, air-dried and then 120 °C dried under vacuum overnight before testing at 35 °C and 2 bar*
[d] *Methanol treated for 8 hours and air-dried before testing at 25 °C and 1 bar*
[e] *Methanol treated for 24 hours, air-dried for another 24 hours and then degassed under vacuum for 24 hours before testing at 35 °C and 2 bar*

(g) Contortion Sites in the Diamine

Many choices of sterically hindered building blocks have been also explored on the diamines in PIM-PIs, including SBF, SBI, TB and Trip units [74-82]. The structures of various PIM-PIs containing contorted diamines or both contorted dianhydrides and diamines are shown in Fig. 18, 19 and 20. Generally, the gas separation performance of PIM-PIs with contorted diamines depended strongly on the type of dianhydride used, their overall separation performance fell below those containing the contorted dianhydrides mentioned before. The hydrogen separation performances of several representative PIM-PIs containing sterically contorted diamines are summarized in Table 6.

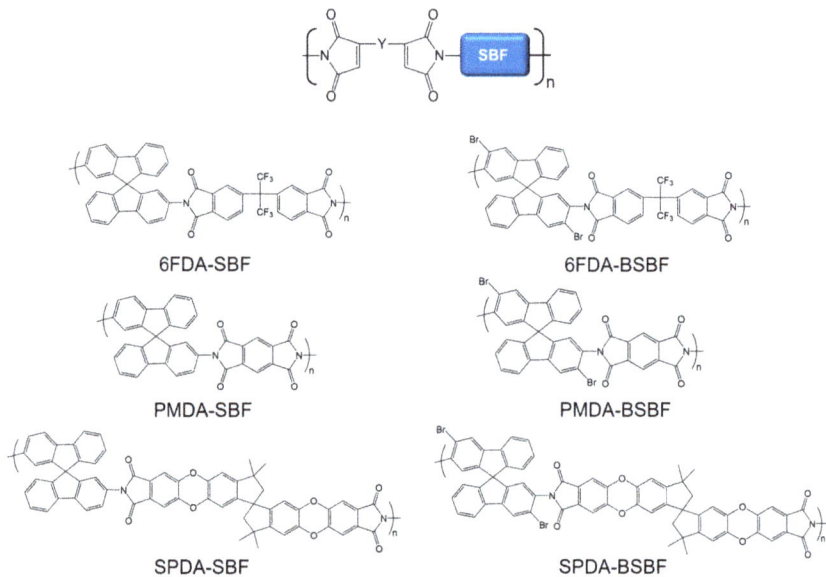

Figure 18. Structures of representative PIM-PIs containing SBF diamines

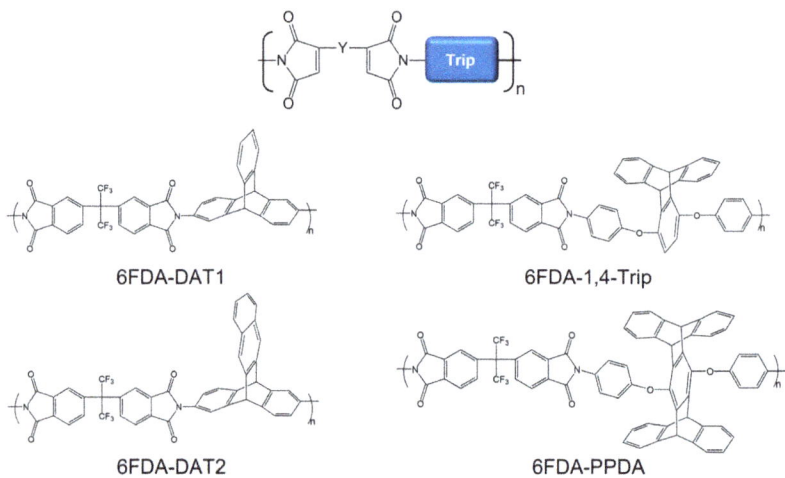

Figure 19. Structures of representative PIM-PIs containing Trip diamines

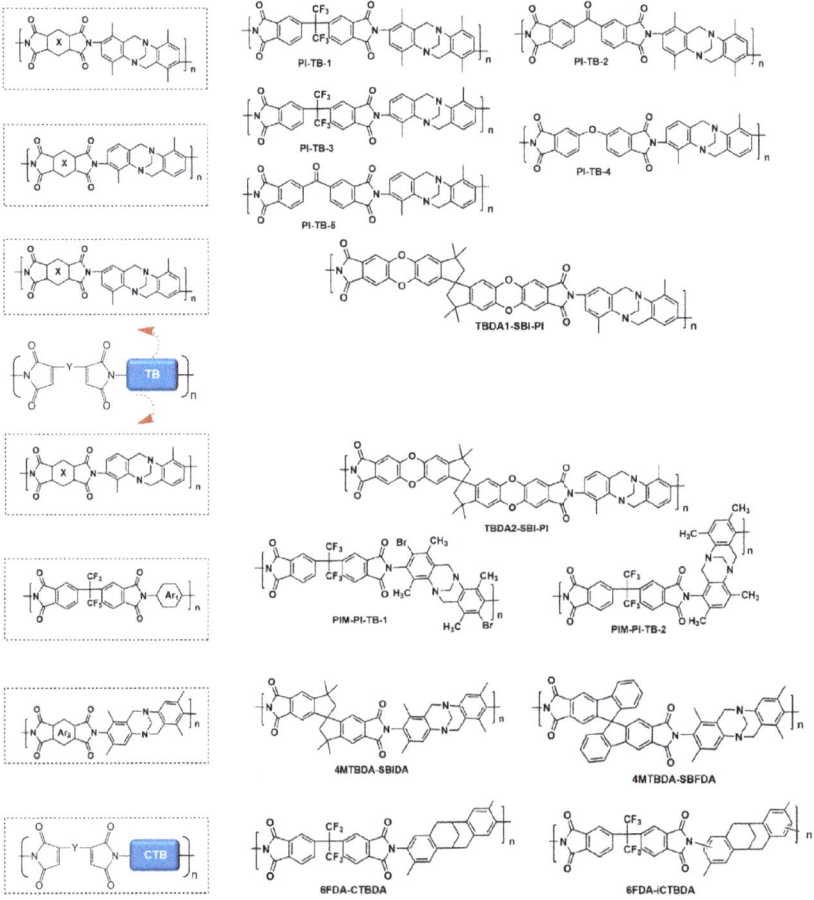

Figure 20. Structures of representative PIM-PIs containing TB diamines. Reproduced with permission [10]. Copyright 2018, Elsevier Ltd.

Table 6. Summary of hydrogen separation performance of PIM-PIs with contorted diamines[10, 75-82]

PIM-PIs with Contorted Diamine	Permeability (Barrer)			Selectivity	
	H_2	N_2	CH_4	H_2/N_2	H_2/CH_4
6FDA-SBF[a]	234	7.8	6.4	30	36.6
PMDA-SBF[a]	230	8.5	9.1	27.1	25.3
SPDA-SBF[a]	501	28.6	41.1	17.5	12.2
6FDA-BSBF[a]	531	27	24.9	19.7	21.3
PMDA-BSBF[a]	560	28.8	36.5	19.4	15.3
SPDA-BSBF[a]	919	69	102	13.3	9
PI-TB-1[b]	607	31	27	19.6	22.5
PI-TB-2[b]	134	2.5	2.1	53.6	63.8
PI-TB-3[b]	299	9.5	6.7	31.5	44.6
PI-TB-4[b]	40	1.3	1	30.8	40
PI-TB-5[b]	53.8	1.9	1.7	28.3	31.6
TBDA1-SBI-PI[d]	915	35	45	26.1	20.3
TBDA2-SBI-PI[d]	1155	49	65	23.6	17.8
PIM-PI-TB-1[a]	612	42	44	14.6	13.9
PIM-PI-TB-2[a]	582	34	31	17.1	18.8
4MTBDA-SBIDA[c]	3200	372	591	8.6	5.4
4MTBDA-SBFDA[c]	2901	264	371	11	7.8
6FDA-DAT1[a]	198	4.7	3.2	42.1	61.9
6FDA-DAT2[a]	281	9	7.1	31.2	39.6
6FDA-1,4-Trip[e]	49	0.66	0.37	74.2	132
6FDA-PPDA[f]	131	3.2	2.5	40.9	52.4

[a] *Methanol treated and then 120 °C dried under vacuum for 24 hours before testing at 35 °C and 2 bar*
[b] *Methanol treated, 120 °C dried under vacuum for 2 hours and then stored under ambient condition before testing at 35 °C and 1 bar*
[c] *Methanol treated and air-dried overnight before testing at 25 °C and 1 bar*
[d] *Dried under high vacuum for 24 hours, methanol treated, and then 120 °C dried again under vacuum for 24 hours before testing at 35 °C and 1 bar*
[e] *Dried at 200 °C for 24 hours, methanol treated, and then dried again at 200 °C for another 24 hours before testing*
[f] *Methanol treated for 24 hours and then 140 °C dried for another 24 hours before testing at 35 °C*

(h) Functionalized PIMs

While ladder PIMs have successfully realized attractive performances and good processability for hydrogen separation, most of them still lack the competitiveness for more challenging separations that involve condensable species, like CO_2/N_2 separation for CO_2 capture from flue gas. Therefore, proper functionalization on PIM structures to optimize their affinity is a facile and effective design strategy to enhance the solubility of condensable gases like CO_2 [10]. While more details will be given in a later section where polymer chain-functionalization is to be discussed, the structures of some representative functionalized ladder PIMs are shown here in Fig. 21, which includes amidoxime-functionalized PIM-1 (AOPIM-1), tetrazole-functionalized PIMs (TZPIMs), thioamide-functionalized PIM-1 (ThioPIM-1) and carboxylated PIM-1 (PIM-1-COOH) [83-86]. Some of these functionalized ladder PIMs also demonstrated improved H_2/N_2 and H_2/CH_4 separations due to the increased intersegmental hydrogen bonding between the introduced functional groups. For example, AOPIM-1 exhibited a lower H_2 permeability but much higher H_2/N_2 and H_2/CH_4 selectivities than PIM-1, which allowed AOPIM-1's overall separation performance to be placed on the 2008 upper bounds. The hydrogen separation performances of some representative functionalized ladder PIMs are summarized in Table 7. Similarly, the development of functionalized PIM-PIs, for example, the hydroxyl-functionalized ones, also targets gas separations that involve condensable gases [87].

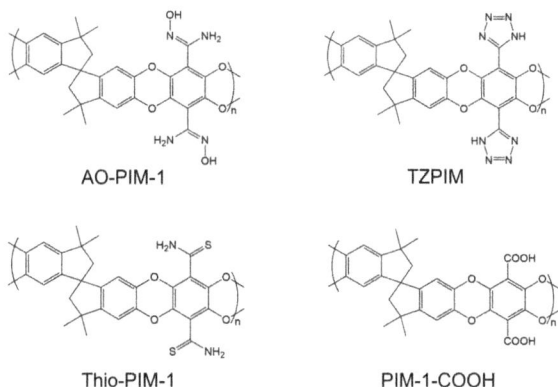

AO-PIM-1   TZPIM

Thio-PIM-1   PIM-1-COOH

Figure 21. Structures of representative functionalized ladder PIMs.

Table 7. Summary of hydrogen separation performance of functionalized ladder PIMs[83, 84, 86]

Functionalized ladder PIMs	Permeability (Barrer)			Selectivity	
	H_2	N_2	CH_4	H_2/N_2	H_2/CH_4
AOPIM-1[a]	912	33	34	27.6	26.8
Thio-PIM-1[b]	610	37	56	16.5	10.9
PIM-1-COOH[c]	408	24	-	17	-

[a] *Methanol treated and then 120 °C dried under vacuum for 24 hours before testing at 35 °C and 2 bar*
[b] *Ethanol treated before testing at 25 °C and 1 bar*
[c] *Hydrolyzed at 120 °C for 5 hours before testing at 35 °C and 4.4 bar*

(i) Updated Separation Performance

The discovery of PIM structure simply opens a new realm of material research on polymeric GS membranes, which has been accelerating the update on the state-of-the-art performance upper bounds. While 17 years lapsed before the 1991 upper bound was first renewed in 2008 by the advent of ladder PIMs like PIM-1 and PIM-7, it only took another 7 years for the 2015 upper bounds for H_2/N_2 and H_2/CH_4 separations to be proposed by Swaiden et al. based on the extraordinary performance of the groundbreaking TPIMs and KAUST-PIs [24, 37, 52]. The 2019 upper bound for CO_2/N_2 performance was also recently published by McKeown's group using a large series of Trip-based ladder PIMs [88]. Most of the novel PIM structures were placed between the 2008 and 2015 upper bounds. As the design principles become more refined with an increasing number of relevant studies, the pool of high-performance PIMs will be further widened. The hydrogen separation performance of numerous representative ladder PIMs and PIM-PIs is compared with the three established upper bounds in Fig. 22. One general design principle widely recognized is to use rigid contorted bicyclic units, such as EA, TB and Trip for achieving superior separation performance [10]. Also, from studies on PIM-PIs, it has been discovered that the gas separation performance tends to follow the order of increasing rigidity of the contortion sites, from SBI, SBF, TB, EA to Trip, in both dianhydride and diamine moieties. However, the best separation performance achieved so far was by having contortion sites in the dianhydride moiety, whereas the performance enhancement was generally less prominent when placing contortion sites in the diamine moiety.

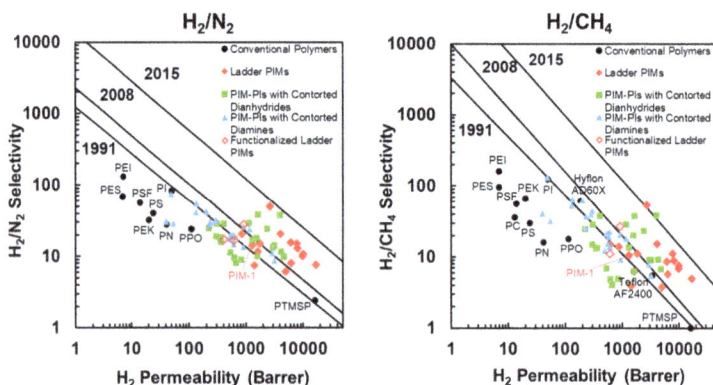

Figure 22. Hydrogen separation performance of representative ladder PIMs and PIM-PIs on 1991, 2008 and 2015 upper bounds

2.2   Polymers for H_2/CO_2 Separation

Compared with H_2/N_2 and H_2/CH_4 separations which only involve gases of low to less than moderate condensabilities, the H_2-selective separation of H_2 and CO_2 is much more challenging because CO_2 tends to exhibit a similar fast permeation rate as H_2 through polymeric membranes, owing to its high solubility in dense polymers, especially those having large free volume that could act as CO_2-favored sorption sites or those that could form polar interactions with CO_2 [49]. Besides, the size difference between H_2 and CO_2 gas molecules is much smaller than that between H_2 and N_2 or CH_4. In order to achieve a high selectivity for H_2 over CO_2, the solubility selectivity which favors the more condensable species needs to be suppressed such that the diffusivity selectivity and hence the size-sieving effect could be the governing factor for the overall gas transport, subsequently leading to the desired high H_2/CO_2 permselectivity [1]. Glassy polymers with relatively more rigid polymer chains, smaller free volume and higher diffusivity selectivity are attractive materials for enabling highly H_2-selective separation of H_2 and CO_2.

In terms of the operation conditions, the H_2/CO_2 separation is favored at high temperatures because the diffusivity of H_2 could be enhanced while the solubility of CO_2 could be suppressed [7, 8]. Operating at elevated temperatures also makes direct industrial sense because the syngas streams containing H_2 and CO_2 are processed or produced at temperatures typically from 150 °C to 300 °C [7]; therefore, no additional heat exchange processes would be required for cooling or heating these gas streams before separation. As such, it is of great industrial interest to find polymeric membrane materials that exhibit superior thermal and mechanical properties to withstand the harsh environments encountered in the H_2/CO_2 separation which aims at the efficient pre-combustion CO_2 capture from syngas. This operation constraint limits the polymer choice to only a few classes of high-performance, heat-resistant glassy polymers, such as polybenzazoles (e.g. polybenzimidazole (PBI) and polybenzoxazoles (PBO)), polyimides, polysulfones, polyamides and polyphosphazenes [89]. Among which, PBI has been showing the greatest promises for efficient membrane H_2/CO_2 separation.

2.2.1 Polybenzimidazole

Polybenzimidazoles (PBI) are a class of heterocyclic glassy polymers from the polybenzazole family and are one of the most widely recognized and studied polymeric material for H_2/CO_2 separation because of their extraordinary thermal stability and attractive separation performance, especially at high temperatures [3, 89]. The structure of the earliest and also the most representative PBI, m-PBI, which has been commercialized by Celanese under the trade name of Celazole®, is shown in Fig. 23

[90]. The *m*-PBI was synthesized from the polycondensation reaction between 3,3-diaminobenzidine and isophthalic acid and it possessed a very high T_g of 425 to 435 °C [91]. Besides its outstanding thermal properties and heat-resistance (decomposition temperature as high as 550 °C), PBI also exhibits high mechanical strength and superior chemical resistance that is uncommon to find in most glassy polymers owing to its rigid aromatic backbone as well as the extensive inter-segmental hydrogen bonding between the imidazole groups [92]. The highly rigid polymer chain also provides higher resistance against CO_2-induced plasticization that could severely deteriorate the size-selectivity of the polymer, which is another advantage of PBI for H_2/CO_2 separation [93]. However, their extremely high chemical resistance results in the low solubility and hence poor solution-processability of bulk PBI polymers in most organic solvents, which limits the use of standard solution-casting method to fabricate thin PBI films [94]. In fact, the poor processability problem is encountered by most thermally and chemically resistant polymers because their extreme stability leads to their low solubility in common organic solvents, rendering their membrane fabrication via the common solution-casting method very difficult. Structural modifications, such as substituting the rigid 3,3-diamine monomer with the 3,3',4,4'-tetraaminodiphenylsulfone (TADPS) monomer that contains a flexible and kinked sulfonyl linkage, are recently found capable of enhancing the solubility of PBI by inhibiting the chain packing and interchain interactions [94].

m-PBI (Celazole®)

TADPS-PBI

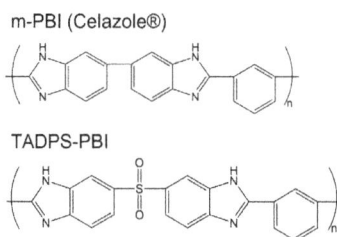

Figure 23. The chemical structures of m-PBI and TADPS-PBI

Many research efforts have also been devoted, since the early 2000s, to the fabrication of 3,3-diamine-based PBIs into flat and fibrous membranes for studying their gas transport properties. Their success has revealed very attractive H_2/CO_2 separation performances. For example, *m*-PBI at 43 °C exhibited a H_2 permeability of around 4 *Barrer* and a H_2/CO_2 selectivity of around 15 which placed its overall performance very close to the 2008 upper bound [95], as shown in Fig. 24. While the H_2 permeability of PBI membranes is small at low temperatures due to the strong π-π stacking effect arising from its rigid, rod-like aromatic chains and strong intermolecular hydrogen bonding, their separation performances in terms of permeability and selectivity could be tremendously enhanced at high operating temperatures. In addition, their H_2 permeability tends to experience a much larger increase than their CO_2 permeability due to the suppressed solubility coefficient of CO_2 at high temperatures [3, 7]. For example, at 250 °C, *m*-PBI could achieve a pure-gas H_2 permeability of around 77 *Barrer* and a H_2/CO_2 selectivity of around 23 which allowed its overall performance to be placed far beyond the 2008 upper bound (Fig. 24) [95]. Structural modifications, such as bulky substitutions on the diacid monomer or using more flexible diacid structures, could also effectively improve the H_2 permeability by hindering the interchain packing and interactions to create a more opened-up pore structure or by increasing the intra-segmental mobility, respectively [92, 95]. As for now, the poor processability of PBI has already been deemed much less of an issue after more than a decade of research efforts, especially after the asymmetric PBI hollow fiber membranes (HFMs) have been successfully fabricated, which marked its fulfilment of the important prerequisite – spinnability – for becoming a commercially practical polymer [96]. Recently, Zhu et al. prepared polyprotic acid-crosslinked PBI membranes via a simple acid doping method, which demonstrated unprecedented size-sieving nature due to their much tightened pore structure after crosslinking (Fig. 25) [97]. The resultant membranes obtained an extraordinarily high H_2/CO_2 selectivity of around 140 at 150 °C which was comparable to some state-of-the-art 2D materials.

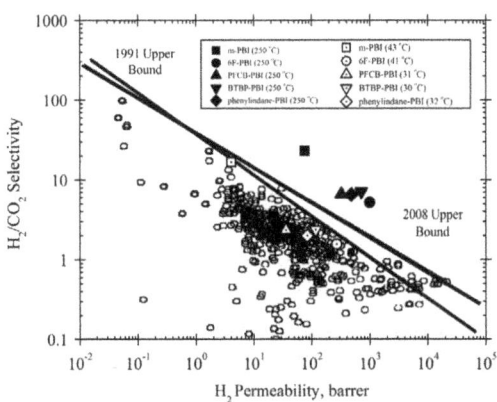

Figure 24. H_2/CO_2 separation performance of m-PBI at low and high temperatures. Reproduced with permission [95]. Copyright 2014, Elsevier Ltd.

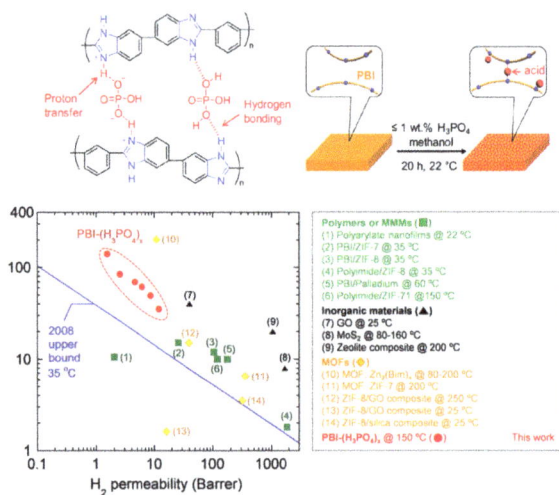

Figure 25. The illustrative schematic and H_2/CO_2 separation performance of acid-crosslinked PBI. Reproduced with permission [97]. Copyright 2018, The Royal Society of Chemistry.

2.2.2 Polyimide

The VTEC™ Polyimide (PI) developed and commercialized by RBI during the mid to late 2000s is also an advanced membrane material for efficient H_2/CO_2 separation [98]. Like PBI, the VTEC™ PI also demonstrated superior thermal stability that is able to withstand > 250 °C operating temperatures [89]. Its chemical resistance and mechanical strength at high temperatures are also impressive. The exact chemical structure of VTEC™ PI is not disclosed given proprietary information, but studies found that it possessed very good film-forming property. The precursor polyamic acid solution could be easily cast into films, and after a high temperature treatment process that initiated the condensation reaction, these polyamic acid prepolymer films then turned into the final VTEC™ PI membranes. This is a similar procedure as the production of Kapton® PI films shown in Fig. 26. Nevertheless, despite the facile precursor-facilitated film-forming protocol, the final VTEC™ PI polymers themselves have poor solubility in organic solvents and hence lack solution-processability, which is typical of all chemically and heat-resistant glassy polymers. The H_2/CO_2 separation performance of VTEC™ PI at high temperatures is very attractive. It demonstrates a mixed-gas H_2 permeability of 83

Barrer and a H_2/CO_2 selectivity of around 9 at 250 °C, which in combination well surpasses the 2008 upper bound and is comparable to the *m*-PBI membranes [89].

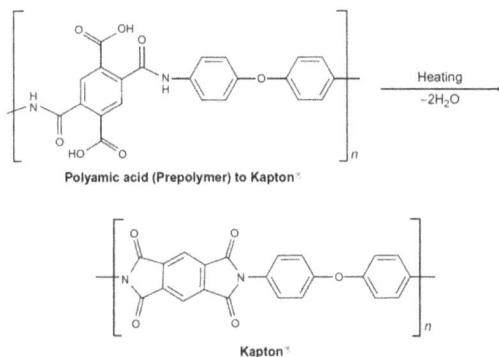

Polyamic acid (Prepolymer) to Kapton®

Kapton®

Figure 26. Film-forming process of Kapton® PI. Reproduced with permission [89]. Copyright 2011, Elsevier Ltd.

2.3   Polymers for Reverse-selective Hydrogen Separations

Given the high intra-segmental mobility and large free volume of rubbery polymers, condensable gases, such as CO_2 and hydrocarbons (HCs), tend to exhibit preferential sorption in them over the light gases [23]. This results in rubbery polymers typically possessing a very high solubility selectivity for CO_2 and HCs that significantly outweighs their diffusivity selectivity for light gases like H_2. Subsequently, rubbery polymers are often intrinsically more permeable to CO_2 and HCs than H_2, leading to the reverse-selective membranes, which are also known as H_2-rejective membranes [2]. Fig. 27 illustrates the difference between size-selective and reverse-selective membrane separations. In addition, the high solubility of CO_2 in rubbery polymers allows CO_2 to occupy the sorption sites in the dense membranes at a faster rate than H_2 via a phenomenon known as competitive sorption, which tends to further enhance the preferential permeation and hence the selectivity of CO_2 under mixed-gas conditions [99]. Despite having rigid backbones, high-free-volume glassy polymers, such as polyacetylenes and PIMs, can be also reverse-selective owing to their possession of large amounts of free volume that acts as sorption sites for condensable gases. For example, PTMSP demonstrates fairly good HCs over H_2 (HCs/H_2) selectivities [100].

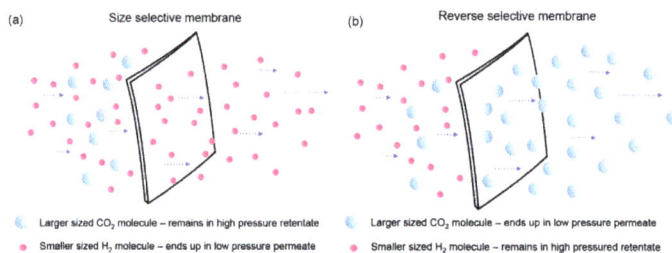

Figure 27. Difference between (a) size-selective and (b) reverse-selective membrane gas separation. Reproduced with permission [23]. Copyright 2013, Elsevier Ltd.

In general, reverse-selective gas separations like CO_2/H_2 and HCs/H_2 favor low operating temperatures because the solubility of CO_2 could be further enhanced, which is opposite to the H_2/CO_2 separation where a high temperature is desired to reduce the solubility of CO_2 and to enlarge the size-sieving effect [2, 7]. As shown in Fig. 28, the CO_2/H_2 upper bound established based on data from the 2008 Robeson's paper displays an upward-sloping line due to its reverse-(size)-selective nature. In contrast, all other separations that preferentially permeate the smaller gas molecules by either the diffusivity selectivity (e.g. H_2/N_2, H_2CH_4) or by the solubility selectivity (e.g. CO_2/N_2, CO_2/CH_4), have downward-sloping upper bounds. In this case, an upward-sloping line no longer indicates a trade-off relationship between the permeability and selectivity, so the interpretation of CO_2/H_2 separation performance could be made by directly comparing the respective magnitudes of permeability and selectivity between polymers.

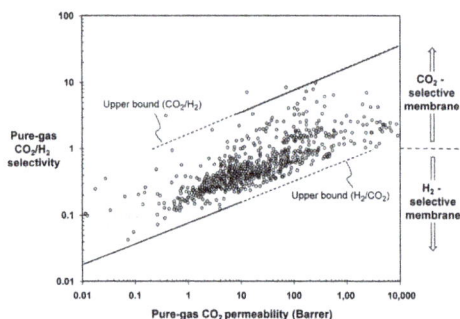

Figure 28. H_2/CO_2 and CO_2/H_2 upper bounds. Reproduced with permission [7]. Copyright 2017, American Chemical Society.

2.3.1 Poly(ethylene oxide) for CO_2/H_2 Separation

Poly(ethylene oxides) (PEO), also known as poly(ethylene glycols) (PEG), are the simplest structured member from the rubbery polyether family, which have been widely used for membrane CO_2/H_2 and CO_2/N_2 separations. The general structure of PEO is shown in Fig. 29. The ether moieties form the characteristic structure of PEO and have high affinity for the CO_2 molecules due to their polar interactions [23]. As compared to other interactive moieties including acetates, nitriles and carbonates, polymers containing ether oxygen (EO) atoms were found by Lin and Freeman to exhibit the highest CO_2 solubility and could provide an attractive combination of a CO_2 permeability of 13 *Barrer* and a CO_2/H_2 selectivity of 7.4 when being tested as semi-crystalline pure PEO membranes [101]. Although polar groups in the polymer matrix generally tend to enhance packing due to the stronger interchain interactions which subsequently promote chain crystallization, EO units actually tend to impede the efficient chain packing due to their bent structure and could actually increase the free volume of polymers and the gas diffusivity when they are in an amorphous state [23].

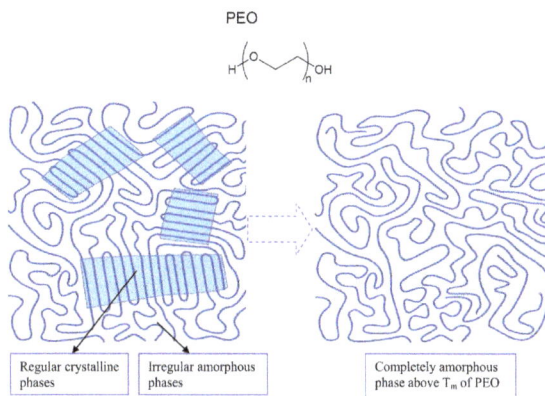

Figure 29. The chemical and matrix structure of PEO. Reproduced with permission [102]. Copyright 2018, Elsevier Ltd.

Nevertheless, despite that a high concentration of EO in the polymer could theoretically improve the permeability for CO_2 by possibly enhancing both its diffusivity and solubility, the high tendency of EO-containing segments to crystallize among themselves remains a challenge for utilizing pure PEO polymers for practical membrane CO_2/H_2 separation. The polymer matrix of semi-crystalline PEO polymers assumes a structure of impermeable, barrier-like crystals being embedded or dispersed in an amorphous phase that is highly permeable to gases (Fig. 29) [102]. As such, the overall gas permeation through PEO membranes is governed by both the volume of amorphous phase and the tortuosity of transport pathways through the amorphous phase which is strongly influenced by the content of PEO crystallites. Reducing the PEO crystallinity to a hypothetical fully amorphous state could significantly enhance the CO_2 permeability, it is the key to design a highly CO_2-permeable PEO membrane. This has been commonly and effectively achieved by branching and/or crosslinking the PEO chains such that the crystallization tendency could be suppressed. Followed by increasing the EO content in the branched or crosslinked chains, both the diffusivity and solubility of CO_2 could then be enhanced [103]. For example, crosslinked polyether polymers based on a mixture of diacrylate terminated PEG monomers could obtain very attractive CO_2/H_2 separation performance with a CO_2 permeability of 100 *Barrer* and a CO_2/H_2 selectivity of 9.5 obtained at 23 °C and 6.8 bar [104]. Lin et al. developed another crosslinked copolymer network based on a poly(ethylene glycol) methyl ether acrylate (PEGMEA) monomer and a PEG-diacrylate crosslinker, which was entirely amorphous and exhibited an outstanding separation performance with a CO_2 permeability of 420 *Barrer* and a CO_2/H_2 selectivity of 10 [105].

While it is known that the solubility of CO_2 could be enhanced at low temperatures, the crystallization tendency of EO-containing segments also favors low operating temperatures. By suppressing the crystallization tendency of PEOs via crosslinking, the operating temperatures could be reduced to improve the CO_2/H_2 separation performance. The increased sorption and solubility of CO_2 at low temperatures could also result in higher plasticization of polymers by the sorbed CO_2, which swells the interchain spacing, increases the chain mobility and FFV, and eventually enhances the CO_2 diffusivity as well [106]. On top of that, this enhanced plasticization also significantly hampers the diffusivity selectivity of H_2 [107]. Therefore, the combined effect of crosslinking PEOs and operating them at lower temperatures could lead to a tremendous enhancement in CO_2/H_2 separation performance. For example, an extraordinary CO_2/H_2 selectivity of 37 could be achieved by PEGMEA membranes at -20 °C [106].

Despite their promising CO_2/H_2 separation performances, pure PEO or polyether membranes are in general mechanically unstable. A common approach to overcome this weakness is to utilize the versatile chemistry of PEO to copolymerize it with mechanically strong glassy polymers like polyimides and polyamides [108, 109]. Their mechanical properties could be significantly enhanced by the harder and stronger glassy segments/micro-domains, while the efficient CO_2/H_2 separation still takes place primarily in the CO_2-favored PEO phase.

2.3.2 PTMSP and PDMS for HCs/H₂ or HCs/N₂ Separation

The separation of organic vapors, mainly light paraffins and olefins like C_{2-4}, from hydrogen- or nitrogen-containing streams is important to the refinery industry because of the high value of these minor components. While such separations share similar gas transport characteristics with the separation of H_2 and CO_2 when using polymeric membranes, both of them can proceed via the diffusivity- or solubility-selective mode. The particularly high condensability of light HCs and the high throughput nature of refinery processes suggest that the use of solubility-selective polymers, especially those with a high free volume, is much preferable and could deliver better separation efficiencies. This is because the high-free-volume polymers have ability to achieve more effective preferential sorption of the large and highly condensable HCs over light gases. Common choice of polymers for HCs/H_2 and HCs/N_2 separations include organosilane-based rubbery polymers like PDMS and ultra-high-free-volume glassy polyacetylenes like PTMSP.

PDMS

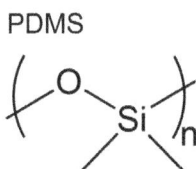

Figure 30. The chemical structure of PDMS.

PDMS is among the most permeable rubbery polymers owing to its highly flexible siloxane linkage and the bulky dimethyl siloxane (DMS) side groups (Fig. 30) such that its permeability for C_{2-4} HCs could be extremely high which is in the order of about 10^3 to 10^4 *Barrer* with good C_{2-4}/H_2 or C_{2-4}/N_2 selectivity [107, 110]. For example, Pinnau et al. reported a C_2H_6 permeability of 3900 *Barrer* and a C_2H_6/H_2 selectivity of 4, and a C_3H_8 permeability of 7400 *Barrer* and a C_3H_8/H_2 selectivity of 7.5 demonstrated by PDMS membranes at 35 °C, which could be further tremendously improved at lower temperatures [107]. Substituting bulkier groups on the side chain or directly on the backbone of PDMS was a common method to alter its transport properties. However, such modifications interestingly tended to decrease the permeability of HCs and increase the size-selectivity of PDMS for H_2 possibly due to the hindered intra-segmental mobility and enhanced chain rigidity from bulky moieties on the polymer chains, which has been reflected in some studies by the increased T_g of polymers when the bulkiness of substitution increased [28]. Apparently, the HCs/light gas separations do not favor such a structural change.

PTMSP, as a rigid-chain, ultra-high-free-volume glassy polymer, possesses even better C_{2+}/H_2 separation performances than PDMS in both the permeability and selectivity for HCs, especially for the relatively larger ones like n-butane [100]. However, the practical use of PTMSP is strongly limited by its weak chemical resistance against aliphatic or aromatic hydrocarbon contaminants in the refinery processes. On the other hand, easily crosslinkable rubbery PDMS exhibits less impressive but still attractive enough performance, while having much more practical chemical resistance against the harsh vapor environments than PTMSP [107].

3.   Polymeric Membranes for N_2 Separation

Among the light gases, nitrogen (N_2) is considered as a slow permeating gas due to its relatively large size and low condensability. However, N_2 is one of the most abundant gases encountered in industrial processes which will be commonly mixed with other valuable gases. Therefore, as put forward in the beginning of this chapter, the industrial meaning of N_2 separation, apart from its utilization as pure N_2 gas, lies in the efficient capture of other components in the gas mixture, such as (1) the capture of CO_2 from flue gas of power plants and (2) the recovery of methane (CH_4) from gas streams in petrochemical or refinery processes for natural gas production and purification. Moreover, both CO_2 and CH_4 are greenhouse gases which need to be reduced or captured before releasing the constituting gas streams into the atmosphere. Several polymers which have been vigorously researched for these applications will be discussed in the following sections.

3.1 CO_2/N_2 Separation

The majority of CO_2 emissions are generated by combusting fossil fuels, such as coal and natural gas, to generate electricity in power plants or petrol engines, and the combustion gas products typically contain a moderate to major amount of N_2 at the atmospheric pressure. Therefore, it is of critical importance to capture CO_2 from these gas products to curb the anthropogenic climate change. Both rubbery and glassy membranes can be utilized for CO_2/N_2 separation depending on the transport characteristics required for the separation. Rubbery polymers favor solubility selectivity for CO_2/N_2 separation due to their flexible chains that selectively permeate condensable gas species more easily while glassy polymers are contingent on the diffusivity selectivity due to their more rigid chains that selectively permeate gas species by size-discrimination. It is also important to note that because of the atmospheric operating pressure which is typically below the onset pressure for CO_2-induced plasticization, the plasticization phenomenon is not a particular issue for glassy polymers when being used for membrane CO_2/N_2 separation. The main challenge for CO_2/N_2 separation is, however, its enormous volume requirement. For example, a typical 600 MWe coal-fired power plant emits 460 tons of CO_2 per hour. It requires not only the use of highly permeable and selective membrane materials but also the fabrication of these materials into useful configurations with a large membrane area at an economically competitive cost [111].

3.1.1 Rubbery Polymers

Considering the differences in the kinetic diameter and condensability which both favor the permeation of CO_2 over N_2 through polymeric membranes, rubbery-type polymers have been widely investigated for the CO_2/N_2 separation targeting CO_2 capture from flue gas. In particular, PEO-based polymers have been regarded as one of the most suitable rubbery membrane materials for this separation owing to the excellent affinity of CO_2 with polyether moieties which increases the solubility of CO_2 and in turn further enhances the permeability of CO_2 and the throughput of the process [111]. Most of the PEO or PEO-containing polymers discussed in Section 2.3.1, which demonstrated promising CO_2/H_2 separation performances, also possessed very attractive CO_2/N_2 separation efficiencies. The chemical structures and CO_2/N_2 separation performances of a number of representative PEO-containing polymeric membranes are shown in Table 8.

To date, the best performing PEO-based membrane was developed by Jiang et al. via the construction of a semi-interpenetrating network (SIPN) by embedding PEGDME into a crosslinked backbone of either PEGMEA, PEGDA or a mixture of both PEGMEA and PEGDA, as provided in Fig. 31 [112]. The SIPN crosslinked PEO-based membranes retained their amorphous structures which thereby enhanced their gas separation performance. Without PEGDME, the crosslinked PEGMEA/PEGDA (7:3) membrane only possessed a CO_2 permeability of 405.7 *Barrer* and a CO_2/N_2 selectivity of 49.8. However, when PEGDME was embedded into the crosslinked PEGMEA/PEGDA (7:3) membranes with the weight percentage of PEGDME being equal to the total weight percentages of PEGMEA and PEGDA, the CO_2 permeability was enhanced to 2980 *Barrer* while the CO_2/N_2 selectivity slightly decreased to 45.7. These unprecedented results further pushed the separation performance of crosslinked PEO membranes to go beyond even the recently updated 2019 upper bound for CO_2/N_2 separation, which will be discussed in a later section.

Table 8. Chemical structures, CO₂ permeability and CO₂/N₂ selectivity of representative PEO-based membranes [111]. Copyright 2016, John Wiley and Sons.

Polymer/monomer	Chemical structure	CO$_2$ permeability [Barrer]	CO$_2$/N$_2$ selectivity
PEO		12	48
Pebax 1074		120	51
PEO-*co*-PBT (Polyactive 1000)		71	44
Pebax+PEG-DME (50 wt %)	(PEGDME)	606	44
Crosslinked poly(ethylene glycol) diacrylate (XLPEGDA)	(PEGDA)	110	50
Crosslinked poly(ethylene glycol) methyl ether acrylate (XLPEGMEA)	(PEGMEA)	570	41

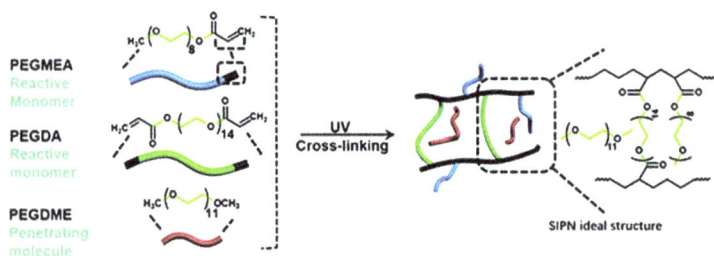

Figure 31. A schematic illustration and reaction of the fabrication of SIPN crosslinked PEO-based membrane. Reproduced with permission [112]. Copyright 2017, The Royal Society of Chemistry.

3.1.2 Glassy Polymers

The separation of CO_2 from N_2 by glassy polymers, such as polyacetylenes, poly(arylene ether)s, polyimides, and polysulfones, which benefit from the smaller kinetic diameter of CO_2 over N_2, has also been extensively investigated [4]. The main advantage of utilizing glassy polymers instead of rubbery polymers stems from the relatively higher thermal and mechanical stability. However, prior to the invention of PIMs and the introduction of any modifications to the polymer chains, traditional glassy polymers have relatively average separation performance because of their relatively small free volume and often their lack of sufficient CO_2 affinity. For example, 6FDA-durene, a widely studied research polyimide, displays a CO_2 permeability of 456 *Barrer* and a CO_2/N_2 selectivity of 12.8, which translate to an overall CO_2/N_2 separation performance point that falls far below the 2008 upper bound [113]. In order to boost the efficiency of glassy polymers for CO_2/N_2 separation, the CO_2 diffusivity has to be increased by introducing more fractional free volume (FFV) while maintaining or improving the chain stiffness of the polymer backbones so that the CO_2/N_2 selectivity will not be compromised. As a result, both PIMs and thermally rearranged (TR) polymers were invented and extensively explored for CO_2 capture from flue gas [57, 63, 114].

3.1.2.1 Polymers of Intrinsic Microporosity (PIMs)

As discussed in Section 2.1.2.2, PIMs were originally synthesized via the polycondensation reactions of aromatic tetrahydroxy- and tetrafluoro-containing monomers resulting in benzodioxane-bridged ladder structures that are sterically contorted by a spiro-carbon center like the spirobisindane (SBI) unit [58]. As a result of their rigid and contorted backbone structures, PIMs pack their chains highly inefficiently, which gives rise to their high FFV and BET surface area that in turn endow their membranes with a high gas diffusivity and a moderate CO_2/N_2 selectivity (e.g. ~ 25 for PIM-1) [63]. However, extending the contorted spiro-carbon centers by substituting tetrafluoroterephthalonitrile (TFTPN) in PIM-1 with phenazine-containing tetrachlorospirobisindane monomers in PIM-7 lowers the BET surface area of PIM-7 and results in a lower CO_2 diffusivity but a higher CO_2/N_2 diffusivity selectivity [63, 115]. Nevertheless, given their intrinsic microporosity granted by the frustrated chain packing in PIMs, these early PIMs still possess much more attractive CO_2/N_2 separation performance than the traditional glassy polymers. It is important to note that their relatively higher CO_2/N_2 permselectivity stems from their high solubility selectivity for CO_2 over N_2 instead of their diffusivity selectivity [111]. Besides, PIMs possess one of the largest free volumes among known glassy polymers, which allows them to demonstrate a high CO_2 solubility which, coupled with their high diffusivity, translates to an unusually high CO_2 permeability (in the range of 10^3 to 10^4 *Barrer*).

The high backbone rigidity of PIMs that arises from the presence of a variety of sites of contortion also helps them retain a sufficient size-sieving effect. As a result, numerous ladder PIMs and PIM-PIs have been developed that exhibit promising CO_2/N_2 separation performances, which include many that have been discussed before in Section 2.1.2.2. Recent development of PIMs was concentrated at increasing the chain rigidity to improve the diffusivity selectivity without compromising their CO_2/N_2 selectivity by substituting the SBI contorted structure into the more rigid bridged-bicyclic-ring system, such as the ethanoantracene (EA), Troger's base (TB) and triptycene (Trip) moieties. Carta et al. demonstrated via molecular modelling that the SBI structure was more flexible than EA and TB [66]. Instead of a direct combination of Trip and TB structures that aims at maximizing the structural contortedness by having as many contortion sites as possible in the backbone, a more distinct innovation was created via the benzodioxane linkage of benzotriptycene (BTrip) moieties, which resulted in unique solution-processable, ribbon-shaped porous 2D polymers, such as PIM-TMN-Trip and PIM-BTrip as shown in Fig. 32 [62, 88]. Because of the highly stiffened BTrip structures and the 2D co-planar arrangement of each BTrip group in the polymer chains, a freshly methanol treated PIM-TMN-Trip membrane could obtain an ultra-high CO_2 permeability of 52,800 *Barrer* which is on par with the most permeable polyacetylene-based membranes, such as PTMSP. Meanwhile, the CO_2/N_2 selectivity of PIM-TMN-Trip (14.9) is also higher than that of PTMSP (10.7) [88]. As a result of this new innovation in PIM design, the upper bound of CO_2/N_2 separation was recently revised by the gas separation performances of several series of BTrip-based PIMs as shown in Fig. 32.

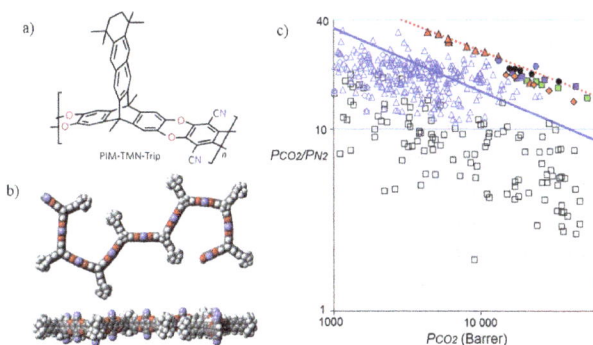

Figure 32. (a) PIM-TMN-Trip chemical structure, (b) two 3D conformational views of PIM-TMN-Trip fragment, and (c) the upper-bound relationships for the CO_2/N_2 gas pair with the recently proposed upper bound shown in a red dotted line and the 2008 upper bound shown in solid blue line. The CO_2/N_2 gas separation performance of BTrip-based PIMs are provided in solid symbols and open symbols indicate the data of other PIMs and non-PIM polymers. (a) and (b) are reproduced with permission [62]. Copyright 2017, Nature Publishing Group. (c) is reproduced with permission [88]. Copyright 2019, The Royal Society of Chemistry.

3.1.2.2 Thermally Rearranged (TR) Polymers

TR polymers are generally derived from *ortho*-functionalized polyimides (*o*-PIs) or polyamides (*o*-PAs) via a solid-state thermal post-treatment process that can generate a rigid and microporous polymer/membrane structure by releasing CO_2 or H_2O depending on the polymer/membrane precursor [114, 116]. As a result, higher free volume membranes can be fabricated with narrowly distributed micropores that are interconnected, which are beneficial to induce rapid diffusion of small gases while maintaining good size-sieving properties [117]. However, conventional TR polymers, which are obtained from *o*-PIs, face serious energy penalty because of their high temperature requirement (> 400°C) for initiating the thermal conversion. On the other hand, *o*-PAs require a lower energy for their thermal conversion which can be attributed to the more flexible secondary amine in the polyamide backbones as compared to the more rigid tertiary amine in polyimides. However, the gas separation performance of *o*-PAs and their corresponding TR polymers are generally lower than that of *o*-PIs and their corresponding TR polymers. Therefore, reducing the energy penalty whilst improving their gas separation performance has become one of the key foci for TR polymer research. Reducing the glass transition temperature (T_g) of the polymer precursor which subsequently decreases the thermal conversion temperature has been achieved by introducing more flexible macromolecular chains onto the *ortho*-hydroxyl groups, such as acetate, pivalate or allyl ether groups [118, 119]. As compared to the pristine 6FDA-HAB polymers which were also thermally treated at the same temperature (e.g., 350°C), the allyl-functionalized 6FDA-HAB polymers demonstrated a much higher CO_2 permeability at the expense of a slightly lowered CO_2/N_2 selectivity, because of their higher TR conversion ratio at a lower temperature. Similarly, a reduced thermal conversion temperature has also been observed via allylation of *ortho*-hydroxyl polyamides [120].

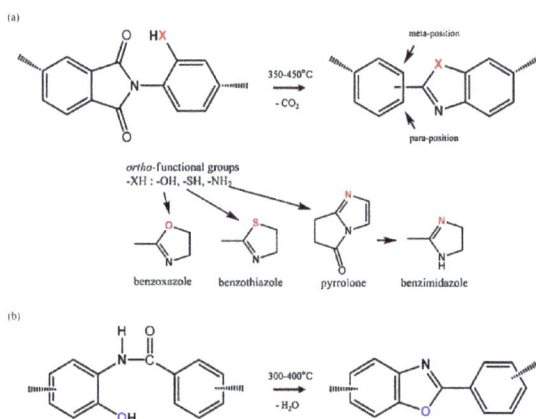

Figure 33. Thermal rearrangement of (a) ortho-functionalized polyimide and (b) ortho-functionalized polyamide to generate TR polymers. Reproduced with permission [116]. Copyright 2015, Elsevier Ltd

3.2 N₂/CH₄ Separation

Nitrogen removal from methane in refineries for natural gas purification to meet pipeline specifications is a huge industrial process due to the high nitrogen content in natural gas reserves [121, 122]. However, nitrogen removal from methane has been considered one of the most challenging separation processes by membrane technology owing to the almost similar kinetic diameters and condensability between N_2 and CH_4 (Table 1). In light of this challenge, there are limited polymer materials that could demonstrate promising separation performance for N_2/CH_4 separation.

Perfluoropolymers are currently the best known material for this application, which have defined the N_2/CH_4 separation upper bound in 2008 and are commercially available under the trade names of Teflon AF, Hyflon AD and Cytop [52, 123, 124]. Among these commercial perfluoropolymers, both Teflon AF and Hyflon AD, which are based on perfluorodioxolane structures, display the best N_2/CH_4 separation performance. Therefore, numerous dioxolane-based perfluoropolymers and copolymers have been recently synthesized, studied and reported to possess N_2/CH_4 separation performance located above the corresponding 2008 upper bound. Yavari et al. reported that poly(perfluoro-2-methylene-4-methyl-1,3-dioxolane) (poly(PFMMD)) membranes obtained an N_2 permeability of 7.7 *Barrer* and a N_2/CH_4 selectivity of 3.85 [125]. Fang et al. copolymerized PFMMD with perfluoro-(2-methylene-1,3-dioxolane) (PFMD) and showed that the 50/50 PFMMD-co-PFMD membranes displayed an extraordinary N_2/CH_4 selectivity of 6 [126]. In addition to the perfluorodioxolane-based polymers, Belov et al. discovered that poly(hexafluoropropylene) (poly(HFP)) membranes could achieve better N_2/CH_4 separation properties than the commercial Hyflon AD80 membranes, although they were still located below the 2008 upper bound [127]. In accordance with these new developments, Wu et al. proposed that there existed a so-called perfluoropolymer upper bound for N_2/CH_4 separation which emphasized the outstanding diffusivity selectivity for N_2 over CH_4 of dioxolane-based perfluoropolymers [128].

4.   Material Strategies for Performance and Properties Enhancement

4.1   Grafting

In order to allow a larger degree of control over the gas separation performance of the fabricated membranes, the grafting technique has been widely investigated whereby a specific functional moiety is chemically transferred onto the polymer backbones. Suzuki et al. grafted amine-terminated dimethylsiloxane (DMS) onto poly(styrene-co-maleic anhydride) (poly(St-co-MAH)), which increased the CO_2 permeability tremendously (more than 100-fold), although the CO_2/N_2 selectivity decreased by 44% [129]. This phenomenon was attributed to the larger free volume of the poly(St-co-MAH-graft-DMS) as compared to the pristine poly(St-co-MAH). An increase in free volume of the membranes was also observed when thermally labile macromolecules, for example, cyclodextrins (CDs) and saccharides, were grafted onto 3,5-diaminobenzoic acid (DABA) containing 6FDA-durene-DABA copolyimide backbones [130-132]. The additional free volume insignificantly altered the CO_2/N_2 separation performance of the fabricated 6FDA-durene/DABA(9:1)-g-CD membranes when the membranes were annealed at 200°C, as provided in Table 9. When the membranes were annealed at 400°C, the CO_2 permeability could be improved by 79% by grafting γ-CD onto the 6FDA-durene/DABA(9:1) copolyimide at the expense of only a 13% reduction in the CO_2/N_2 selectivity, which was ascribed to the formation of micropores because of the decomposed CDs [132]. A similar micropore formation was also observed by thermally decomposing the grafted saccharides (such as glucose, sucrose and raffinose) in the 6FDA-durene/DABA(1:1)-g-saccharide membranes [131]. Therefore, both lower increment in CO_2 permeability and lower reduction in CO_2/N_2 selectivity can be achieved by varying the size of grafted CDs and saccharides in the 6FDA-durene/DABA copolyimides.

Table 9. Summary of the CO_2/N_2 separation performance of grafted polyimide membranes [113, 114]

Membrane	Annealing Temperature (°C)	Permeability (Barrer)		CO_2/N_2 Selectivity
		CO_2	N_2	
6FDA-durene-DABA (9:1)	200	235	13.4	17.6
6FDA-durene-DABA (9:1)-g-αCD	200	241	14.5	16.6
6FDA-durene-DABA (9:1)-g-βCD	200	239	14.4	16.6
6FDA-durene-DABA (9:1)-g-γCD	200	251	15.5	16.3
6FDA-durene-DABA (9:1)	400	520	38.0	13.7
6FDA-durene-DABA (9:1)-g-αCD	400	569	48.5	11.7
6FDA-durene-DABA (9:1)-g-βCD	400	772	62.9	12.3
6FDA-durene-DABA (9:1)-g-γCD	400	930	77.8	11.9
6FDA-durene-DABA (5:5)	400	366	23.0	16.0
6FDA-durene-DABA (5:5)-g-glucose	400	533	33.7	15.8
6FDA-durene-DABA (5:5)-g-sucrose	400	370	21.7	17.0
6FDA-durene-DABA (5:5)-g-raffinose	400	407	27.8	14.7

4.2 Polymer Chain-Functionalization

Polymer chain-functionalization is commonly utilized to modify the polymer chains prior to the membrane fabrication. For both hydrogen and nitrogen separation, PIM-1 is the most widely functionalized polymer owing to its nitrile (–CN) groups. These reactive nitrile groups can be easily converted into a variety of polar functional groups, including the N- and S- containing ones, such as those in AOPIM-1, TZPIMs and ThioPIM-1, or the hydroxyl or carboxyl groups, such as those in PIM-1-COOH, which all are capable of modifying the polar interaction between the polymers and the permeating CO_2 gases [83-

86]. These polar functional groups are capable of not only inducing higher affinity of CO_2 with the polymers, but also altering the packing density of the polymer chains in the membranes, which can consequently tune the gas separation performance. Mason et al. showed that reducing the nitrile groups into amine groups enhanced both CO_2 sorption uptake and the ideal CO_2/N_2 sorption/solubility selectivity [133]. However, the amine groups also introduced the formation of inter- and intra-chain hydrogen bonds which imparted additional chain stiffness to the polymer network in the membrane. As a result, amine-PIM-1 was H_2-selective instead of CO_2-selective as compared to other functionalized PIM-1, and the ethanol-treated membrane demonstrated H_2/CO_2 separation performance above the Robeson 2008 upper bound. On the other hand, TZPIM-1 exhibited the highest CO_2 permeability, while AOPIM-1 displayed the highest ideal CO_2/N_2 selectivity among the functionalized PIM-1 membranes [83]. Further methylation of TZPIM-1 (referred to as MTZ100PIM-1) had also been investigated by Du et al. [134]. It improved the processability, solubility in polar protic and aprotic solvents and thermal stability of TZPIM-1, and the resultant MTZ100PIM-1 had lower CO_2 permeability and CO_2/N_2 selectivity than TZPIM-1.

4.3　Crosslinking

Crosslinking is considered as a common post-treatment on polymeric membranes to enhance their gas separation performance, particularly to improve their diffusivity selectivity or molecular-sieving properties. In addition, crosslinking is also widely employed as a method to improve the stability of the membranes, especially to mitigate the plasticization phenomenon often observed in separations involving highly condensable gases at high operating pressures. Generally, there are three crosslinking methods which vary by the source for bond formation or cleavage during the crosslinking reaction; namely, chemical, thermal and UV (also known as photo-oxidative) crosslinking.

4.3.1　Chemical Crosslinking

Diamine crosslinking is a very popular chemical crosslinking method which was originally patented by Hayes in 1989 to improve the O_2/N_2 separation performance of polyimide membranes [135]. About a decade later in 2001, Liu et al. cross-linked the 6FDA-durene membranes with p-xylenediamine (pXDA) in methanol solutions to enhance He/N_2 selectivity by more than 8-fold, which was the first reported literature of diamine crosslinking with polyimide membranes [136]. A wide variety of diamine molecules has thereafter been utilized to tune the molecular-sieving properties of polyimide membranes, particularly for H_2/CO_2, H_2/N_2, H_2/CH_4 and, in certain cases, CO_2/N_2 and CO_2/H_2 separation [137]. Yong et al. showed that by changing the molecular structures of the diamine molecules, the gas separation performance of PIM-1/Matrimid (90:10) polymeric blending membranes could be tuned from CO_2-selective into H_2-selective, which depended on not only the reactivity but also the spatial structure of the diamine molecules [138]. Meanwhile, Zhao et al. demonstrated that H_2-selective Matrimid® membranes could be modified into CO_2-selective membranes by crosslinking Matrimid® with diamino-based polymers, such as Jeffamine® or O,O′-Bis(2-aminopropyl) polypropylene glycol-block-polyethylene glycol-block-polypropylene glycol [139]. On one hand, diamine crosslinking generally enhanced the gas separation performance by decreasing the diffusivity of gases and increasing the molecular-sieving properties because of the tightened interchain spacing of the membranes. On the other hand, when being crosslinked with Jeffamine, the Matrimid® membranes exhibited both higher CO_2 permeability and CO_2/N_2 and CO_2/H_2 selectivities despite their tightened interchain spacing. This was ascribed to the rubbery nature of Jeffamine or the PEO-containing phase that commonly exhibits small d-spacings [140] but possesses mobile polymer chains to allow easy gas passages.

Although diamine crosslinking has been proven effective in enhancing the gas transport properties of polyimide membranes, the polyamide network generated by the crosslinking process could possibly revert back to its corresponding polyimide network at elevated temperatures depending on the combination of the polymer and the diamine molecule [141]. The release of diamine molecules and the polymer chain scission have also been observed when the reverse-imidization of the polyamide network occurred [142, 143]. Besides, the polymer chain scission issue could also occur in the conventional solution-phase diamine crosslinking [144], which consequently inspired the development of vapor-phase crosslinking method by Shao et al. to minimize such issue [145]. The advantages of vapor-phase crosslinking were further reported by Wang et al. when they crosslinked 6FDA-durene membranes with diethylenetriamine (DETA) [146]. The vapor-phase crosslinked membrane exhibited a much higher increment of H_2/CO_2 selectivity (130-fold) than the solution-phase one (13-fold) albeit its lower gel content. Japip et al. demonstrated that the vapor-phase crosslinking of 6FDA-durene-ZIF71 mixed matrix membranes (MMMs) by tris(2-aminoethyl)amine (TAEA) could significantly improve the H_2/CO_2 separation performance as compared to the TAEA-crosslinked pristine 6FDA-durene membrane because of the additional crosslinked network formed at the interface between the polyimide matrix and ZIF-71 nanoparticles [147]. Meanwhile, the *in-situ* free volume, gas permeability and diffusivity measurements of the crosslinked membranes at 150 °C revealed the thermal motion of the crosslinked network and also its dependency on the molecular structure of the di/triamine crosslinkers [148]. A larger tri-podal crosslinker (TAEA) displayed superior H_2/CO_2 separation performance at lower temperatures (e.g. 35 °C), while a shorter di-podal crosslinker (EDA) presented superior H_2/CO_2 separation performance at higher temperatures (150 °C) because of the restricted thermal motion owing to its shorter chain length.

Diol crosslinking is an alternative chemical crosslinking strategy, but is limited to polymers containing carboxylic acid or sulfonic acid groups only [137]. As opposed to diamine crosslinking, diol crosslinking has been generally reported not to reduce the gas permeability. Instead, gas permeabilities generally tend to increase on polyimide membranes after diol crosslinking. Staudt-Bickel and Koros showed that 6FDA-mPD/DABA copolyimide crosslinked with ethylene glycol displayed higher CO_2 permeability and CO_2/N_2 selectivity by 45% and 35%, respectively, owing to the increased free volume because of the insertion of ethylene glycol [149]. However, if the uncrosslinked polymer membrane has a particularly large free volume because of its loose packing structure, the gas permeabilities would still decrease notwithstanding the improved selectivity. Zhang et al. demonstrated that crosslinking PIM-TB copolymers that contained carboxylic acid groups (CoPIM-TB) with glycidol could reduce the permeability of all light gases due to the tightened d-spacing. As a result, the selectivity for a number of gas pairs, such as CO_2/N_2 and H_2/N_2, was greatly enhanced [150].

4.3.2 Thermal Crosslinking

Thermal crosslinking can be considered as one of the oldest methods to improve gas separation performance and the stability of polymeric membranes. Diacetylene and benzenecyclobutane groups are two of the earliest molecular structures utilized to generate thermally crosslinked membranes with improved gas separation performance [151-153]. In 2008, Kratochvil and Koros presented a new thermal crosslinking technique for carboxylic acid containing polyimides by decarboxylation of the acid pendant groups at high temperatures [154]. Similar to the diol crosslinking, the decarboxylation-induced thermal crosslinking increased the d-spacing which consequently enhanced the gas permeability [155-157]. More specifically, the increased d-spacing originated from the breaking of hydrogen bonding and the subsequent formation of anhydride intermediate during the thermal annealing process. However, during the decarboxylation step which formed phenyl radicals, the resultant d-spacing would be determined by the original polymer structure and free volume. Zhang et al. demonstrated that 6FDA-CADA1 (CADA1 refers to 2-(3,6-bis(4-amino-2-(trifluoromethyl) phenoxy)-9H-xanthen-9-yl) benzoic acid) polymers displayed an increasing d-spacing trend upon thermal crosslinking even after the temperature (425 °C) was above their T_g [156]. In contrast, both Askari et al. and Xiao et al. reported that the d-spacing of 6FDA-Durene/DABA polymers esterified with CDs increased as a function of annealing temperature up to 400 °C, but the *d*-spacing decreased when the membranes were heated at 425 °C, as schematically drawn in **Error! Reference source not found.** [130, 132]. A similar decarboxylation-induced thermal crosslinking modification was also observed for DC-PIM-1 (DC refers to decarboxylated) to increase CO_2 permeability and CO_2/N_2 selectivity when being compared to its non-thermally-crosslinked counterpart cPIM-1 [84, 158].

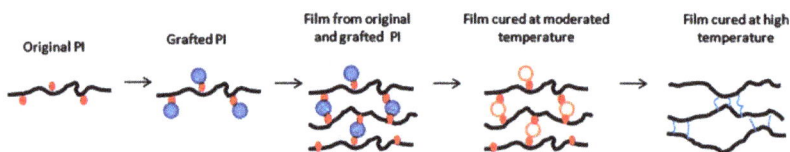

Figure 34. Schematic evolution of polyimides bearing pendant carboxylic acid groups through grafting with CDs and thermal crosslinking process. Reproduced with permission [130]. Copyright 2011, The Royal Society of Chemistry.

During the thermal crosslinking process, an inert environment (i.e., vacuum or nitrogen atmosphere) is typically used to minimize the effect of oxygen which may accelerate the degradation of polymer chains. Nonetheless, Song et al. introduced thermal oxidative crosslinking on PIM-1 membranes (TOX-PIM-1) with improved molecular-sieving properties at the expense of gas permeability [159]. The reduced gas permeability was attributed to the reduced ultramicroporosity upon thermal oxidative crosslinking due to the narrower bottleneck of the interconnected micropores. In contrast to the high free volume PIM-1, Zhang et al. demonstrated that it was possible to enhance both gas permeability and selectivity of a lower free volume cardo-polyimide containing the phenolphthalein structure via thermal oxidative crosslinking [160]. The *d*-spacing was increased as a function of the crosslinking temperature.

4.3.3 UV Crosslinking

UV/photo-induced crosslinking of polymers containing benzophenone units is the most well-known classic organic photochemistry reaction which takes place via radical coupling after the hydrogen removal from benzylic hydrogen donor groups by an excited benzophenone unit [161]. One of the first reports on UV crosslinking which could enhance the gas separation performance of polyimide membranes was found in the patent by R.A. Hayes [162]. Afterwards, Kita et al. demonstrated the enhanced selectivity with a reduced gas permeability as a function of UV irradiation time on polyimide membranes based on benzophenone tetracarboxylic dianhydride (BTDA) and 2,4-diaminomesitylene (DAM) [163]. Because of the radical coupling that results in a bond formation, UV crosslinking reduces the diffusivity coefficients of the permeating gases and is more effective in improving the selectivity for gas pairs that features a relatively big size difference such as the H_2/CH_4 separation. Besides polyimide, polyarylate and polyaryletherketone, which both possess benzophenone groups, can also be crosslinked via UV irradiation [164-166]. However, polyarylate polymers that contain aromatic esters and a free *ortho* or *para* position next to the ester moiety may undergo the photo-Fries rearrangement instead of crosslinking. Meanwhile, polyimide membranes, which do not contain benzophenone groups, will not undergo UV crosslinking. Rather, they experience densification as a result of photo-oxidative chain scission of the polymers [167-169]. Nevertheless, the photo-oxidative effect is generally not as significant for polymers with low free volumes as for those with high free volumes such as PIMs.

4.4 Thin-film Composite (TFC) Membranes

Practical membrane configurations usually consist of a thin selective layer on top of a porous mechanically stable support to improve the throughput (flux) of membranes for gas separation. There are in general two different practical membrane configurations, i.e., integrally skinned asymmetric membranes made from a single polymer and thin film composite (TFC) membranes made from a combination of different polymers [170]. Because of the additional polymeric materials used in TFC membranes, they provide a higher degree of control over the gas separation performance. In addition, it becomes now possible to utilize a highly crosslinked polymer which may not be soluble after its polymerization reaction, via a bottom-up approach, thereby broadening the material selection. In contrast, a more general top-down approach uses the soluble or solution-processable polymers which are coated onto a porous substrate to create the desired TFC membranes. While the gas separation properties of TFC membranes are mostly constrained by the polymers chosen as the selective layer, the presence of an intermediate gutter layer and/or a sealing layer and the structural properties of the porous support will also inadvertently influence the overall separation properties [171].

One of the bottom-up methods to fabricate TFC membranes is based on the interfacial polymerization (IP) technique which takes place on the interface of two immiscible liquids such as water and oil. Jimenez-Solomon et al. reported the synthesis of highly crosslinked polyarylates (PAR) TFC membranes via the interfacial reaction between aqueous phenoxide and trimesoylchloride (TMC) in the hexane solution, as illustrated in

Fig. 35 [172]. Because of the ultra-thinness of the highly crosslinked PAR nanofilms, the permeation of small gas molecules such as He and H_2 was not affected by the contorted molecular structures. In contrast, the degree of contortion in PAR nanofilms exerts a greater effect on the permeation of larger gas molecules such as N_2 and CH_4. Thus, the molecular-sieving properties for hydrogen purification could be enhanced by increasing the degree of contortion as observed in PAR-RES (resorcinol) nanofilms. In addition to PAR, TFC membranes made of polyamide (PA) or hybrid inorganic-organic PI have demonstrated outstanding hydrogen purification performances [173, 174].

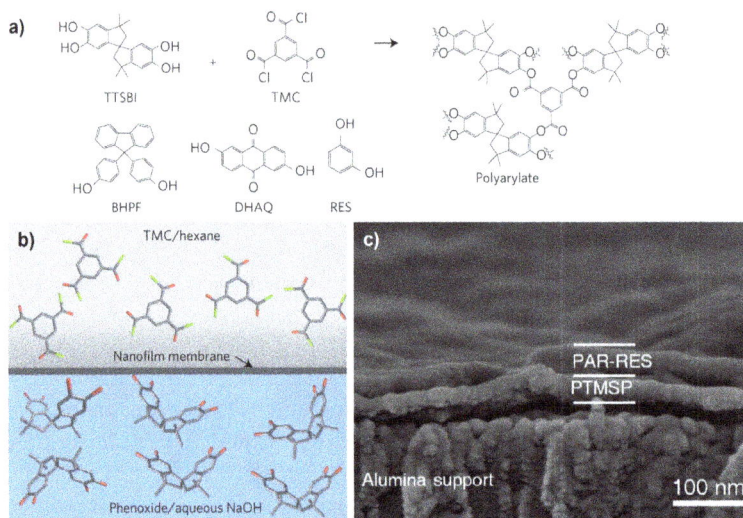

*Figure 35. The fabrication method of PAR TFC membranes (**a**) chemical reaction and monomers, (**b**) schematic interfacial polymerization, and (**c**) SEM image of a representative PAR-RES TFC membrane. Reproduced with permission [172]. Copyright 2016, Nature Publishing Group.*

4.5   Mixed Matrix Membranes (MMMs)

Among the many advanced membrane modification methods, incorporating inorganic, organic or hybrid nanoparticles/nanomaterials (NPs/NMs) into the polymer matrix gains phenomenal popularity among the membrane community, especially for gas separation [175]. Such strategy utilizes the unique topologies, sub-nanometer molecular-sieving pores (targeting the diffusivity selectivity) or special functionalities (targeting the solubility selectivity) of the incorporated NPs/NMs to potentially enhance the overall gas transport properties of the composite membrane materials. This membrane design is widely known as the mixed matrix membranes (MMMs) which are capable of simultaneously harnessing the polymer processability and tapping the superior gas separation properties of nano-fillers for overcoming the inherent trade-off limitation on polymeric membranes [7].

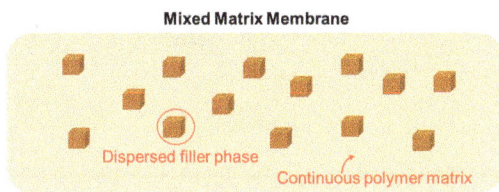

Figure 36. Illustration of MMM structure

As shown in the illustrative diagram in Fig. 36, the MMM structure is characterized by the discrete nano-filler phase dispersed within a continuous polymer phase/matrix. Ever since the pioneering work on the first MMM fabricated by silicone rubbers being incorporated with zeolite 5A in the 1970s by Paul and Kemp et al., there have been a substantial number of filler choices being extensively studied during the following few decades, such as zeolites, activated carbons, carbon molecular sieves, metal oxides, mesoporous silica, carbon nanotubes and polyhedral oligomeric silsesquioxane (POSS), which are generally considered as the early conventional fillers [175, 176].

The central challenge encountered in designing MMMs is the formation of nano-defects at the polymer-filler interface, which could significantly shoot up the permeability but deteriorate the selectivity to an unusable level, due to the poor compatibility or interaction between polymers and fillers [8]. Common strategies deployed to solve this issue are to strengthen the interfacial interaction between the filler and polymer by modifying the polymer functionalities or the filler surface in order to improve their adhesion [7, 177, 178]. The success of MMM design also depends on the choice of polymer as well as the particle size and dispersion because both the intrinsic gas transport properties of the polymer and the size-dependent dispersibility of fillers can play a critical role in determining the resultant composite properties [179].

4.5.1 Advanced Filler Choices

The poor interfacial problem in early MMMs arose from the strong inorganic nature of early fillers as well as their limited tunability in size, shape, morphology or functionality. A variety of new hybrid or organic nanomaterials with intrinsic porosity, tunable 2D/3D architect and better polymer interactions, such as metal organic frameworks (MOFs), covalent organic frameworks (COFs), porous organic cages (POCs) and organic macrocyclic molecules (OMMs), have been substantially explored as more advanced fillers for MMM designs and have created MMMs with much better gas transport characteristics that could surpass the trade-off limitations.

4.5.1.1 Metal Organic Frameworks (MOFs)

MOFs were first incorporated into polymeric membranes for gas separation in 2004 and have since gained explosive popularity because the highly tunable properties, outstanding gas transport characteristics as well as much improved interfacial interactions demonstrated by MOF-incorporated membranes simply presented a major breakthrough in the MMM research. MOFs are porous hybrid materials possessing three- or two-dimensional (3D or 2D) porous networks of transition metal complexes [180]. Owing to the abundance of transition metal nodes and the functionalizability of the polytopic organic linkers, MOFs are considered a powerful toolbox to design MMMs with improved polymer-fillers compatibility and ultimately gas separation performance. Among thousands of MOFs synthesized to date, zeolitic imidazolate frameworks (ZIFs), Cu-based MOFs, MOF-74, Materials Institute Lavoisier (MIL), and University of Oslo-66 (UiO-66) series are the most typical series of MOFs utilized for fabricating MMMs for gas separation. Depending on the applications or types of gases to be separated, a judicious selection of MOFs is required mainly based on their aperture size, functional groups of the polytopic organic linkers and the transition metal nodes. This would lead to MMM designs with either enhanced diffusivity or solubility in addition to having better dispersion of fillers with superior polymer-filler interactions [181, 182].

Spatial distribution of the embedded MOFs with micro-, 3D nano- or 2D nano-crystals/sheets inside the polymer matrices is another important aspect to fabricate MMMs with an enhanced separation performance. Rodenas et al. utilized a tomographic focused-ion-beam scanning electron microscope (FIB-SEM) to prove the better spatial occupancy of 2D nano-sheets Cu-based MOFs (CuBDC) inside the Matrimid® matrix than their 3D nano- and micro-crystals, as shown in Fig. 37 [183]. Subsequently, MMMs embedded with 2D nano-sheets demonstrated a superior selectivity owing to the improved interaction of gas molecules with CuBDC nano-sheets that resulted in an enhanced molecular-sieving capability. Meanwhile, Kang et al. verified this observation using another Cu-based MOF ($[Cu_2(ndc)_2(dabco)]_n$) and discovered that MMMs containing 2D nano-sheets showed the best H_2/CO_2 separation performance among all of their prepared MMMs including those containing 3D micro- and nano-crystals [184].

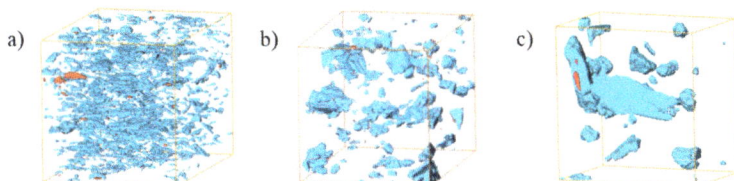

*Figure 37. Spatial occupancy of 8 wt% CuBDC (**a**) 2D nano-sheets, (**b**) 3D nano-crystals, and (**c**) 3D sub-micrometer crystals in Matrimid® 5218 MMMs from 3D surface-rendered view of the segmented FIB-SEM tomograms. Reproduced with permission [183]. Copyright 2015, Nature Publishing Group.*

4.5.1.2 Covalent Organic Frameworks (COFs)

COFs were first discovered in 2005 and only found its applicability as an MMM filler choice very recently [185]. They comprise purely organic building blocks joined together by covalent bonds and possess a rigid 2D or 3D extended structure with a high surface area, widely tunable chemistry and pore structure, and excellent affinity for organic polymers [181]. As an emerging filler choice, the first COF-incorporated MMM (Fig. 38) for gas separation application was developed in 2016 by Kang et al. using exfoliated water-stable, 2D COF-sheets, which obtained much higher H_2/CO_2 separation performance arising from the favorable gas transport characteristics of 2D COF and the good polymer-filler compatibility [186]. Following the success of Kang's work, Biswal et al. later also prepared two 2D-COF-incorporated MMMs based on a tert-butylpolybenzimidazole (PBI-BuI) polymer matrix which displayed good mechanical and gas transport properties [187].

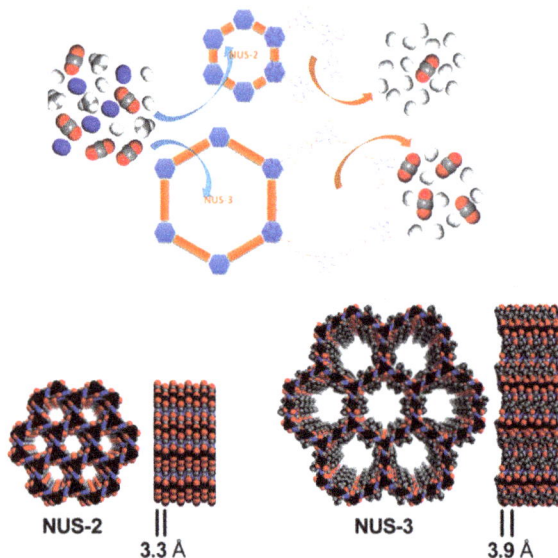

Figure 38. Illustration of gas transport through 2D COF-sheets (NUS-2 and NUS-3) and their simulated crystal structures. Reproduced with permission [186]. Copyright 2016, American Chemical Society.

4.5.1.3 Porous Organic Cages (POCs)

POCs are a relatively new class of nanomaterials first created in 2009 [188]. Like COFs, POCs are also completely organically built, but instead of having an extended framework-type structure, POCs exhibits a confined, molecular structure that also has highly tunable pore architectures and diverse choices of functionalities [181]. POCs possess excellent solubility in a variety of common solvents, making them a versatile filler choice for fabricating MMMs via the standard solution-mixing procedure. The pioneering POC-based MMM work using the classic CC3 cage molecules (Fig. 39) was not successful in realizing superior gas separation performance due to the high crystallization tendency of the dissolved CC3 cages during and after solvent evaporation [188, 189]. Later, Zhu et al. synthesized a new POC structure, called ASPOC, that had less stronger inter-cage interactions and hence better solubility in solvents and interactions with polymers, such that more homogeneously blended Matrimid®-based MMMs were obtained with greatly enhanced CO_2/N_2 performance [190].

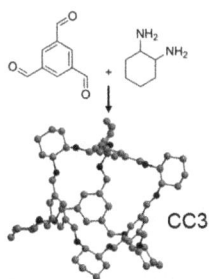

Figure 39. The molecular structure of CC3 cage used in GS MMMs. Reproduced with permission [189]. Copyright 2013, John Wiley & Sons.

4.5.1.4 Organic Macrocyclic Molecules (OMMs)

OMMs, such as cyclodextrins (CD) or calixarenes (CA), are an interesting class of supramolecules that have been discovered for a long time and have found many good applications in analytical chemistry, host-guest recognition, and biochemistry [191]. They generally possess a unique 3D molecular structure with a single open cavity that varies in size and rim functionalities. These cavities in OMMs could possibly act as either ultra-permeable or highly size-sieving gas transport channels for enhancing the separation performance, but such an idea is still quite underexplored at present. Despite showing interesting gas sorption properties, the application of OMMs in gas separation is relatively much rarer as compared to other fillers choices as mentioned before. Liu et al. incorporated β-CD into PIM-1 polymers via a copolymerization scheme to develop a series of PIM-CD membranes (Fig. 40) which achieved significantly enhanced H_2 and CO_2 permeabilities without sacrificing the selectivity, surpassing the 2008 upper bounds for H_2/N_2, H_2/CH_4 and CO_2/N_2 separations at only small loadings of β-CD [192]. However, the β-CD pore size was too large to maintain the selectivity at higher loadings. Wu et al. later prepared a series of 4-tert-butylcalix[4]arene (CA)-incorporated PIM-1 membranes (PIM-CA) using a similar copolymerization scheme (Fig. 41). As the size of CA4 was much smaller than CD and fell within the range of gas molecule sizes, their PIM-CA membranes were able to achieve a much higher selectivity at higher loadings and hence demonstrated greatly enhanced gas separation performances, especially for CO_2/N_2, which were well beyond the 2008 upper bounds [193]. CA molecules were also found capable of favorably tuning the polymer structure towards more efficient gas transport due to their rigid cup-shape conformation and their possession of bulky t-butyl pendant groups.

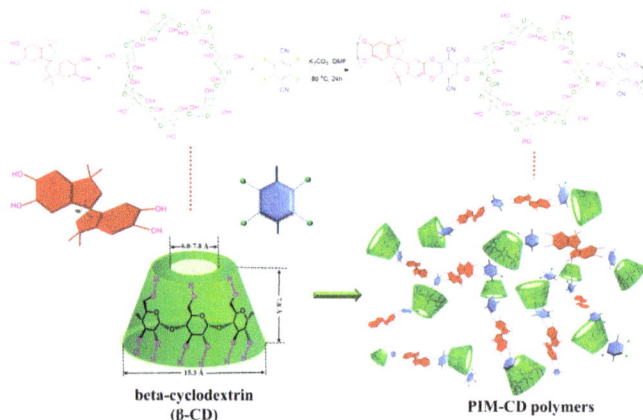

Figure 40. Illustration of the synthesis and polymer structure of PIM-CD copolymers. Reproduced with permission [192]. Copyright 2017, Elsevier Ltd.

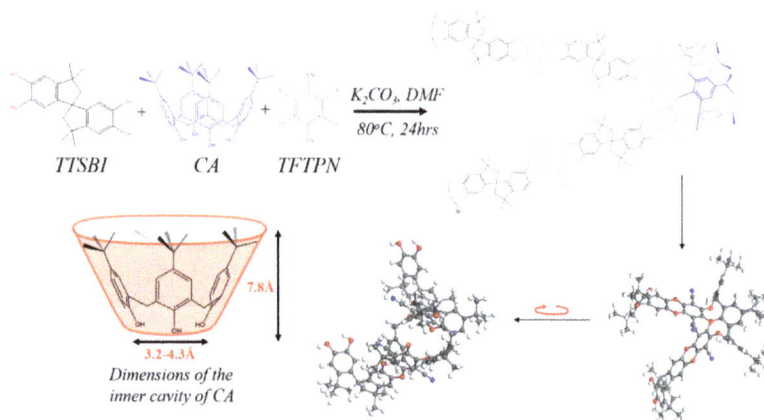

Figure 41. Illustration of the synthesis and polymer structure of PIM-CA copolymers. Reproduced with permission [193]. Copyright 2018, John Wiley & Sons.

In a very recent study by Wu et al., an unconventional post-fabrication infiltration method for incorporating OMMs into MMMs after the dense polymer films have been formed was proposed. It enabled the infiltration of a water-soluble member of CA, 4-sulfocalix[4]arene (SCA4) into already fabricated dense membranes (Fig. 42) [194]. The post-fabrication infiltration method sidestepped the problem of interfacial nano-defects or filler agglomeration tendency, while the SCA4 molecules could also act as strongly size-sieving molecular gatekeepers that drastically boosted the H_2/N_2 and H_2/CH_4 separation performance of the membrane. Just like framework-type fillers, OMMs also possess high tunability on their cavity size, functionality and physical properties.

4.3.1.5 2D Nanosheets

2D nanosheets have also attracted growing attention as an efficient MMM filler. 2D nanomaterials, such as graphene oxide (GO) sheets, 2D MOFs, and 2D COFs, generally have large aspect ratios and exhibit outstanding size-sieving capability using their sub-nano-sized interlayer spacing [184, 186, 195, 196]. The presence of large aspect ratios results in 2D nanosheets in the polymer matrix with a higher tortuosity for gas transport. This favors the passage of smaller gas molecules over bulkier ones and further enhances the size-selectivity. In addition to the interlayer spacing, surface pores/defects can be created on the nanosheets that act as angstrom-level, size-sieving openings in connection with the interlayer spacing so as to further enhance the overall size-selective effect. For example, thermally reduced GO (rGO) nanosheets could possess size-sieving surface defects due to the removal of thermally labile O-containing functionalities on the nanosheets. Similarly, 2D COF-sheets could possess large and highly permeable intrinsic surface pores arising from their porous architectures [197, 198].

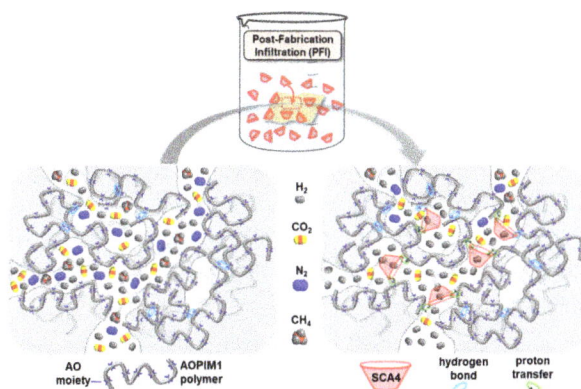

Figure 42. Illustration of SCA4 molecules acting as gatekeepers inside microporous membranes via the PFI method. Reproduced with permission [194]. Copyright 2020, The Royal Society of Chemistry.

4.5.2 Thin-film Nanocomposite (TFN) Membranes

The emergence of finely sized, highly polymer-compatible and easily dispersible organic or hybrid fillers prompts the advancement of TFC membranes into thin-film nanocomposite (TFN) membranes [199, 200]. The key characteristic of the TFN membranes is that their ultra-thin selective layer is incorporated with high-performance, functional nanofillers (i.e., the selective layer is an MMM) such that their gas separation could be greatly enhanced while only a small amount of nanomaterials is consumed due to the ultra-thinness of the selective layer. An illustrative diagram of TFN membranes was shown in Fig. 43. The size and interfacial chemistry of the nanofillers are critical to the successful design of TFN membranes because the minimal thinness as well as the mechanical stability are determined by these factors.

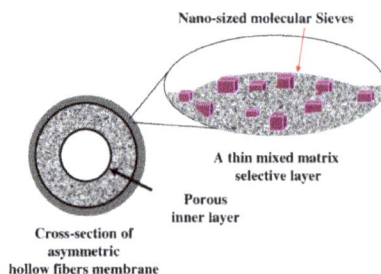

Figure 43. Illustration of TFN membrane structure. Reproduced with permission [175]. Copyright 2007, Elsevier Ltd.

5. Summary

H_2- and N_2-related gas separations are of paramount importance to the ongoing clean energy and environmental development. Great promises of practical applications have been found in polymeric membranes given their high energy-efficiency, small footprint and ease of operation. Since their first successful commercial implementation in the 1970s, polymeric membranes have been playing a growing role in the GS industry, especially when an increasing number of new gas pairs to be separated are being put on the table.

Early successes have been witnessed on relatively simpler separations that involved primarily a size-discrimination process, such as H_2/N_2 and H_2/CH_4 separations, and quite a number of commercially viable polymeric membrane products have been developed over the past several decades. Despite the inherent trade-off relationship between the permeability and selectivity of polymeric membranes, with the advent of the ultra-efficient new generation of intrinsically microporous polymers, like PIMs and PIM-PIs, these limits have been constantly transcended or prompted for updates. As such, the use of polymeric membranes for these so-called straightforward separations is gaining a greater competitive edge over the traditional thermally driven, energy-intensive technologies. While these new-generation polymers indeed produce undeniably extraordinary performances for not only hydrogen separations but also other gas separations, the physical aging problem which gradually deteriorates the gas permeability with time remains a key challenge against their applicability for actual industrial processes. Nevertheless, with more studies and research efforts being devoted to overcome this constraint, the membrane community are accumulating more and more tools and strategies against the aging problem, including, but not limited to, controlled crosslinking, advanced thermal treatment, and the incorporation of functional nanomaterials

On the other hand, when dealing with the more challenging gas separations that involve condensable species, which present a fast-growing market of gas separation these days due to the rising industrial demand for CO_2 capture and hydrocarbon treatment/recycling/separation, the membrane design could be much more complicated and the gas transport properties and the commercial practicality of the resultant polymeric membranes are less predictable. For instance, in such separations, the final gas transport properties of membranes are often a result of the competition between the size-selectivity and the solubility selectivity of the polymers. Therefore, a good membrane design not only needs to consider the diffusivity-related factors, such as FFV and chain rigidity, but also solubility-related ones, such as affinities and polar functionalities. Besides, the plasticization effect induced by condensable gases which severely compromises the separation efficiency and possibly mechanical stability of polymeric membranes remains a critical issue yet to be satisfactorily addressed. This is especially true for very thin polymeric films (either in flat or HFM mode) which are the most preferred but also the most plasticization-affected membrane configuration. While in this chapter a variety of polymeric membrane materials have been extensively discussed largely on their gas separation performances, in the actual design of polymeric membranes involving condensable gases, their intrinsic separation performance needs to be always considered in conjunction with their durability against the strongly plasticizing environment. This multifaceted design principle is also the future direction for new polymer development that aims at delivering easily processable, high-performance, and aging- and plasticization-resistant membranes.

Apart from designing new polymer backbones, the advent of the mixed matrix membrane (MMM) concept also opens a new door to gas separation membrane design given its high versatility and the virtually countless filler choices. Many of which have already demonstrated extraordinary composite properties previously unperceived in the neat polymers. With the development of thin film nanocomposite (TFN) membranes, the community has been one big step closer to the eventual commercial use of MMMs for actual industrial processes which could be another effective catalyst that will speed up the takeover of the gas separation industry by membrane technology.

References

[1] J.D. Perry, K. Nagai, W.J. Koros, Polymer membranes for hydrogen separations, MRS Bull., 31 (2006) 745-749. https://doi.org/10.1557/mrs2006.187

[2] N.W. Ockwig, T.M. Nenoff, Membranes for hydrogen separation, Chem. Rev., 107 (2007) 4078-4110. https://doi.org/10.1021/cr0501792

[3] X. Huang, H. Yao, Z. Cheng, Hydrogen separation membranes of polymeric materials, in: Y.-P. Chen, S. Bashir, J.L. Liu (Eds.) Nanostructured Materials for Next-Generation Energy Storage and Conversion: Hydrogen Production, Storage, and Utilization, Springer Berlin Heidelberg, Berlin, Heidelberg, 2017, pp. 85-116. https://doi.org/10.1007/978-3-662-53514-1_3

[4] C.E. Powell, G.G. Qiao, Polymeric CO_2/N_2 gas separation membranes for the capture of carbon dioxide from power plant flue gases, J. Membr. Sci., 279 (2006) 1-49. https://doi.org/10.1016/j.memsci.2005.12.062

[5] M.A. Carreon, Molecular sieve membranes for N_2/CH_4 separation, J. Mater. Res., 33 (2018) 32-43. https://doi.org/10.1557/jmr.2017.297

[6] L. Liu, A. Chakma, X. Feng, D. Lawless, Separation of VOCs from N_2 using poly(ether block amide) membranes, Can. J. Chem. Eng., 87 (2009) 456-465. https://doi.org/10.1002/cjce.20181

[7] M. Galizia, W.S. Chi, Z.P. Smith, T.C. Merkel, R.W. Baker, B.D. Freeman, 50th anniversary perspective: Polymers and mixed matrix membranes for gas and vapor separation: A review and prospective opportunities, Macromolecules, 50 (2017) 7809-7843. https://doi.org/10.1021/acs.macromol.7b01718

[8] P. Bernardo, E. Drioli, G. Golemme, Membrane gas separation: A review/state of the art, Ind. Eng. Chem. Res., 48 (2009) 4638-4663. https://doi.org/10.1021/ie8019032

[9] (a) S. Loeb, S. Sourirajan, Sea water demineralization by means of an osmotic membrane, in: R.F. Gould (Ed.) Saline Water Conversion—II, American Chemical Society, Washington, D.C., 1963, pp. 117-132; https://doi.org/10.1021/ba-1963-

0038.ch009 (b) A. Giwa, M. Ahmed, S.W. Hasan, Polymers for membrane filtration in water purification, in: R. Das (Ed.), Polymeric Materials for Clean Water, Springer, Cham, 2019, pp. 167-190. https://doi.org/10.1007/978-3-030-00743-0_8

[10] Y. Wang, X. Ma, B.S. Ghanem, F. Alghunaimi, I. Pinnau, Y. Han, Polymers of intrinsic microporosity for energy-intensive membrane-based gas separations, Mater. Today Nano, 3 (2018) 69-95. https://doi.org/10.1016/j.mtnano.2018.11.003

[11] A.F. Ismail, K.C. Khulbe, T. Matsuura, Fundamentals of gas permeation through membranes, in: Gas Separation Membranes: Polymeric and Inorganic, Springer International Publishing, Cham, 2015, pp. 11-35. https://doi.org/10.1007/978-3-319-01095-3_2

[12] R.J. Gardner, R.A. Crane, J.F. Hannan, Hollow fiber permeator for separating gases, Chem. Eng. Prog., 73 (1977) 76-78.

[13] R.R. Zolandz, G.K. Fleming, Gas permeation applications, in: W.S.W. Ho, K.K. Sirkar (Eds.) Membrane Handbook, Chapman and Hall, New York, 1992, pp. 78. https://doi.org/10.1007/978-1-4615-3548-5_5

[14] J.M.S. Henis, M.K. Tripodi, Multicomponent membranes for gas separations, U.S. Patent 4,230,463A, 1980

[15] W.J. Schell, C.D. Houston, Spiral-wound permeators for purifications and recovery, Chem. Eng. Prog., 78 (1982).

[16] B. Zornoza, C. Casado, A. Navajas, Advances in hydrogen separation and purification with membrane technology, in: L.M. Gandía, G. Arzamendi, P.M. Diéguez (Eds.) Renewable Hydrogen Technologies, Elsevier, Amsterdam, 2013, pp. 245-268. https://doi.org/10.1016/B978-0-444-56352-1.00011-8

[17] H. Makino, Y. Kusuki, H. Yoshida, A. Nakamaura, Process for preparing aromatic polyimide semipermeable membranes, U.S. Patent 4,378,324, 1983

[18] P. Luis, T. Van Gerven, B. Van der Bruggen, Recent developments in membrane-based technologies for CO_2 capture, Prog. Energy Combust. Sci., 38 (2012) 419-448. https://doi.org/10.1016/j.pecs.2012.01.004

[19] S. Adhikari, S. Fernando, Hydrogen membrane separation techniques, Ind. Eng. Chem. Res., 45 (2006) 875-881. https://doi.org/10.1021/ie050644l

[20] M. Mulder, Membrane processes, in: M. Mulder (Ed.) Basic Principles of Membrane Technology, Springer Netherlands, Dordrecht, 1991, pp. 198-280. https://doi.org/10.1007/978-94-017-0835-7_6

[21] J.G. Wijmans, R.W. Baker, The solution-diffusion model: A review, J. Membr. Sci., 107 (1995) 1-21. https://doi.org/10.1016/0376-7388(95)00102-I

[22] S. Matteucci, Y. Yampolskii, B.D. Freeman, I. Pinnau, Transport of gases and vapors in glassy and rubbery polymers, in: Y. Yampolskii, B.D. Freeman, I. Pinnau (Eds.) Materials Science of Membranes for Gas and Vapor Separation, John Wiley & Sons, Chichester, 2006, pp. 1-47. https://doi.org/10.1002/047002903X.ch1

[23] C.H. Lau, P. Li, F.Y. Li, T.S. Chung, D.R. Paul, Reverse-selective polymeric membranes for gas separations, Prog. Polym. Sci., 38 (2013) 740-766. https://doi.org/10.1016/j.progpolymsci.2012.09.006

[24] L.M. Robeson, Correlation of separation factor versus permeability for polymeric membranes, J. Membr. Sci., 62 (1991) 165-185. https://doi.org/10.1016/0376-7388(91)80060-J

[25] B.D. Freeman, Basis of permeability/selectivity tradeoff relations in polymeric gas separation membranes, Macromolecules, 32 (1999) 375-380. https://doi.org/10.1021/ma9814548

[26] C.F. Tien, A.C. Savoca, A.D. Surnamer, M. Langsam, Chemical structure/permeation relationship for polysilylpropynes, Polym. Mater. Sci. Eng. Prepr., 61 (1989) 507-511.

[27] Y. Alqaheem, A. Alomair, M. Vinoba, A. Pérez, Polymeric gas-separation membranes for petroleum refining, Int. J. Polym. Sci., 2017 (2017). https://doi.org/10.1155/2017/4250927

[28] S.A. Stern, Polymers for gas separations: The next decade, J. Membr. Sci., 94 (1994) 1-65. https://doi.org/10.1016/0376-7388(94)00141-3

[29] D.F. Sanders, Z.P. Smith, R. Guo, L.M. Robeson, J.E. McGrath, D.R. Paul, B.D. Freeman, Energy-efficient polymeric gas separation membranes for a sustainable future: A review, Polymer, 54 (2013) 4729-4761. https://doi.org/10.1016/j.polymer.2013.05.075

[30] R.W. Baker, Gas separation, in: Membrane Technology and Applications, Third Edition, John Wiley & Sons, United Kingdom, 2012, pp. 325-378. https://doi.org/10.1002/9781118359686.ch8

[31] C.L. Aitken, W.J. Koros, D.R. Paul, Gas transport properties of biphenol polysulfones, Macromolecules, 25 (1992) 3651-3658. https://doi.org/10.1021/ma00040a008

[32] C.L. Aitken, W.J. Koros, D.R. Paul, Effect of structural symmetry on gas transport properties of polysulfones, Macromolecules, 25 (1992) 3424-3434. https://doi.org/10.1021/ma00039a018

[33] M.W. Hellums, W.J. Koros, J.C. Schmidhauser, Gas separation properties of spirobiindane polycarbonate, J. Membr. Sci., 67 (1992) 75-81. https://doi.org/10.1016/0376-7388(92)87041-U

[34] W.J. Koros, G.K. Fleming, Membrane-based gas separation, J. Membr. Sci., 83 (1993) 1-80. https://doi.org/10.1016/0376-7388(93)80013-N

[35] T. Heinze, T. Liebert, Chemical characteristics of cellulose acetate, Macromol. Symp., 208 (2004) 167-238. https://doi.org/10.1002/masy.200450408

[36] A.C. Puleo, D.R. Paul, S.S. Kelley, The effect of degree of acetylation on gas sorption and transport behavior in cellulose acetate, J. Membr. Sci., 47 (1989) 301-332. https://doi.org/10.1016/S0376-7388(00)83083-5

[37] R. Swaidan, B. Ghanem, I. Pinnau, Fine-tuned intrinsically ultramicroporous polymers redefine the permeability/selectivity upper bounds of membrane-based air and hydrogen separations, ACS Macro Lett., 4 (2015) 947-951. https://doi.org/10.1021/acsmacrolett.5b00512

[38] K. Toi, G. Morel, D.R. Paul, Gas sorption and transport in poly(phenylene oxide) and comparisons with other glassy polymers, J. Appl. Polym. Sci., 27 (1982) 2997-3005. https://doi.org/10.1002/app.1982.070270823

[39] A.S. Hay, Oxidation of phenols and resulting products, U.S. Patent 3,306,875A, 1967

[40] G.A. Polotskaya, A.V. Penkova, A.M. Toikka, Z. Pientka, L. Brozova, M. Bleha, Transport of small molecules through polyphenylene oxide membranes modified by fullerene, Sep. Sci. Technol., 42 (2007) 333-347. https://doi.org/10.1080/01496390600997963

[41] B.J. Story, W.J. Koros, Sorption and transport of CO_2 and CH_4 in chemically modified poly(phenylene oxide), J. Membr. Sci., 67 (1992) 191-210. https://doi.org/10.1016/0376-7388(92)80025-F

[42] O.M. Ekiner, G. Vassilatos, Polyaramide hollow fibers for H_2/CH_4 separation: II. Spinning and properties, J. Membr. Sci., 186 (2001) 71-84. https://doi.org/10.1016/S0376-7388(00)00665-7

[43] J.M. García, F.C. García, F. Serna, J.L. de la Peña, High-performance aromatic polyamides, Prog. Polym. Sci., 35 (2010) 623-686. https://doi.org/10.1016/j.progpolymsci.2009.09.002

[44] O.M. Ekiner, G. Vassilatos, Polyaramide hollow fibers for hydrogen/methane separation — spinning and properties, J. Membr. Sci., 53 (1990) 259-273. https://doi.org/10.1016/0376-7388(90)80018-H

[45] P. Falcigno, M. Masola, D. Williams, S. Jasne, Comparison of properties of polyimides containing DAPI isomers and various dianhydrides, in: C. Feger, M.M. Khohasteh, J.E. McGrath (Eds.) Polyimides: Materials, Chemistry and Characterization, Elsevier, New York, 1989, pp. 497-512.

[46] Y. Sasaki, H. Inoue, H. Itatani, M. Kashima, Process for preparing polyimide solution, U.S. Patent 4,290,936A, 1981

[47] Z. Xu, C. Dannenberg, J. Springer, S. Banerjee, G. Maier, Gas separation properties of polymers containing fluorene moieties, Chem. Mater., 14 (2002) 3271-3276. https://doi.org/10.1021/cm0112789

[48] M. Aguilar-Vega, D.R. Paul, Gas transport properties of polycarbonates and polysulfones with aromatic substitutions on the bisphenol connector group, J. Polym. Sci. Part B: Polym. Phys., 31 (1993) 1599-1610. https://doi.org/10.1002/polb.1993.090311116

[49] Y. Yampolskii, Polymeric gas separation membranes, Macromolecules, 45 (2012) 3298-3311. https://doi.org/10.1021/ma300213b

[50] I. Pinnau, L.G. Toy, Gas and vapor transport properties of amorphous perfluorinated copolymer membranes based on 2,2-bistrifluoromethyl-4,5-difluoro-1,3-dioxole/tetrafluoroethylene, J. Membr. Sci., 109 (1996) 125-133. https://doi.org/10.1016/0376-7388(95)00193-X

[51] M. Macchione, J.C. Jansen, G. De Luca, E. Tocci, M. Longeri, E. Drioli, Experimental analysis and simulation of the gas transport in dense Hyflon® AD60X membranes: Influence of residual solvent, Polymer, 48 (2007) 2619-2635. https://doi.org/10.1016/j.polymer.2007.02.068

[52] L.M. Robeson, The upper bound revisited, J. Membr. Sci., 320 (2008) 390-400. https://doi.org/10.1016/j.memsci.2008.04.030

[53] K. Nagai, T. Masuda, T. Nakagawa, B.D. Freeman, I. Pinnau, Poly[1-(trimethylsilyl)-1-propyne] and related polymers: synthesis, properties and functions, Prog. Polym. Sci., 26 (2001) 721-798. https://doi.org/10.1016/S0079-6700(01)00008-9

[54] T. Masuda, E. Isobe, T. Higashimura, K. Takada, Poly[1-(trimethylsilyl)-1-propyne]: A new high polymer synthesized with transition-metal catalysts and characterized by extremely high gas permeability, J. Am. Chem. Soc., 105 (1983) 7473-7474. https://doi.org/10.1021/ja00363a061

[55] Y. Ichiraku, S.A. Stern, T. Nakagawa, An investigation of the high gas permeability of poly (1-trimethylsilyl-1-propyne), J. Membr. Sci., 34 (1987) 5-18. https://doi.org/10.1016/S0376-7388(00)80017-4

[56] I. Pinnau, L.G. Toy, Transport of organic vapors through poly(1-trimethylsilyl-1-propyne), J. Membr. Sci., 116 (1996) 199-209. https://doi.org/10.1016/0376-7388(96)00041-5

[57] P.M. Budd, E.S. Elabas, B.S. Ghanem, S. Makhseed, N.B. McKeown, K.J. Msayib, C.E. Tattershall, D. Wang, Solution-processed, organophilic membrane derived from a polymer of intrinsic microporosity, Adv. Mater., 16 (2004) 456-459. https://doi.org/10.1002/adma.200306053

[58] P.M. Budd, B.S. Ghanem, S. Makhseed, N.B. McKeown, K.J. Msayib, C.E. Tattershall, Polymers of intrinsic microporosity (PIMs): Robust, solution-processable, organic nanoporous materials, Chem. Commun., (2004) 230-231. https://doi.org/10.1039/b311764b

[59] P.M. Budd, Polymer with intrinsic microporosity (PIM), in: E. Drioli, L. Giorno (Eds.) Encyclopedia of Membranes, Springer Berlin Heidelberg, Berlin, Heidelberg, 2016, pp. 1606-1607.

[60] B.S. Ghanem, N.B. McKeown, P.M. Budd, J.D. Selbie, D. Fritsch, High-performance membranes from polyimides with intrinsic microporosity, Adv. Mater., 20 (2008) 2766-2771. https://doi.org/10.1002/adma.200702400

[61] Z.X. Low, P.M. Budd, N.B. McKeown, D.A. Patterson, Gas permeation properties, physical aging, and its mitigation in high free volume glassy polymers, Chem. Rev., 118 (2018) 5871-5911. https://doi.org/10.1021/acs.chemrev.7b00629

[62] I. Rose, C.G. Bezzu, M. Carta, B. Comesaña-Gándara, E. Lasseuguette, M.C. Ferrari, P. Bernardo, G. Clarizia, A. Fuoco, J.C. Jansen, Kyle E. Hart, T.P. Liyana-Arachchi, C.M. Colina, N.B. McKeown, Polymer ultrapermeability from the inefficient packing of 2D chains, Nat. Mater., 16 (2017) 932-937. https://doi.org/10.1038/nmat4939

[63] P.M. Budd, K.J. Msayib, C.E. Tattershall, B.S. Ghanem, K.J. Reynolds, N.B. McKeown, D. Fritsch, Gas separation membranes from polymers of intrinsic

microporosity, J. Membr. Sci., 251 (2005) 263-269.
https://doi.org/10.1016/j.memsci.2005.01.009

[64] B.S. Ghanem, R. Swaidan, X. Ma, E. Litwiller, I. Pinnau, Energy-efficient hydrogen separation by AB-type ladder-polymer molecular sieves, Adv. Mater., 26 (2014) 6696-6700. https://doi.org/10.1002/adma.201401328

[65] B. Ghanem, F. Alghunaimi, N. Alaslai, X. Ma, I. Pinnau, New phenazine-containing ladder polymer of intrinsic microporosity from a spirobisindane-based AB-type monomer, RSC Adv., 6 (2016) 79625-79630. https://doi.org/10.1039/C6RA16393A

[66] M. Carta, R. Malpass-Evans, M. Croad, Y. Rogan, J.C. Jansen, P. Bernardo, F. Bazzarelli, N.B. McKeown, An efficient polymer molecular sieve for membrane gas separations, Science, 339 (2013) 303-307. https://doi.org/10.1126/science.1228032

[67] C.G. Bezzu, M. Carta, A. Tonkins, J.C. Jansen, P. Bernardo, F. Bazzarelli, N.B. McKeown, A spirobifluorene-based polymer of intrinsic microporosity with improved performance for gas separation, Adv. Mater., 24 (2012) 5930-5933. https://doi.org/10.1002/adma.201202393

[68] B.S. Ghanem, N.B. McKeown, P.M. Budd, N.M. Al-Harbi, D. Fritsch, K. Heinrich, L. Starannikova, A. Tokarev, Y. Yampolskii, Synthesis, characterization, and gas permeation properties of a novel group of polymers with intrinsic microporosity: PIM-polyimides, Macromolecules, 42 (2009) 7881-7888. https://doi.org/10.1021/ma901430q

[69] Y. Rogan, L. Starannikova, V. Ryzhikh, Y. Yampolskii, P. Bernardo, F. Bazzarelli, J.C. Jansen, N.B. McKeown, Synthesis and gas permeation properties of novel spirobisindane-based polyimides of intrinsic microporosity, Polym. Chem., 4 (2013) 3813-3820. https://doi.org/10.1039/c3py00451a

[70] Y. Rogan, R. Malpass-Evans, M. Carta, M. Lee, J.C. Jansen, P. Bernardo, G. Clarizia, E. Tocci, K. Friess, M. Lanč, N.B. McKeown, A highly permeable polyimide with enhanced selectivity for membrane gas separations, J. Mater. Chem. A, 2 (2014) 4874-4877. https://doi.org/10.1039/C4TA00564C

[71] X. Ma, M.A. Abdulhamid, I. Pinnau, Design and synthesis of polyimides based on carbocyclic pseudo-Tröger's Base-derived dianhydrides for membrane gas separation applications, Macromolecules, 50 (2017) 5850-5857. https://doi.org/10.1021/acs.macromol.7b01054

[72] B.S. Ghanem, R. Swaidan, E. Litwiller, I. Pinnau, Ultra-microporous triptycene-based polyimide membranes for high-performance gas separation, Adv. Mater., 26 (2014) 3688-3692. https://doi.org/10.1002/adma.201306229

[73] R. Swaidan, M. Al-Saeedi, B. Ghanem, E. Litwiller, I. Pinnau, Rational design of intrinsically ultramicroporous polyimides containing bridgehead-substituted triptycene for highly selective and permeable gas separation membranes, Macromolecules, 47 (2014) 5104-5114. https://doi.org/10.1021/ma5009226

[74] X. Ma, O. Salinas, E. Litwiller, I. Pinnau, Novel spirobifluorene- and dibromospirobifluorene-based polyimides of intrinsic microporosity for gas separation applications, Macromolecules, 46 (2013) 9618-9624. https://doi.org/10.1021/ma402033z

[75] Y. Zhuang, J.G. Seong, Y.S. Do, H.J. Jo, Z. Cui, J. Lee, Y.M. Lee, M.D. Guiver, Intrinsically microporous soluble polyimides incorporating Tröger's base for membrane gas separation, Macromolecules, 47 (2014) 3254-3262. https://doi.org/10.1021/ma5007073

[76] Y. Zhuang, J.G. Seong, Y.S. Do, W.H. Lee, M.J. Lee, M.D. Guiver, Y.M. Lee, High-strength, soluble polyimide membranes incorporating Tröger's Base for gas separation, J. Membr. Sci., 504 (2016) 55-65. https://doi.org/10.1016/j.memsci.2015.12.057

[77] B. Ghanem, N. Alaslai, X. Miao, I. Pinnau, Novel 6FDA-based polyimides derived from sterically hindered Tröger's base diamines: Synthesis and gas permeation properties, Polymer, 96 (2016) 13-19. https://doi.org/10.1016/j.polymer.2016.04.068

[78] M. Lee, C.G. Bezzu, M. Carta, P. Bernardo, G. Clarizia, J.C. Jansen, N.B. McKeown, Enhancing the gas permeability of Tröger's Base derived polyimides of intrinsic microporosity, Macromolecules, 49 (2016) 4147-4154. https://doi.org/10.1021/acs.macromol.6b00351

[79] Z. Wang, D. Wang, J. Jin, Microporous polyimides with rationally designed chain structure achieving high performance for gas separation, Macromolecules, 47 (2014) 7477-7483. https://doi.org/10.1021/ma5017506

[80] F. Alghunaimi, B. Ghanem, N. Alaslai, R. Swaidan, E. Litwiller, I. Pinnau, Gas permeation and physical aging properties of iptycene diamine-based microporous polyimides, J. Membr. Sci., 490 (2015) 321-327. https://doi.org/10.1016/j.memsci.2015.05.010

[81] J.R. Wiegand, Z.P. Smith, Q. Liu, C.T. Patterson, B.D. Freeman, R. Guo, Synthesis and characterization of triptycene-based polyimides with tunable high fractional free volume for gas separation membranes, J. Mater. Chem. A, 2 (2014) 13309-13320. https://doi.org/10.1039/C4TA02303J

[82] S. Luo, Q. Liu, B. Zhang, J.R. Wiegand, B.D. Freeman, R. Guo, Pentiptycene-based polyimides with hierarchically controlled molecular cavity architecture for efficient membrane gas separation, J. Membr. Sci., 480 (2015) 20-30. https://doi.org/10.1016/j.memsci.2015.01.043

[83] R. Swaidan, B.S. Ghanem, E. Litwiller, I. Pinnau, Pure- and mixed-gas CO_2/CH_4 separation properties of PIM-1 and an amidoxime-functionalized PIM-1, J. Membr. Sci., 457 (2014) 95-102. https://doi.org/10.1016/j.memsci.2014.01.055

[84] N. Du, G.P. Robertson, J. Song, I. Pinnau, M.D. Guiver, High-performance carboxylated polymers of intrinsic microporosity (PIMs) with tunable gas transport properties, Macromolecules, 42 (2009) 6038-6043. https://doi.org/10.1021/ma9009017

[85] N. Du, H.B. Park, G.P. Robertson, M.M. Dal-Cin, T. Visser, L. Scoles, M.D. Guiver, Polymer nanosieve membranes for CO_2-capture applications, Nat. Mater., 10 (2011) 372-375. https://doi.org/10.1038/nmat2989

[86] C.R. Mason, L. Maynard-Atem, N.M. Al-Harbi, P.M. Budd, P. Bernardo, F. Bazzarelli, G. Clarizia, J.C. Jansen, Polymer of intrinsic microporosity incorporating thioamide functionality: Preparation and gas transport properties, Macromolecules, 44 (2011) 6471-6479. https://doi.org/10.1021/ma200918h

[87] X. Ma, R. Swaidan, Y. Belmabkhout, Y. Zhu, E. Litwiller, M. Jouiad, I. Pinnau, Y. Han, Synthesis and gas transport properties of hydroxyl-functionalized polyimides with intrinsic microporosity, Macromolecules, 45 (2012) 3841-3849. https://doi.org/10.1021/ma300549m

[88] B. Comesaña-Gándara, J. Chen, C.G. Bezzu, M. Carta, I. Rose, M.-C. Ferrari, E. Esposito, A. Fuoco, J.C. Jansen, N.B. McKeown, Redefining the Robeson upper bounds for CO_2/CH_4 and CO_2/N_2 separations using a series of ultrapermeable benzotriptycene-based polymers of intrinsic microporosity, Energy Environ. Sci., 12 (2019) 2733-2740. https://doi.org/10.1039/C9EE01384A

[89] J.R. Klaehn, C.J. Orme, E.S. Peterson, F.F. Stewart, J.M. Urban-Klaehn, High temperature gas separations using high performance polymers, in: S.T. Oyama, S.M. Stagg-Williams (Eds.) Membrane Science and Technology, Elsevier, Great Britain, 2011, pp. 295-307. https://doi.org/10.1016/B978-0-444-53728-7.00013-6

[90] H. Vogel, C.S. Marvel, Polybenzimidazoles, new thermally stable polymers, J. Polym. Sci., 50 (1961) 511-539. https://doi.org/10.1002/pol.1961.1205015419

[91] Y. Wang, T.X. Yang, K. Fishel, B. Benicewicz, T.S. Chung, Polybenzimidazoles, in: O. Olabisi, K. Adewale (Eds.) Handbook of Thermoplastics Second Edition, CRC Press, Boca Raton, 2015, pp. 617-667.

[92] S.C. Kumbharkar, P.B. Karadkar, U.K. Kharul, Enhancement of gas permeation properties of polybenzimidazoles by systematic structure architecture, J. Membr. Sci., 286 (2006) 161-169. https://doi.org/10.1016/j.memsci.2006.09.030

[93] S.D. Kenarsari, D. Yang, G. Jiang, S. Zhang, J. Wang, A.G. Russell, Q. Wei, M. Fan, Review of recent advances in carbon dioxide separation and capture, RSC Adv., 3 (2013) 22739-22773. https://doi.org/10.1039/c3ra43965h

[94] H. Borjigin, K.A. Stevens, R. Liu, J.D. Moon, A.T. Shaver, S. Swinnea, B.D. Freeman, J.S. Riffle, J.E. McGrath, Synthesis and characterization of polybenzimidazoles derived from tetraaminodiphenylsulfone for high temperature gas separation membranes, Polymer, 71 (2015) 135-142. https://doi.org/10.1016/j.polymer.2015.06.021

[95] X. Li, R.P. Singh, K.W. Dudeck, K.A. Berchtold, B.C. Benicewicz, Influence of polybenzimidazole main chain structure on H_2/CO_2 separation at elevated temperatures, J. Membr. Sci., 461 (2014) 59-68. https://doi.org/10.1016/j.memsci.2014.03.008

[96] S.C. Kumbharkar, Y. Liu, K. Li, High performance polybenzimidazole based asymmetric hollow fibre membranes for H_2/CO_2 separation, J. Membr. Sci., 375 (2011) 231-240. https://doi.org/10.1016/j.memsci.2011.03.049

[97] L. Zhu, M.T. Swihart, H. Lin, Unprecedented size-sieving ability in polybenzimidazole doped with polyprotic acids for membrane H_2/CO_2 separation, Energy Environ. Sci., 11 (2018) 94-100. https://doi.org/10.1039/C7EE02865B

[98] J.R. Klaehn, C.J. Orme, E.S. Peterson, T.A. Luther, M.G. Jones, A.K. Wertsching, J.M. Urban-Klaehn: CO_2 separation using thermally optimized membranes: A comprehensive project report (2000 - 2007) (Idaho National Laboratory, Idaho 2008). https://doi.org/10.2172/928082

[99] W.J. Koros, R.T. Chern, V. Stannett, H.B. Hopfenberg, A model for permeation of mixed gases and vapors in glassy polymers, J. Polym. Sci. Polym. Phys. Ed., 19 (1981) 1513-1530. https://doi.org/10.1002/pol.1981.180191004

[100] I. Pinnau, C.G. Casillas, A. Morisato, B.D. Freeman, Hydrocarbon/hydrogen mixed gas permeation in poly(1-trimethylsilyl-1-propyne) (PTMSP), poly(1-phenyl-1-propyne) (PPP), and PTMSP/PPP blends, J. Polym. Sci. Part B: Polym. Phys., 34 (1996) 2613-2621. https://doi.org/10.1002/(SICI)1099-0488(19961115)34:15<2613::AID-POLB9>3.0.CO;2-T

[101] H. Lin, B.D. Freeman, Gas solubility, diffusivity and permeability in poly(ethylene oxide), J. Membr. Sci., 239 (2004) 105-117. https://doi.org/10.1016/j.memsci.2003.08.031

[102] S.B. Aziz, R.M. Abdullah, Crystalline and amorphous phase identification from the tanδ relaxation peaks and impedance plots in polymer blend electrolytes based on [CS:AgNt]x:PEO(x-1) (10 ≤ x ≤ 50), Electrochim. Acta, 285 (2018) 30-46. https://doi.org/10.1016/j.electacta.2018.07.233

[103] Y. Hirayama, Y. Kase, N. Tanihara, Y. Sumiyama, Y. Kusuki, K. Haraya, Permeation properties to CO_2 and N_2 of poly(ethylene oxide)-containing and crosslinked polymer films, J. Membr. Sci., 160 (1999) 87-99. https://doi.org/10.1016/S0376-7388(99)00080-0

[104] N.P. Patel, A.C. Miller, R.J. Spontak, Highly CO_2-permeable and selective polymer nanocomposite membranes, Adv. Mater., 15 (2003) 729-733. https://doi.org/10.1002/adma.200304712

[105] H. Lin, E. Van Wagner, B.D. Freeman, L.G. Toy, R.P. Gupta, Plasticization-enhanced hydrogen purification using polymeric membranes, Science, 311 (2006) 639-642. https://doi.org/10.1126/science.1118079

[106] M. Wessling, S. Schoeman, T. van der Boomgaard, C.A. Smolders, Plasticization of gas separation membranes, Gas Sep. Purif., 5 (1991) 222-228. https://doi.org/10.1016/0950-4214(91)80028-4

[107] I. Pinnau, Z. He, Pure- and mixed-gas permeation properties of polydimethylsiloxane for hydrocarbon/methane and hydrocarbon/hydrogen separation, J. Membr. Sci., 244 (2004) 227-233. https://doi.org/10.1016/j.memsci.2004.06.055

[108] V.I. Bondar, B.D. Freeman, I. Pinnau, Gas sorption and characterization of poly(ether-b-amide) segmented block copolymers, J. Polym. Sci. Part B: Polym. Phys., 37 (1999) 2463-2475. https://doi.org/10.1002/(SICI)1099-0488(19990901)37:17<2463::AID-POLB18>3.0.CO;2-H

[109] H.Z. Chen, Y.C. Xiao, T.S. Chung, Synthesis and characterization of poly (ethylene oxide) containing copolyimides for hydrogen purification, Polymer, 51 (2010) 4077-4086. https://doi.org/10.1016/j.polymer.2010.06.046

[110] S.H. Choi, J.H. Kim, S.B. Lee, Sorption and permeation behaviors of a series of olefins and nitrogen through PDMS membranes, J. Membr. Sci., 299 (2007) 54-62. https://doi.org/10.1016/j.memsci.2007.04.022

[111] J. Liu, X. Hou, H.B. Park, H. Lin, High-performance polymers for membrane CO_2/N_2 Separation, Chem. Eur. J., 22 (2016) 15980-15990. https://doi.org/10.1002/chem.201603002

[112] X. Jiang, S. Li, L. Shao, Pushing CO_2-philic membrane performance to the limit by designing semi-interpenetrating networks (SIPN) for sustainable CO_2 separations, Energy Environ. Sci., 10 (2017) 1339-1344. https://doi.org/10.1039/C6EE03566C

[113] S.L. Liu, R. Wang, T.S. Chung, M.L. Chng, Y. Liu, R.H. Vora, Effect of diamine composition on the gas transport properties in 6FDA-durene/3,3'-diaminodiphenyl sulfone copolyimides, J. Membr. Sci., 202 (2002) 165-176. https://doi.org/10.1016/S0376-7388(01)00754-2

[114] H.B. Park, C.H. Jung, Y.M. Lee, A.J. Hill, S.J. Pas, S.T. Mudie, E. Van Wagner, B.D. Freeman, D.J. Cookson, Polymers with cavities tuned for fast selective transport of small molecules and ions, Science, 318 (2007) 254-258. https://doi.org/10.1126/science.1146744

[115] N.B. McKeown, P.M. Budd, K.J. Msayib, B.S. Ghanem, H.J. Kingston, C.E. Tattershall, S. Makhseed, K.J. Reynolds, D. Fritsch, Polymers of intrinsic microporosity (PIMs): Bridging the void between microporous and polymeric materials, Chem. Eur. J., 11 (2005) 2610-2620. https://doi.org/10.1002/chem.200400860

[116] S. Kim, Y.M. Lee, Rigid and microporous polymers for gas separation membranes, Prog. Polym. Sci., 43 (2015) 1-32. https://doi.org/10.1016/j.progpolymsci.2014.10.005

[117] H.B. Park, S.H. Han, C.H. Jung, Y.M. Lee, A.J. Hill, Thermally rearranged (TR) polymer membranes for CO_2 separation, J. Membr. Sci., 359 (2010) 11-24. https://doi.org/10.1016/j.memsci.2009.09.037

[118] R. Guo, D.F. Sanders, Z.P. Smith, B.D. Freeman, D.R. Paul, J.E. McGrath, Synthesis and characterization of thermally rearranged (TR) polymers: Influence of ortho-positioned functional groups of polyimide precursors on TR process and gas

transport properties, J. Mater. Chem. A, 1 (2013) 262-272.
https://doi.org/10.1039/C2TA00799A

[119] A. Tena, S. Rangou, S. Shishatskiy, V. Filiz, V. Abetz, Claisen thermally rearranged (CTR) polymers, Sci. Adv., 2 (2016) e1501859.
https://doi.org/10.1126/sciadv.1501859

[120] M.R. de la Viuda, A. Tena, S. Neumann, S. Willruth, V. Filiz, V. Abetz, Novel functionalized polyamides prone to undergo thermal Claisen rearrangement in the solid state, Polym. Chem., 9 (2018) 4007-4016. https://doi.org/10.1039/C8PY00467F

[121] R.W. Baker, K. Lokhandwala, Natural gas processing with membranes: An overview, Ind. Eng. Chem. Res., 47 (2008) 2109-2121.
https://doi.org/10.1021/ie071083w

[122] A. Tena, A. Marcos-Fernández, A.E. Lozano, J.G. de la Campa, J. de Abajo, L. Palacio, P. Prádanos, A. Hernández, Thermally segregated copolymers with PPO blocks for nitrogen removal from natural gas, Ind. Eng. Chem. Res., 52 (2013) 4312-4322. https://doi.org/10.1021/ie303378k

[123] Z. Cui, E. Drioli, Y.M. Lee, Recent progress in fluoropolymers for membranes, Prog. Polym. Sci., 39 (2014) 164-198.
https://doi.org/10.1016/j.progpolymsci.2013.07.008

[124] Y. Yampolskii, N. Belov, A. Alentiev, Perfluorinated polymers as materials of membranes for gas and vapor separation, J. Membr. Sci., 598 (2020) 117779.
https://doi.org/10.1016/j.memsci.2019.117779

[125] M. Yavari, M. Fang, H. Nguyen, T.C. Merkel, H. Lin, Y. Okamoto, Dioxolane-based perfluoropolymers with superior membrane gas separation properties, Macromolecules, 51 (2018) 2489-2497.
https://doi.org/10.1021/acs.macromol.8b00273

[126] M. Fang, Z. He, T.C. Merkel, Y. Okamoto, High-performance perfluorodioxolane copolymer membranes for gas separation with tailored selectivity enhancement, J. Mater. Chem. A, 6 (2018) 652-658. https://doi.org/10.1039/C7TA09047A

[127] N.A. Belov, A.A. Zharov, A.V. Shashkin, M.Q. Shaikh, K. Raetzke, Y.P. Yampolskii, Gas transport and free volume in hexafluoropropylene polymers, J. Membr. Sci., 383 (2011) 70-77. https://doi.org/10.1016/j.memsci.2011.08.029

[128] A.X. Wu, J.A. Drayton, Z.P. Smith, The perfluoropolymer upper bound, AIChE J., 65 (2019) e16700. https://doi.org/10.1002/aic.16700

[129] F. Suzuki, K. Nakane, Y. Hata, Grafting of siloxane on poly (styrene-co-maleic acid) and application of this grafting technique to a porous membrane for gas separation, J. Membr. Sci., 104 (1995) 283-290. https://doi.org/10.1016/0376-7388(95)00043-C

[130] Y.C. Xiao, T.S. Chung, Grafting thermally labile molecules on cross-linkable polyimide to design membrane materials for natural gas purification and CO_2 capture, Energy Environ. Sci., 4 (2011) 201-208. https://doi.org/10.1039/C0EE00278J

[131] M.L. Chua, Y.C. Xiao, T.S. Chung, Effects of thermally labile saccharide units on the gas separation performance of highly permeable polyimide membranes, J. Membr. Sci., 415 (2012) 375-382. https://doi.org/10.1016/j.memsci.2012.05.022

[132] M. Askari, Y.C. Xiao, P. Li, T.S. Chung, Natural gas purification and olefin/paraffin separation using cross-linkable 6FDA-Durene/DABA co-polyimides grafted with α, β, and γ-cyclodextrin, J. Membr. Sci., 390–391 (2012) 141-151. https://doi.org/10.1016/j.memsci.2011.11.030

[133] C.R. Mason, L. Maynard-Atem, K.W.J. Heard, B. Satilmis, P.M. Budd, K. Friess, M. Lanč, P. Bernardo, G. Clarizia, J.C. Jansen, Enhancement of CO_2 affinity in a polymer of intrinsic microporosity by amine modification, Macromolecules, 47 (2014) 1021-1029. https://doi.org/10.1021/ma401869p

[134] N. Du, G.P. Robertson, M.M. Dal-Cin, L. Scoles, M.D. Guiver, Polymers of intrinsic microporosity (PIMs) substituted with methyl tetrazole, Polymer, 53 (2012) 4367-4372. https://doi.org/10.1016/j.polymer.2012.07.055

[135] R.A. Hayes, Amine-modified polyimide membranes, U.S. Patent 4,981,497, 1991

[136] Y. Liu, R. Wang, T.S. Chung, Chemical cross-linking modification of polyimide membranes for gas separation, J. Membr. Sci., 189 (2001) 231-239. https://doi.org/10.1016/S0376-7388(01)00415-X

[137] K. Vanherck, G. Koeckelberghs, I.F.J. Vankelecom, Crosslinking polyimides for membrane applications: A review, Prog. Polym. Sci., 38 (2013) 874-896. https://doi.org/10.1016/j.progpolymsci.2012.11.001

[138] W. F. Yong, F.Y. Li, T.S. Chung, Y.W. Tong, Highly permeable chemically modified PIM-1/Matrimid membranes for green hydrogen purification, J. Mater. Chem. A, 1 (2013) 139 14-13925. https://doi.org/10.1039/c3ta13308g

[139] H.Y. Zhao, Y.M. Cao, X.L. Ding, M.Q. Zhou, J.H. Liu, Q. Yuan, Poly(ethylene oxide) induced cross-linking modification of Matrimid membranes for selective

separation of CO_2, J. Membr. Sci., 320 (2008) 179-184.
https://doi.org/10.1016/j.memsci.2008.03.070

[140] H. Lin, T. Kai, B.D. Freeman, S. Kalakkunnath, D.S. Kalika, The effect of cross-linking on gas permeability in cross-linked poly(ethylene glycol diacrylate), Macromolecules, 38 (2005) 8381-8393. https://doi.org/10.1021/ma0510136

[141] H.K. Yun, K. Cho, J.K. Kim, C.E. Park, S.M. Sim, S.Y. Oh, J.M. Park, Adhesion improvement of epoxy resin/polyimide joints by amine treatment of polyimide surface, Polymer, 38 (1997) 827-834. https://doi.org/10.1016/S0032-3861(96)00592-7

[142] L. Shao, T.S. Chung, S.H. Goh, K.P. Pramoda, Polyimide modification by a linear aliphatic diamine to enhance transport performance and plasticization resistance, J. Membr. Sci., 256 (2005) 46-56. https://doi.org/10.1016/j.memsci.2005.02.030

[143] C.E. Powell, X.J. Duthie, S.E. Kentish, G.G. Qiao, G.W. Stevens, Reversible diamine cross-linking of polyimide membranes, J. Membr. Sci., 291 (2007) 199-209. https://doi.org/10.1016/j.memsci.2007.01.016

[144] B.T. Low, Y.C. Xiao, T.S. Chung, Y. Liu, Simultaneous occurrence of chemical grafting, cross-linking, and etching on the surface of polyimide membranes and their Impact on H_2/CO_2 separation, Macromolecules, 41 (2008) 1297-1309. https://doi.org/10.1021/ma702360p

[145] L. Shao, B.T. Low, T.S. Chung, A.R. Greenberg, Polymeric membranes for the hydrogen economy: Contemporary approaches and prospects for the future, J. Membr. Sci., 327 (2009) 18-31. https://doi.org/10.1016/j.memsci.2008.11.019

[146] H. Wang, D.R. Paul, T.S. Chung, Surface modification of polyimide membranes by diethylenetriamine (DETA) vapor for H_2 purification and moisture effect on gas permeation, J. Membr. Sci., 430 (2013) 223-233. https://doi.org/10.1016/j.memsci.2012.12.008

[147] S. Japip, K.S. Liao, Y.C. Xiao, T.S. Chung, Enhancement of molecular-sieving properties by constructing surface nano-metric layer via vapor cross-linking, J. Membr. Sci., 497 (2016) 248-258. https://doi.org/10.1016/j.memsci.2015.09.045

[148] S. Japip, K.S. Liao, T.S. Chung, Molecularly tuned free volume of vapor cross-linked 6FDA-Durene/ZIF-71 MMMs for H_2/CO_2 separation at 150 °C, Adv. Mater., 29 (2017) 1603833. https://doi.org/10.1002/adma.201603833

[149] C. Staudt-Bickel, W. J. Koros, Improvement of CO_2/CH_4 separation characteristics of polyimides by chemical crosslinking, J. Membr. Sci., 155 (1999) 145-154. https://doi.org/10.1016/S0376-7388(98)00306-8

[150] C. Zhang, L. Fu, Z. Tian, B. Cao, P. Li, Post-crosslinking of triptycene-based Tröger's base polymers with enhanced natural gas separation performance, J. Membr. Sci., 556 (2018) 277-284. https://doi.org/10.1016/j.memsci.2018.04.013

[151] M.E. Rezac, E. Todd Sorensen, H.W. Beckham, Transport properties of crosslinkable polyimide blends, J. Membr. Sci., 136 (1997) 249-259. https://doi.org/10.1016/S0376-7388(97)00170-1

[152] A.C. Puleo, D.R. Paul, P.K. Wong, Gas sorption and transport in semicrystalline poly(4-methyl-1-pentene), Polymer, 30 (1989) 1357-1366. https://doi.org/10.1016/0032-3861(89)90060-8

[153] C.T. Wright, D.R. Paul, Feasibility of thermal crosslinking of polyarylate gas-separation membranes using benzocyclobutene-based monomers, J. Membr. Sci., 129 (1997) 47-53. https://doi.org/10.1016/S0376-7388(96)00327-4

[154] A.M. Kratochvil, W.J. Koros, Decarboxylation-induced cross-linking of a polyimide for enhanced CO_2 plasticization resistance, Macromolecules, 41 (2008) 7920-7927. https://doi.org/10.1021/ma801586f

[155] W. Qiu, C.C. Chen, L. Xu, L. Cui, D.R. Paul, W.J. Koros, Sub-T_g cross-linking of a polyimide membrane for enhanced CO_2 plasticization resistance for natural gas separation, Macromolecules, 44 (2011) 6046-6056. https://doi.org/10.1021/ma201033j

[156] C. Zhang, P. Li, B. Cao, Decarboxylation crosslinking of polyimides with high CO_2/CH_4 separation performance and plasticization resistance, J. Membr. Sci., 528 (2017) 206-216. https://doi.org/10.1016/j.memsci.2017.01.008

[157] R. Xu, L. Li, X. Jin, M. Hou, L. He, Y. Lu, C. Song, T. Wang, Thermal crosslinking of a novel membrane derived from phenolphthalein-based cardo poly(arylene ether ketone) to enhance CO_2/CH_4 separation performance and plasticization resistance, J. Membr. Sci., 586 (2019) 306-317. https://doi.org/10.1016/j.memsci.2019.05.084

[158] N. Du, M.M. Dal-Cin, G.P. Robertson, M.D. Guiver, Decarboxylation-induced cross-linking of polymers of intrinsic microporosity (PIMs) for membrane gas separation, Macromolecules, 45 (2012) 5134-5139. https://doi.org/10.1021/ma300751s

[159] Q. Song, S. Cao, R.H. Pritchard, B. Ghalei, S.A. Al-Muhtaseb, E.M. Terentjev, A.K. Cheetham, E. Sivaniah, Controlled thermal oxidative crosslinking of polymers of intrinsic microporosity towards tunable molecular sieve membranes, Nat. Commun., 5 (2014) 4813. https://doi.org/10.1038/ncomms5813

[160] C. Zhang, B. Cao, P. Li, Thermal oxidative crosslinking of phenolphthalein-based cardo polyimides with enhanced gas permeability and selectivity, J. Membr. Sci., 546 (2018) 90-99. https://doi.org/10.1016/j.memsci.2017.10.015

[161] A.A. Lin, V.R. Sastri, G. Tesoro, A. Reiser, R. Eachus, On the crosslinking mechanism of benzophenone-containing polyimides, Macromolecules, 21 (1988) 1165-1169. https://doi.org/10.1021/ma00182a052

[162] R.A. Hayes, Polyimide gas separation membranes, U.S. Patent 4,717,393, 1988

[163] H. Kita, T. Inada, K. Tanaka, K.-i. Okamoto, Effect of photocrosslinking on permeability and permselectivity of gases through benzophenone- containing polyimide, J. Membr. Sci., 87 (1994) 139-147. https://doi.org/10.1016/0376-7388(93)E0098-X

[164] M.S. McCaig, D.R. Paul, Effect of UV crosslinking and physical aging on the gas permeability of thin glassy polyarylate films, Polymer, 40 (1999) 7209-7225. https://doi.org/10.1016/S0032-3861(99)00125-1

[165] C.T. Wright, D.R. Paul, Gas sorption and transport in UV-irradiated polyarylate copolymers based on tetramethyl bisphenol-A and dihydroxybenzophenone, J. Membr. Sci., 124 (1997) 161-174. https://doi.org/10.1016/S0376-7388(96)00215-3

[166] J.R. Rowlett, Q. Liu, W. Zhang, J.D. Moon, M.E. Dose, J.S. Riffle, B.D. Freeman, J.E. McGrath, Gas transport properties and characterization of UV crosslinked poly(phenylene oxide-co-arylene ether ketone) copolymers, J. Mater. Chem. A, 4 (2016) 16047-16056. https://doi.org/10.1039/C6TA05320C

[167] S. Matsui, T. Ishiguro, A. Higuchi, T. Nakagawa, Effect of ultraviolet light irradiation on gas permeability in polyimide membranes. 1. Irradiation with low pressure mercury lamp on photosensitive and nonphotosensitive membranes, J. Polym. Sci. Part B: Polym. Phys., 35 (1997) 2259-2269. https://doi.org/10.1002/(SICI)1099-0488(199710)35:14<2259::AID-POLB6>3.0.CO;2-R

[168] I.K. Meier, M. Langsam, H.C. Klotz, Selectivity enhancement via photooxidative surface modification of polyimide air separation membranes, J. Membr. Sci., 94 (1994) 195-212. https://doi.org/10.1016/0376-7388(93)E0174-I

[169] Q. Song, S. Cao, P. Zavala-Rivera, L. Ping Lu, W. Li, Y. Ji, S.A. Al-Muhtaseb, A.K. Cheetham, E. Sivaniah, Photo-oxidative enhancement of polymeric molecular sieve membranes, Nat. Commun., 4 (2013) 1918. https://doi.org/10.1038/ncomms2942

[170] C.Z. Liang, T.S. Chung, J.-Y. Lai, A review of polymeric composite membranes for gas separation and energy production, Prog. Polym. Sci., 97 (2019) 101141. https://doi.org/10.1016/j.progpolymsci.2019.06.001

[171] K. Xie, Q. Fu, G.G. Qiao, P.A. Webley, Recent progress on fabrication methods of polymeric thin film gas separation membranes for CO_2 capture, J. Membr. Sci., 572 (2019) 38-60. https://doi.org/10.1016/j.memsci.2018.10.049

[172] M.F. Jimenez-Solomon, Q. Song, K.E. Jelfs, M. Munoz-Ibanez, A.G. Livingston, Polymer nanofilms with enhanced microporosity by interfacial polymerization, Nat. Mater., 15 (2016) 760-767. https://doi.org/10.1038/nmat4638

[173] Z. Ali, F. Pacheco, E. Litwiller, Y. Wang, Y. Han, I. Pinnau, Ultra-selective defect-free interfacially polymerized molecular sieve thin-film composite membranes for H_2 purification, J. Mater. Chem. A, 6 (2018) 30-35. https://doi.org/10.1039/C7TA07819F

[174] M.J.T. Raaijmakers, M.A. Hempenius, P.M. Schön, G.J. Vancso, A. Nijmeijer, M. Wessling, N.E. Benes, Sieving of hot gases by hyper-cross-linked nanoscale-hybrid membranes, J. Am. Chem. Soc., 136 (2014) 330-335. https://doi.org/10.1021/ja410047u

[175] (a) T. S. Chung, L.Y. Jiang, Y. Li, S. Kulprathipanja, Mixed matrix membranes (MMMs) comprising organic polymers with dispersed inorganic fillers for gas separation, Prog. Polym. Sci., 32 (2007) 483-507; https://doi.org/10.1016/j.progpolymsci.2007.01.008 (b) P. Banerjee, R. Das, P. Das, A. Mukhopadhyay, Membrane technology, in: R. Das (Ed.), Carbon Nanotubes for Clean Water, Springer, Cham, 2018, pp. 127-150. https://doi.org/10.1007/978-3-319-95603-9_6

[176] Y.C. Xiao, K.Y. Wang, T.S. Chung, J. Tan, Evolution of nano-particle distribution during the fabrication of mixed matrix TiO_2-polyimide hollow fiber membranes, Chem. Eng. Sci., 61 (2006) 6228-6233. https://doi.org/10.1016/j.ces.2006.05.040

[177] M.D. Guiver, G.P. Robertson, Y. Dai, F. Bilodeau, Y.S. Kang, K.J. Lee, J.Y. Jho, J. Won, Structural characterization and gas-transport properties of brominated Matrimid polyimide, J. Polym. Sci. A Polym. Chem., 40 (2002) 4193-4204.

[178] R. Mahajan, R. Burns, M. Schaeffer, W.J. Koros, Challenges in forming successful mixed matrix membranes with rigid polymeric materials, J. Appl. Polym. Sci., 86 (2002) 881-890.

[179] R. Mahajan,W.J. Koros, Factors controlling successful formation of mixed-matrix gas separation materials, Ind. Eng. Chem. Res., 39 (2000) 2692-2696.

[180] J.L.C. Rowsell, O.M. Yaghi, Metal–organic frameworks: A new class of porous materials, Microporous Mesoporous Mater., 73 (2004) 3-14. https://doi.org/10.1016/j.micromeso.2004.03.034

[181] Y. Cheng, Y. Ying, S. Japip, S.D. Jiang, T.S. Chung, S. Zhang, D. Zhao, Advanced porous materials in mixed matrix membranes, Adv. Mater., 30 (2018) 1802401. https://doi.org/10.1002/adma.201802401

[182] B. Ghalei, K. Sakurai, Y. Kinoshita, K. Wakimoto, A.P. Isfahani, Q. Song, K. Doitomi, S. Furukawa, H. Hirao, H. Kusuda, S. Kitagawa, E. Sivaniah, Enhanced selectivity in mixed matrix membranes for CO_2 capture through efficient dispersion of amine-functionalized MOF nanoparticles, Nat. Energy, 2 (2017) 17086. https://doi.org/10.1038/nenergy.2017.86

[183] T. Rodenas, I. Luz, G. Prieto, B. Seoane, H. Miro, A. Corma, F. Kapteijn, F.X. Llabrés i Xamena, J. Gascon, Metal–organic framework nanosheets in polymer composite materials for gas separation, Nat. Mater., 14 (2015) 48-55. https://doi.org/10.1038/nmat4113

[184] (a) Z. Kang, Y. Peng, Z. Hu, Y. Qian, C. Chi, L.Y. Yeo, L. Tee, D. Zhao, Mixed matrix membranes composed of two-dimensional metal–organic framework nanosheets for pre-combustion CO_2 capture: A relationship study of filler morphology versus membrane performance, J. Mater. Chem. A, 3 (2015) 20801-20810; https://doi.org/10.1039/C5TA03739E (b) G.R. Xu, J.M. Xu, H.C. Su, X.Y. Liu, L. Li, H.L. Zhao, H.J. Feng, R. Das, Two-dimensional (2D) nanoporous membranes with sub-nanopores in reverse osmosis desalination: Latest developments and future directions, Desalination, 451 (2019) 18-34. https://doi.org/10.1016/j.desal.2017.09.024

[185] A.P. Côté, A.I. Benin, N.W. Ockwig, M. Keeffe, A.J. Matzger, O.M. Yaghi, Porous, crystalline, covalent organic frameworks, Science, 310 (2005) 1166. https://doi.org/10.1126/science.1120411

[186] Z. Kang, Y. Peng, Y. Qian, D. Yuan, M.A. Addicoat, T. Heine, Z. Hu, L. Tee, Z. Guo, D. Zhao, Mixed matrix membranes (MMMs) comprising exfoliated 2D covalent organic frameworks (COFs) for efficient CO_2 separation, Chem. Mater., 28 (2016) 1277-1285. https://doi.org/10.1021/acs.chemmater.5b02902

[187] B.P. Biswal, H.D. Chaudhari, R. Banerjee, U.K. Kharul, Chemically stable covalent organic framework (COF)-polybenzimidazole hybrid membranes: Enhanced gas separation through pore modulation, Chem. Eur. J., 22 (2016) 4695-4699. https://doi.org/10.1002/chem.201504836

[188] T. Tozawa, J.T.A. Jones, S.I. Swamy, S. Jiang, D.J. Adams, S. Shakespeare, R. Clowes, D. Bradshaw, T. Hasell, S.Y. Chong, C. Tang, S. Thompson, J. Parker, A. Trewin, J. Bacsa, A.M.Z. Slawin, A. Steiner, A.I. Cooper, Porous organic cages, Nat. Mater., 8 (2009) 973-978. https://doi.org/10.1038/nmat2545

[189] A.F. Bushell, P.M. Budd, M.P. Attfield, J.T.A. Jones, T. Hasell, A.I. Cooper, P. Bernardo, F. Bazzarelli, G. Clarizia, J.C. Jansen, Nanoporous organic polymer/cage composite membranes, Angew. Chem. Int. Ed., 52 (2013) 1253-1256. https://doi.org/10.1002/anie.201206339

[190] G. Zhu, F. Zhang, M.P. Rivera, X. Hu, G. Zhang, C.W. Jones, R.P. Lively, Molecularly mixed composite membranes for advanced separation processes, Angew. Chem. Int. Ed., 58 (2019) 2638-2643. https://doi.org/10.1002/anie.201811341

[191] C.D. Gutsche, Using the baskets: Calixarenes in action, in: Calixarenes: An Introduction, Second Edition, The Royal Society of Chemistry, Cambridge, 2008, pp. 208-237. https://doi.org/10.1039/9781847558190-00208

[192] J.T. Liu, Y.C. Xiao, K.S. Liao, T.S. Chung, Highly permeable and aging resistant 3D architecture from polymers of intrinsic microporosity incorporated with beta-cyclodextrin, J. Membr. Sci., 523 (2017) 92-102. https://doi.org/10.1016/j.memsci.2016.10.001

[193] J. Wu, J.T. Liu, T.S. Chung, Structural tuning of polymers of intrinsic microporosity via the copolymerization with macrocyclic 4-tert-butylcalix[4]arene for enhanced gas separation performance, Adv. Sustainable Syst., 2 (2018) 1800044. https://doi.org/10.1002/adsu.201800044

[194] J. Wu, S. Japip, T.S. Chung, Infiltrating molecular gatekeepers with coexisting molecular solubility and 3D-intrinsic porosity into a microporous polymer scaffold for gas separation, J. Mater. Chem. A, 8 (2020) 6196-6209. https://doi.org/10.1039/C9TA12028A

[195] J. Shen, M. Zhang, G. Liu, K. Guan, W. Jin, Size effects of graphene oxide on mixed matrix membranes for CO_2 separation, AIChE J., 62 (2016) 2843-2852. https://doi.org/10.1002/aic.15260

[196] M. Chen, F. Soyekwo, Q. Zhang, C. Hu, A. Zhu, Q. Liu, Graphene oxide nanosheets to improve permeability and selectivity of PIM-1 membrane for carbon dioxide separation, J. Ind. Eng. Chem., 63 (2018) 296-302. https://doi.org/10.1016/j.jiec.2018.02.030

[197] G. Dong, J. Hou, J. Wang, Y. Zhang, V. Chen, J. Liu, Enhanced CO_2/N_2 separation by porous reduced graphene oxide/Pebax mixed matrix membranes, J. Membr. Sci., 520 (2016) 860-868. https://doi.org/10.1016/j.memsci.2016.08.059

[198] Y. Wang, X. Wang, J. Guan, L. Yang, Y. Ren, N. Nasir, H. Wu, Z. Chen, Z. Jiang, 110th anniversary: Mixed matrix membranes with fillers of intrinsic nanopores for gas separation, Ind. Eng. Chem. Res., 58 (2019) 7706-7724. https://doi.org/10.1021/acs.iecr.9b01568

[199] B.H. Jeong, E.M.V. Hoek, Y. Yan, A. Subramani, X. Huang, G. Hurwitz, A.K. Ghosh, A. Jawor, Interfacial polymerization of thin film nanocomposites: A new concept for reverse osmosis membranes, J. Membr. Sci., 294 (2007) 1-7. https://doi.org/10.1016/j.memsci.2007.02.025

[200] S. Yu, S. Li, S. Huang, Z. Zeng, S. Cui, Y. Liu, Covalently bonded zeolitic imidazolate frameworks and polymers with enhanced compatibility in thin film nanocomposite membranes for gas separation, J. Membr. Sci., 540 (2017) 155-164. https://doi.org/10.1016/j.memsci.2017.06.047

Keyword Index

About the Editor

Dr. Rasel Das is working as a Postdoctoral Associate and Senior Scientist in the Department of Chemistry, Stony Brook University, NY, USA. He was a Visiting Researcher and JSPS Postdoctoral Fellow at the Global Innovation Center, Kyushu University, Japan from 2018 to 2020. He finished DAAD Postdoctoral Fellowship at the Leibniz-Institute of Surface Engineering, Germany in 2017. He was awarded for Endeavour Research Fellowship at Australian Institute for Bioengineering and Nanotechnology, The University of Queensland, Australia in 2018. He is leading the research and product development team of Water Desalination and Transport Division, MHD Technology Corp. Florida, USA. He has published >60 journal articles, books and editorials, examining and defining 'Functional Nanomaterials for Water Purification'. He has been invited to discuss his research at many prestigious institutions like SEAS, Harvard University; and National Institute of Health (NIH), USA. He has been awarded and honored by Elsevier-Atlas and BSN Academic Award in 2015. He is on the editorial board member of the Journal of Environmental & Analytical Toxicology, Lead Guest Editor of Desalination Journal, Topic Editor of Membrane; and also reviewer of many prestigious journals including ACS Environmental Science & Technology, Journal of Membrane Science, Desalination, Nanoscale, etc. He received his Ph.D. from the University of Malaya, Malaysia under the prestigious BrightSparks scholarship after being awarded with distinctions in both undergraduate and postgraduate studies from the University of Science and Technology Chittagong, Bangladesh. His doctoral thesis was on the Functionalization of Carbon Nanotube and Development of NanoBiohybrid for Water Purification.